12th International Conference Quality Production Improvement (QPI 2018)

MATEC Web of Conferences Volume 183 (2018)

Zaborze, Poland
18 - 20 June 2018

Editors:

R. Ulewicz
B. Hadzima

ISBN: 978-1-5108-6857-1

Printed from e-media with permission by:

Curran Associates, Inc.
57 Morehouse Lane
Red Hook, NY 12571

Some format issues inherent in the e-media version may also appear in this print version.

Printed by Curran Associates, Inc. (2018)

For additional information, please contact EDP Sciences – Web of Conferences at the address below.

EDP Sciences – Web of Conferences
17, Avenue du Hoggar
Parc d'Activité de Courtabœuf
BP 112
F-91944 Les Ulis Cedex A
France

Phone: +33 (0) 1 69 18 75 75
Fax: +33 (0) 1 69 28 84 91

contact-edps@webofconferences.org

Additional copies of this publication are available from:

Curran Associates, Inc.
57 Morehouse Lane
Red Hook, NY 12571 USA
Phone: 845-758-0400
Fax: 845-758-2633
Email: curran@proceedings.com
Web: www.proceedings.com

TABLE OF CONTENTS

Preface

12th International Conference Quality Production Improvement - QPI 2018 was successfully held in Zaborze near Myszków, Poland on 18th - 20th June 2018. The objective of the conference is to provide a platform for researchers, engineers, academicians as well as industrial professionals from all over the world to present their research results and development activities in production engineering, quality of materials, technology and management. The success and prosperity of the conference is reflected high level of the papers received.

The proceedings are a compilation of the accepted papers and represent an interesting outcome of the conference. This book covers 4 chapters: Production engineering; Material engineering; Quality management; Management.

We would like to thank speakers, patronages and all the conference for making this year's QPI 2018 conference another unforgettable event. We sincerely appreciate all the authors who have contributed to this volume and also to the reviewers for their strong support to improve all papers.

Especially we would like to thank The Rector of the Czestochowa University of Technology Prof. Norbert Sczygiol and Mayor of the municipality of Olsztyn Tomasz Kucharski.

We sincerely hope that QPI 2018 will be a forum for excellent discussions that will put forward new ideas and promote collaborative researches. We are looking forward to meeting you 2019!

Organizing Team

Conference QPI

12th International Conference Quality Production Improvement - QPI 2018
18-20.06.2018 ZABORZE near MYSZKÓW, POLAND

MAIN ORGANIZER
Quality and Production Managers Association

PARTNERS
Department of Production Engineering and Safety

Faculty of Management, Czestochowa University of Technology

Assocation of Polish Inventors and Rationalizers

MAIN PATRONAGE:
The conference was held under the patronage of Prof. Norbert Sczygiol
The Rector of the Czestochowa University of Technology

OTHER PATRONAGE:

SCIENTIFIC COMMITEE

CHAIRMAN
Prof. nzw. dr hab. inż. Robert Ulewicz - Czestochowa University of Technology, Poland

DEPUTY CHAIRMAN
Prof. Branislav Hadzima - University of Žilina, Slovak Republic

MEMBERS
Prof. Witold Biały - Silesian University of Technology, Poland
Prof. Martina Blašková - University of Žilina, Slovak Republic
Prof. Tatianá Čorejova - University of Žilina, Slovak Republic
Prof. Nicoletto Gianni - University of Parma, Italy
Prof. Janusz Grabara - Czestochowa University of Technology, Poland
Prof. Otakar Bokůvka - University of Žilina, Slovak Republic
Prof. Miloš Hitka - Technical University in Zvolen, Slovak Republic
Prof. František Holešovský - J. E. Purkyně University in Ústí nad Labem, Czech Republic
Prof. Jozef Hrubec - Slovak University of Agriculture in Nitra, Slovak Republic
Prof. Denis Jelačić - University of Zagreb, Croatia
Prof. Dorota Jelonek - Czestochowa University of Technology, Poland
Prof. Jiří Kliber - VSB-TU Ostrava, Czech Republic
Prof. Ludvík Kunz - Czech Academy of Sciences, Czech Republic
Prof. Tomasz Lipiński -- University Of Warmia and Mazury, Poland
Prof. Ružica R. Nikolić - University of Kragujevac, Serbia
Prof. František Nový - University of Žilina, Slovakia
Prof. Jacek Pietraszek - Cracow University of Technology Poland
Prof. Norbert Radek - Kielce University of Technology, Poland
Prof. Anna Šatanova - Technical University in Zvolen, Slovak Republic
Prof. Jacek Selejdak - Czestochowa University of Technology, Poland
Prof. Michał Szota - Czestochowa University of Technology, Poland
Prof. Augustín Sládek - University of Žilina, Slovak Republic
Prof. János Takács - Budapest University of Technology and Economics, Hungary
Prof. Richard Vlosky - Louisiana State University, United States of America
Atul B. Borade, PhD. (IN) - Jawaharlal Darda Institute of Engineering & Technology, India
Juhani Anttila, M.Sc. - Quality Integration, Helsinki, Finland

ORGANIZATION COMMITTEE
Robert Ulewicz - CHAIRMAN - Czestochowa University of Technology, Poland
Jacek Selejdak - Vice-CHAIRMAN - Czestochowa University of Technology, Poland
Magdalena Mazur - Czestochowa University of Technology, Poland
Renata Stasiak-Betlejewska - Czestochowa University of Technology, Poland
Dorota Klimecka-Tatar - Czestochowa University of Technology, Poland
Manuela Ingaldi - Czestochowa University of Technology, Poland

Marta Niciejewska - Czestochowa University of Technology, Poland
Joanna Rosak-Szyrocka - Czestochowa University of Technology, Poland
Krzysztof Knop - Czestochowa University of Technology, Poland
Krzysztof Mielczarek - Czestochowa University of Technology, Poland
Piotr Czaja - Pedagogical University of Cracow, Poland

edp sciences

MATEC Web of Conferences

Statement of Peer review

In submitting conference proceedings to MATEC Web of Conferences, I certify to the Publisher that I adhere to the **Policy on Publishing Integrity** of the journal in order to safeguard good scientific practice in publishing.

1. All articles have been subjected to peer review administered by the proceedings editors.
2. Reviews have been conducted by expert referees, who have been requested to provide unbiased and constructive comments aimed, whenever possible, at improving the work.
3. Proceedings editors have taken all reasonable steps to ensure the quality of the materials they publish and their decision to accept or reject a paper for publication has been based only on the merits of the work and the relevance to the journal.

Title, date and place of the conference

12th International Conference Quality Production Improvement - QPI 2018
18 - 20 June 2018 Zaborze near Myszków
Poland

Proceedings editor(s):

Prof. nzw. dr hab. inż. Robert Ulewicz
Częstochowa University of Technology

Date and editor's signature 09.07.2018

17, avenue du Hoggar - PA de Courtabœuf – BP 112 - 91944 Les Ulis Cedex A (France)
Tél. : 33 (0)1 69 18 75 75 - Fax : 33(0)1 69 07 45 17 – www.edpsciences.org

Identification of bottlenecks and analysis of the state before applying lean management

Radosław Wolniak[1,*], *Bożena* Skotnicka-Zasadzień[2], and *Anna* Gębalska-Kwiecień[3]

[1]Silesian University of Technology, Organisation and Management Faculty, 41-800 Zabrze Roosvelta str., Poland
[2]Silesian University of Technology, Organisation and Management Faculty, 41-800 Zabrze Roosvelta str., Poland
[3]Silesian University of Technology, Organisation and Management Faculty, 41-800 Zabrze Roosvelta str., Poland

Abstract. The analysis of the production process from the time operations point of view is very important method use in Lean management to improve the whole process. In this paper we describe example of this type analysis on the example of industrial organization from automotive industry. The aim of presented paper is to identify the bottleneck in the production process and to analyse operations times in the production process. On the basis of the carried out preliminary analysis it was stated that the lack of proper organization of the place and time of work, lack of proper preparation in terms of content and equipment, lack of proper work standards, and what unfortunately in the case of the RZ machine are very important factors of losses.

1 Theoretical introduction

The problem analysed in this paper concentrate on analysis of the production process to identify the bottlenecks and to analyse the state of the process. This problem is very important from industrial organization point of view, because bottlenecks are very dangerous from production time point of view. We need to identify them and next analyse all the process with the times of operation. The next stage could be to use lean management methods to reduce this time [1]. For example the SMEAD method can be useful in this situation.

The aim of presented paper is to identify the bottleneck in the production process and to analyse operations times in the production process. The SMED analysis will be presented in another paper.

The paper was prepared as a result of project 13/030/BK_18/0039 Development of intelligent production methods as well as work and life environments in the context of production engineering challenges.

* Corresponding author: rwolniak@polsl.pl

According to Cambridge English dictionary we can define the bottleneck term as a problem that delays progress [2]. Other definitions of this term are for example as follows:

- department, facility, machine, or resource already working at its full capacity and which, therefore, cannot handle any additional demand placed on it. Also called critical resource, a bottleneck limits the throughput of associated resources [3],
- bottleneck in production is a point where an operation meets or exceeds the capacity of the facility. In other words, the factory or department can't produce enough units fast enough to keep the rest of the production schedule or other daily operations flowing at the same rate [4].

According to the theory of constraints (TOC), the central task of an effective production management is to find and eliminate the impact of a bottleneck in a company [5]. We can divide the production management into three types from the occurrence of the bottleneck point of view [6]:

- management in a situation with no bottleneck in the production process,
- management in a situation with one bottleneck in the production process,
- management in a situation with many bottlenecks in the production process.

The absence of bottlenecks in the production process means that the production productivity is enough to achieve forecast sales. If one bottleneck occurs, it needs to be determined whether there is only one production process or whether alternative production processes are possible. Contribution margin analysis for individual combinations of products and processes is a useful tool in this case. If it turns out that there are several parallel bottlenecks in the production process, the decision becomes more complex. In such an event, solutions to these problems should be founded with the use of linear programming methods [7].

The Theory of Constraits (TOC) is an organizational methodology of change that focuses on improving profit. The basic premise of a TOC is that each organization has at least one limitation, one factor not allowing more than anything to be the purpose of the organization. This is usually a profit [8, 9].

TOC focuses on process constraints, production organization and also on „Bottlenecks" in different production products. However, the TOC does not focus exclusively on production constraints. Factors such as market demand are taken into account, or the ability to turn this demand into an on-demand sales force. So not only production, but also distribution, finance and accounting, sales and marketing, and HR (human resources) are taken into account [10-13].

After solving the problem, the system capacity limits increase, but this does not mean that they become infinite, because most often, another resource becomes a new cause of further limitations. It is worth pointing out that the proper implementation of the various steps of the TOC cycle will make the existing constraints an opportunity for enterprise development. The key to the success of an enterprise is to understand what limitations exist in its system. We should in this case ask some question. At the strategic level, which limits the creation of exceptional value for our customers? And at operational level, which limits the flow velocity increase? The term „constraint" has in common everyday negative connotations. However, the Theory of Constraints suggests a different perspective. The limitation of the system is, in fact, the point of leverage of the whole system. Once the goals of the system are well defined, understanding its limitations is key to building a game plan - strategies and tactics that guide us to achieve the goal [14-24].

2 Identification an characteristic of bottleneck

After the initial substantive preparation, a brainstorming tool was used to develop a checklist before introducing the SMED method. This list was intended to illustrate the actual state of the situation and to present all aspects to which the team should pay attention during the visit to the production hall during changeover. After making the checklist, the team equipped with a video camera went to the production line and hall and proceeded to the re-documenting and all operations carried out during the change. The culmination of this stage was the development of a preliminary division of duties and a detailed determination of the project's objective.

On the base of this analysis and discussion of the team the main bottleneck was identified in the research. It was reduction of the changeover time to 18 minutes. All the times values in this paper were multiplied by a constant factor.

The characteristic of the process identified as a bottleneck was subsequently made. The analysed production process takes place at the pipe production department. The specification of this line, depending on the reference produced, may require the use of a maximum of seven machines. On of this machines is so called RZ - is responsible for more complex calibration operations, such as the creation of outlets (exhaust system ends).

The necessity of using so many machines requires very good organizational preparation before it starts. Also communication and experience from the rudders and foremen is needed. However, the analysis of the film showed that not everything works as it should, because the total changeover time was 3h 52min. This is a very big result, especially since the line overwhelms an average of 1.5 times during one shift, which during the whole day (the company is working on 3 shifts) gives 16h and 14min of line stoppage. This is a huge loss for the company, ignoring the fact that the main assumption of SMED is to achieve the changeover time with the value of a one-digit unit of time.

After the initial analysis of the film, it was noticed that the main problem in the retooling of the above-mentioned machines lies on the side of the bending machine, whose total changeover time amounted to 2h 14min and partly on the RZ side.

For all other machines, the changeover time is adequate to the time specified in the instructions and is relatively short, as it is within 10 - 18 minutes. In addition, the specification of machines is so simple that the possibility of introducing any improvements is impossible. The only machine, as already mentioned, which in addition to the bending machine requires special attention is RZ. This is due to the fact that the retooling process of this machine in a very blatant manner contradicts the principles of ergonomics, which is why this issue certainly needs to be analysed and implemented. In addition, the time of its retooling also exceeds the limit according to the methodology, as it amounted to as much as 2h 28min [25].

3. Analysis of the current state before the introduction of the SMED method

The activities presented in the publication concern the shortening of the changeover time of the multitask bending machine. This is conditioned by the fact that the machine belongs to the bottleneck and significantly affects the much extended time of the entire retooling. In addition, reducing the changeover time of this machine is very important, because it is the most complex machine in the entire machine park, characterized by the largest number of parts for conversion and the largest number of retooling's.

After analysing the video, it was necessary to analyse the current state. First of all, the specification of all the activities performed during the conversion of the machine along with determining the responsibility and time of their execution - this division is presented in the table below (table 1), so that all aspects of the conversion can be analysed in a very clear and transparent way.

The data presented in the table shows a list of all activities performed by the setter during the conversion of a bending machine. Of all activities, all those that result from poor organization of the workplace, improper preparation for retooling or lack of appropriate conversion standards are bolded. As a result, the time wasted to perform unnecessary activities was 490 seconds, which out of a total re-set time of 8150 seconds, gives 6% of the time [25].

Table 1. A list of operations during the retooling of the bending machine before the implementation of SMED [25].

Number	Action name	Time [s]
1.	Preparation of the bending machine for conversion (manual)	20
2.	Securing the bending machine and setting the position „0"	540
3.	Preparation of the safety field	40
4.	**Searching for tools**	**30**
5.	**Importing tools for retooling the bending machine**	**40**
6.	Unscrewing the tension	100
7.	Removing the spigot nut	80
8.	**Organize the trolley with tools**	**60**
9.	Disassembly of 4 hooves	220
10.	Disassembly of the clamping jaw (release)	80
11.	Disassembly of the stem	80
12.	Disassembly of the roller using the traverse • search for traverse tools • access by traverse to the machine.	500 **(240)** (70)
13.	Disassembly of the clamping jaw	30
14.	Setting the height of the clamping jaw	240
14.	Disassembly of the shaft from the clamping jaw	100
15.	Disassembly of the smoothing jaw	140
16.	Assembly of the smoothing jaw	200
17.	Roll assembly	260
18.	**Consultations**	**60**
19.	Mounting the spigot cap	160
20.	**Placing tools on the trolley**	**60**
21.	Mounting the roller to the clamping jaw	90
22.	Mounting of the clamping jaw	30
23.	Assembly and disassembly of jaws from transport	600
24.	Installation of 1 hoof	190
25.	Pipe radius setting	120
26.	Mounting tension	240
27.	Setting the clamping jaw (using the ratchet)	360
28.	Tensioning and pre-setting	240
29.	Unlocking security	240
30.	Mounting and twisting of the spindle	480
31.	Setting the spindle	120
32.	Setting of the bending machine	2400
	Suma	**8150**

Another aspect requiring improvement concerned the time of screwing in and removing the screws. With this machine, the setter has 22 screws to screw in, where it uses a large heavy wrench, a rattle or other screwdrivers, sometimes also assisting with a hammer. The time needed to screw / unscrew the screws in this way was 40 seconds, which in total gives 440 seconds, which in turn makes up almost 11% of the total retooling time of the bending machine.

In this way, after a very preliminary analysis of the operations, it was possible to identify operations that generated as much as 17% of losses in the overall retooling time of the bending machine. It was very important to find a golden mean, so as to be able to reduce the losses to a minimum.

However, at this stage of the analysis it was very important to also retool the machine RZ. As mentioned earlier in the analyses, the time needed for the reconstruction of RZ was 2h 28min, therefore it was very important to analyse all the activities performed during this operation, so as to be able to identify those that we are able to change.

In this case, the analysis of the activities showed that the situation looked very similar to the bending machine, so the issues related to poor organization of the workplace and improper work preparation resulted in the following results [25]:

- 400 seconds of loss for consultations and documentation review, which is more than 4% of total reconstruction time of RZ,
- 2072 seconds of loss to go to the tool magazine for missing parts, results in over 23% of the total changeover time of the RZ.

Issues related to the organization of work resulted in losses in the form of 27% of total changeover time. However, in the case of the RZ machine, the issue of tightening / unscrewing the screws is also of key importance. As with the bending machine, the time needed to screw / unscrew 1 screw was 40 seconds, which with 36 screws gives a total of 1440 seconds, representing in turn more than 16% of the total retooling time RZ.

4. Conclusion

On the basis of the carried out preliminary analysis it was stated that the lack of proper organization of the place and time of work, lack of proper preparation in terms of content and equipment, lack of proper work standards, and what unfortunately in the case of the RZ machine are very important factors of losses. Those factors play a very important role in the analysed changeover. The amount of time lost on the process of screwing in and removing the screws is also important. A positive aspect of the changes that should be carried out is the fact that they are not issues that require large investments.

References

1. R. Wolniak, *Zeszyty Naukowe Politechniki Śląskiej. Seria Org. i Zarz.*, **75**, 157 (2014)
2. Bottleneck, https://dictionary.cambridge.org/dictionary/english/bottleneck
3. Bottleneck, http://www.businessdictionary.com/definition/bottleneck.html
4. Bottleneck, https://www.myaccountingcourse.com/accounting-dictionary/bottleneck
5. A. Koliński, *Efficiency in Sustainable Supply Chain* (Springer International Publishing, 2017).
6. A. Koliński, A. Tomkowiak, *Gosp. Mat. i Logist.*, **9**, 16 (2010)
7. J. Kolińska, R. Domański, *17th international scientific conference Business Logistic in Modern Management Proceedings* (Osijek, Croatia, 2017).

8. J. Cox, J. Schleier, *Theory of Constraints. Handbook* (McGraw Hill Profesional, New York, 2010)
9. E. Goldrat, *Theory of Constraints / Lean / Six Sigma Integration* (McGraw Hill Profesional, New York, 2010)
10. R. Wolniak, Business and management, **5**, 1093 (2017)
11. R. Wolniak, B. Skotnicka-Zasadzień, M. Zasadzień, *3rd International Conference on Social, Education and Management Engineering* (SEME, Shanghai, 2017)
12. M.M. Srinivasan, *Building Lean Supply Chains with the Theory of Constraints* (McGraw Hill Profesional, New York, 2011)
13. R. Wolniak, B. Skotnicka-Zasadzień, Metalurgija, **4**, 709 (2014)
14. R. Wolniak, Quality & Quantity, **1**, 515 (2013)
15. M. Zasadzień, *6th International Conference on Operations Research and Enterprise Systems* (ICORES, Porto, Portugal 2017)
16. H.W. Dettmer, *Goldrst.s Theory of Constraints: A Systems Approach to Continuous Inmrovement* (ASQ Quality Press, Milwaukee, Wsconsin, 1999)
17. T. Hirts, P. Guernaccini, *Studies in Mana. and Fin. Acc.*, **25**, 117 (2012)
18. T.L. Pereira, D.L. Pacheco, L.R. Henrique, D.V. Rafael, *Business Process Man. Jour.*, **6**, 922 (2014)
19. A. Ray, B. Sarkar, S.K. Sanyal, *International Journal of Accounting & Information Management*, **16**, 155 (2008)
20. A. Zivaljevic, *Man. of Envir. Qual.: An Int. Jour.*, **4**, 505 (2015)
21. J. Brodny, S. Alszer, K. Jolanta, T. Magdalena, *Arch. Contr. Sci.*, **2**, 197 (2017)
22. J. Brodny, K. Stecuła, M. Tutak, Science and technologies in geology, **2**, 65 (2016)
23. M. Wojtaszek, W. Biały, *Manage. Syst. Prod. Eng.*, **3**, 133 (2015)
24. J. Sitko, Syst. Wspomag. Inż. Prod., **4**, 90 (2016)
25. M. Ziółkowska, *Doskonalenie procesu produkcyjnego przy wykorzystaniu metody SMED w wybranym przedsiębiorstwie produkcyjnym* (master thesis on supervising B. Skotnicka-Zasadzień, Zabrze, 2018)

Improving the efficiency of the production process using SMED

Bożena Skotnicka-Zasadzień[1], *Radosław* Wolniak[2], and *Anna* Gębalska-Kwiecień[3]

[1]Silesian University of Technology, Organisation and Management Faculty, 41-800 Zabrze Roosvelta Str., Poland
[2]Silesian University of Technology, Organisation and Management Faculty, 41-800 Zabrze Roosvelta Str., Poland
[3]Silesian University of Technology, Organisation and Management Faculty, 41-800 Zabrze Roosvelta Str., Poland

Abstract. The article presents one of the methods of management used in quality engineering, i.e. the SMED, which is applied to streamline the functioning of machines used on a production line. The research problem that was analysed on one of the production lines was too long a retooling time which resulted in too high a cost of the operation which, consequently, resulted in excessive overall production costs. The main aim of the actions undertaken when applying the SMED method was to identify potentially problematic areas and reduce the retooling time. All those actions allowed to minimise production costs, and improved the functioning of the company as well as the efficiency of the production process.

1 Theoretical introduction

Due to market requirements and the need to reduce constantly rising production costs, enterprises more and more often introduce solutions of the Lean methodology. The Lean concept, which originates from Japan, also referred to as „lean management", consists mainly of getting rid of all the wastefulness in a company. Wastefulness refers to activities such as overproduction, unnecessary movements, stocks, unnecessary transport, any errors in the production process, as well as waiting times and shortages. To achieve the intended benefits, it is necessary to implement many Lean tools and methods [1, 2, 7].

For the proper functioning of the company and a quick response to the needs of customers, it is necessary to introduce shorter production series, which in turn leads to shortening retooling times and reducing production costs. To achieve such goals, the Single Minute Exchange of Die method, or SMED, is used [3, 8, 10, 11, 22].

The SMED method developed by a Japanese engineer Shigeo Shingo, was first implemented in TOYOTA's factories. The SMED methodology was established to reduce the time needed for retooling of the press and setting tools in machines, but its philosophy can be applied to all production processes. It allows to reduce some of the retooling times from several hours down to several minutes, which it is very important when it comes to reducing production costs. Sometimes shortening even one operation allows for the reduction of costs on the whole production line. It is known that achieving a result below

10 minutes is not always possible for all retoolings, but the very application of the method reduces the retooling time in virtually any case. Any reduction of retooling time will consequently reduce costs [4, 9, 12].

The retooling processes of machines and devices have a characteristic common feature, namely they all consist of four stages that are the same for each type of equipment and operation. The first stage is preparation, inspection of parts and tools as well as post-process regulations [5, 6, 13, 15]. The second stage consists of assembly and disassembly of blades, parts and tools. Then settings, measurements and calibrations are made. The final stage is trial series and corrections of the machine settings [14, 16, 18, 19, 21]. The scientific goal in this article was to show how the application of the SMED method and the introduction of improvement measures have an impact on improving the efficiency of the production process.

2 Description of the research problem

In the company surveyed, the main problem was related to the retooling of machines used for pipe bending. The total retooling time of this machine was 116 minutes, which is much too long, which resulted in generating large losses and reducing the efficiency of the production process. In addition to the very long retooling time of the machine, the following factors were also important: the machine is heavily used and plays a very important role in the whole production process. It is the most expensive and most complex machine in the whole production hall, with the largest number of retooling, having the most complicated retooling process due to the largest number of parts to be replaced and has the longest downtimes.

All these aspects caused the management to make the decision to deal with the retooling process in the first place with the goal to shorten this time to about 10 minutes. All the activities that were undertaken were consistent with the SMED terminology [17, 20].

3 Measures to improve the efficiency of the production process

In order to improve the efficiency of the production process, the following actions were taken in accordance with the SMED terminology:

- preparation and analysis of a video recording, the checklist presented in Table 1 was helpful in the analysis
- specifying all the activities performed during the retooling that generate the greatest losses of time. These included: searching for tools, importing tools for retooling the machine, organizing the trolley with tools, consulting, loading tools onto the trolley. The total time of all activities is 125 [17]
- division of internal to external activities
- transforming internal activities into external ones
- proposing and implementing improvement measures.

Table 1. Control checklist prior to implementing SMED [17].

<table>
<tr><td colspan="2" rowspan="2">Control Checklist Prior to SMED</td><td>Process</td><td colspan="2">Machine No.</td></tr>
<tr><td>Production line</td><td>Checked by</td><td>Date</td></tr>
<tr><td>#</td><td colspan="2">Checkpoint</td><td colspan="2">Notes</td></tr>
<tr><td>1</td><td colspan="2">Is there a full set of the necessary tools and equipment available at the station ?</td><td colspan="2"></td></tr>
<tr><td>2</td><td colspan="2">.Are the tools and equipment easily accessible to the operator during the entire retooling process ?</td><td colspan="2"></td></tr>
<tr><td>3</td><td colspan="2">Are the tools and equipment arranged in the order in which they are</td><td colspan="2"></td></tr>
</table>

	used during retooling ?	
4	Is putting the tools and equipment back in their place a common practice ?	
5	Are special carts used and is there a designated location for them ?	
6	Is placing thing on the special carts a common practice ?	
7	Does every machine have its retooling instructions ?	
8	Are the operators who perform the retooling trained according to the retooling instructions ?	
9	Have quality standards been defined for every product ?	
10	Are the standards placed on every machine ?	
11	Are green standards prepared for every reference?	
12	Are green standards on display near the relevant machine?	
13	Are measurement tools located in the right place near the machine that is being used?	
14	Are measurement tools easily accessible to the user?	
15	Is the proper placement of measurement tools common practice?	
16	Is the location in which tools and equipment should be stored after retooling clearly specified?	
17	Is the location in which tools and equipment should be stored prior to retooling clearly specified?	
18	Is there a designated location for tools for the next reference?	

Among the stages described above, the most important was the preparation and analysis of the video recording, as it was used to identify further activities related to the identification of activities, division into internal and external activities and the process of transforming internal activities into external ones. These were key activities because they allowed to identify the bottleneck which, undoubtedly, was the production machine analysed. Meanwhile, in the case of other machines, no action was taken, as they were carried out in accordance with the standards and were within the limits of 5 - 9 minutes [17].

As it turned out, the introduction of appropriate improvement measures in the form of replacing the manual key with a pneumatic wrench made it possible to shorten the screwing and unscrewing time by as much as 75%. All these aspects caused the retooling time of the line to be reduced by 50%, and the machine which was causing the greatest time losses by almost 60%, Table 2 [17].

Table 2. Retooling control checklist [17].

RETOOLING CONTROL CHECKLIST					
Setter's name					
Before starting work in my area I have checked the following elements			OK	NOK	NOTES
1	TOOLS	Check the completeness of the toolset			
1.1		Roll/Roll + roller			
1.2		Pressure clamp/Clamp with adapter + roller/rollers			
1.3		Smoothening clamp with handles and hooks			
1.4		Hooves with sliders			
1.5		Transport clamps			
1.6		Mandrel (assembled, complete, in standard size)			
1.7		Roller pin			
1.9		Calibrating tool (with pins and rings)			

1.10		Screws with washers			
2.1	LUBRICATING	Check for presence of the "Prolube amber" lubricant			
2.2		Check the quality of calibrating tool lubrication			
3.1	TECHNICAL DOCUMENTATION	Check if there is technical documentation			
3.2		Familiarise yourself with the technical documentation			

4 Conclusion

The presented analysis, consistent with the SMED methodology, allowed to improve the efficiency of the production process by:
- shortening the retooling time of the line by 50%,
- shortening the retooling time of the bending machine by less than 60%,
- reducing the screwing and unscrewing time by 75%,
- shortening the machine part replacement time by almost 32%,
- planned return on investment in the purchase of electric impact wrenches at the level of almost 7%.

In addition to shortening the retooling time, the following actions have been proposed to improve the retooling process [17]:
- permanently affixing handles to the smoothening clamp and temporarily mounting the handles on the roll - external operation,
- mounting of the positioning screw to the pressure clamp and setting the clamp height - external operation,
- using a pneumatic wrench to eliminate time wastage during the screwing/unscrewing of a large number of screws with multi-start threads,
- development of a new retooling instruction,
- development of a checklist for setters,

Achieving the results presented was possible by introducing changes in the area of improving the retooling process as well as organization and communication in the workplace [17].

References

1. J. Sitko, Syst. Wspomag. Inż. Prod., **4**, 90 (2016)
2. M. Nowicka-Skowron, R. Ulewicz, *25TH Anniversary International Conference of Metallurgy and Materials,* 1962-1997 (2016)
3. A. Zakrzewska-Bielawska, *Metody zarządzania procesami w świetle studiów i badań empirycznych* (Uniwersytet Ekonomiczny w Krakowie, Kraków, 2009)
4. S. Kowal, K. Knop, Zeszyty Naukowe Quality. Production. Improvememt, **2**, 7 (2017)
5. J.P. Womack, D.T. Jones, *Centrum Informacji Menadżera* (Warszawa, 2001)
6. M. Kruczek, Z. Żebrucki, Logistyka, **2** (2012)
7. J. Maciak, *Komputerowo zintegrowane zarządzanie – tom II* (Zakopane (2011)
8. R. Wolniak, Business and management, **5**, 1093 (2017).
9. R. Wolniak, B. Skotnicka-Zasadzień, M. Zasadzień, *3rd International Conference on Social, Education and Management Engineering* (SEME, Shanghai, 2017)
10. S. Shigeo, *Productivity* (Inc,1985)

11. R. Wolniak, B. Skotnicka-Zasadzień, Metalurgija, **4**, 709 (2014)
12. E. Milewska, Business and Management **5**, 599-606 (2017)
13. M. Zasadzień, *6th International Conference on Operations Research and Enterprise Systems* (ICORES, Porto, Portugal (2017)
14. K. Antosz, A. Pacana, D. Stadnicka, W. Zielecki, (Oficyna Wydawnictwa Politechniki Rzeszowskiej, Rzeszów 2015)
15. B. Szczęśniak, Business and management, 5, 565-572 (2017)
16. J. Brodny, K. Stecuła, M. Tutak, Science and technologies in geology, **65** (2016)
17. M. Ziółkowska, *Doskonalenie procesu produkcyjnego przy wykorzystaniu metody SMED w wybranym przedsiębiorstwie produkcyjnym* (master thesis on supervising B. Skotnicka-Zasadzień, Zabrze, 2018)
18. J. Baran, *International Business Information Management Association, Education Excellence and Innovation Management through Vision 2020: From Regional Development Sustainability to Global Economic Growth* (Vienna, Austria 2017)
19. J. Sitko, K. Michalski, *17th International Multidisciplinary Scientific GeoConference. SGEM 2017, Ecology, economics, education and legislation* (Albena, Bulgaria (2017)
20. D. Klimecka-Tatar, Production Engineering Archives, **17**, 40-44 (2018)
21. R. Ulewicz, *International Conference on the Path Forward for Wood Products: A Global Perspective Location* (Baton Rouge, 2016)
22. M. Ingaldi, S.T. Dziuba, *24th International Conference on Metallurgy and Materials*, (Brno, Czech Rep., 2015)

MATEC Web of Conferences **183**, 01003 (2018) https://doi.org/10.1051/matecconf/201818301003
QPI 2018

Probabilistic Assessment of Road Risks for Improving Logistics Processes

Abdelaziz Lakehal[1,*], and *Fouad* Tachi[2]
[1]Department of Mechanical Engineering, Mohamed Chérif Messaadia University, P.O. Box 1553, Souk-Ahras, 41000, Algeria. e-mail: lakehal21@yahoo.fr
[2]Sonatrach, Algerian Petroleum Institute, School of Skikda, Algeria

Abstract. The Intermodal transport represents a solution, which has proved its effectiveness, for the supply of the various logistic platforms. Road transport is also one of the means of transport used in the logistic function and is the most common. This type of transport is especially recommended for medium and short distance journeys. Transport is an important link in the logistical chain. Several constraints accompany this transport function such as: delays, flexibility, diversity of merchandise, and road risks. To identify this last problem of road risk and to minimize its influence, a Bayesian network has been developed in this paper. Through experts' surveys and research in the literature, the various risks were identified. The structure of the Bayesian network is defined on the basis of this census. The network settings vary from one situation to another. The exploitation of statistics and historical files of the transport company has allowed to define the parameters (probabilities) given in the example studied in this paper.
To prevent risks and anticipate failures in the logistics function, while optimizing a utility function, an influence diagram was used. This tool has provided the ability to control actions and make decisions safely. An example of merchandise transport between two port companies has shown promising results and better efficiency in the anticipation of actions.

1 Introduction

The number of logistic platforms continues to grow. Also, demand on merchandise continues to increase. Movement of merchandise within and outside a company is part of the logistic function. Ensuring logistics between multiple platforms is an activity that needs to be well managed and never left to chance. Managing logistics means organizing, executing and controlling the movement of merchandise. Several synthesis works, prepared by experts, have proposed techniques and means whose purpose is to ensure handling movements, transport, supply, deliveries, and management of the production flows, while respecting the deadlines and costs.

In this context, transport and logistics are part of the same activity. To understand and analyze the impact of failures in merchandise transport business on the logistic function,

* Corresponding author: lakehal21@yahoo.fr

 https://doi.org/10.1051/matecconf/201818301003

several studies have been carried out in recent years [1, 2]. Road transport ensures merchandise serving at long distance. Maximum permissible rolling load (from 40 to 60 t according to countries) is regulated [3]. The road mode is today the dominant mode of transport with nearly 80% of market share. Containers are transported by trucks carrying either one or two containers. The transport of bulk merchandise has its specific risks.

Mathematical modeling, operational research and simulation are widely used tools to meet needs for risk control and optimization of the logistic function. Caution should be taken when projecting the results of these studies on real cases. Even if the results are close to reality, real experiences always hide unforeseen events. Road network architecture and drivers behavior have a big influence on the number of accidents that already occurred [4]. Also the lack of visibility favors the occurrence of road accidents, real-time prediction systems have been developed in recent years [5].

Current research works focus on developing means to eliminate these risks. The way managers prefer to manage these risks is another issue, do they prefer to manage them independently, or do they prefer to manage these risks with their industrial partners in a collaborative approach? Here, the permanent concern is to anticipate the impact of taken actions, and before that, what decision must be taken to optimize the logistic function?

To answer these questions, our study will seek to understand, analyze, and control the risks associated with the logistic function. To do this, and as the answer to all these questions is formulated as a decision problem, we will inventory, prioritize, and quantify these risks through a Bayesian network, and treat the decision issue with an influence diagram, while optimizing a utility function measured by the satisfaction.

Through this article, we will see that road safety is a complex subject since the factors defining the risk are: the human factor, the information, the structure and the reliability of the vehicles. This article does not pretend to solve all the difficulties related to determining the probabilities of road risk causes, but provides useful information so that logistics practitioners can identify their risks and avoid methodological errors, if they are asked to conduct an analysis on the subject. From a probabilistic point of view, the Bayesian networks, developed later in this paper, have the effect of reducing the probability of the variable affecting merchandise delivery with conformity and on schedule. Indeed, the most penalizing undesirable phenomenon can occur only if one or several causes take place.

2 Research Tools for Probabilistic Reasoning

Bayesian networks and influence diagrams [6] are probabilistic graphical models; they are also called belief networks and causal networks. These models were initially used in medical field and subsequently in industry, where they gave encouraging results especially for diagnosis, classification, failure prediction, and risk analysis [7-9]. These models are characterized by their ability to process uncertain information and to represent the interdependencies between different variables of a given problem. Their main advantage lies in their simple and understandable graphic representation, even by a non specialist. Moreover, it is known that probabilistic risk analysis has many advantages because it allows evaluating the probability of failure of a complex system, so that its weak points can be identified. Another advantage of probabilistic graphical models concerns the representation of complex and dynamic systems. Object-oriented Bayesian networks and dynamic Bayesian networks are extensions of Bayesian networks that deal with the dynamic behavior of large systems. In this article, Netica software [10] is used to build an analysis and road risk assessment model to anticipate failures in the logistics function.

2.1 Bayesian belief networks tool

Like all artificial intelligence techniques, Bayesian networks require inputs. On the basis of these input data and by inference, it is possible to calculate the outputs. In the form of probabilities, these output data are represented by probability distributions. An a posteriori probability is the conditional probability on the collected data, by combination of a priori probability and likelihood via the Bayes theorem [11]. The relationship between variables is defined by a conditional probability table.

A conditional probability table is always associated with the child node. This table quantifies the effect of parent nodes on the child node. It describes the probabilities associated with the child nodes according to the different values of the parent nodes. Root nodes are defined by a priori probabilities and do not have a conditional probability table.

Child to parent dependencies, are not allowed in Bayesian networks. Therefore, the variables and arcs constituting the network will form an oriented acyclic graph.
Formally, a BN [12] is defined by:

- an acyclic oriented graph G, G = (V, E), where V is the set of nodes of G, and E is the set of arcs of G,
- a probabilistic space (Ω, Z, P), with a non-empty finite set, Z a set of subspaces of Ω, and P a probability measure on Z with P (Ω) = 1,
- a set of random variables associated with the nodes of the graph G and defined on (Ω, Z, P), such that:

$$P(V_1, V_2, \ldots, V_n) = \prod_{i=1}^{n} P(V_i / C(V_i)) \quad (1)$$

Where C (Vi) is the set of parents (or causes) of Vi in the graph G.

2.2 Influence diagrams tool

The solution to a decision problem is to determine an optimal strategy, which optimizes the expected utility for the decision maker, and to calculate the optimal expected utility to adhere to this strategy. An influence diagram is a probabilistic network for reasoning about decision making under uncertainty. It is a graphical representation of a decision problem involving, in an interleaved order, decisions and observations. It also represents a quantification of the strengths and / or weaknesses of the decision-maker's dependency relationships and preferences. An influence diagram can be considered as a Bayesian network to which are added decision variables and utility functions indicating the decision maker's preferences (Fig. 1).

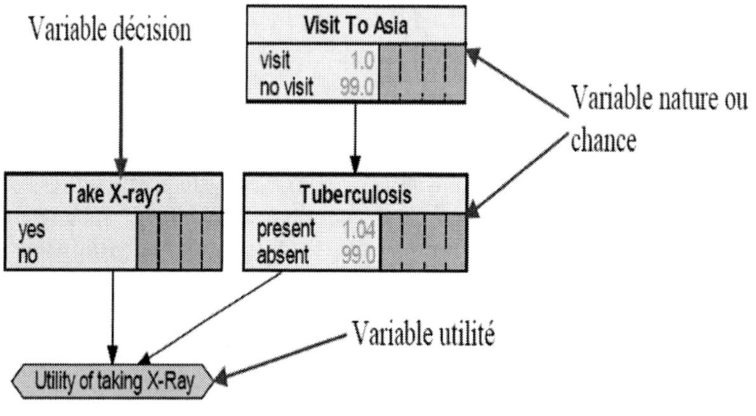

Fig. 1. Nodes types generated with Netica software.

Three types of variables are to be considered for the influence diagrams: chance nodes, decision nodes and utility nodes. Preference is very important in decision making. It can be based on an objective quantity or on a quantity that has no numerical measure, customer satisfaction for example. The probabilistic inference in influence diagrams involves propagating the evidence observed in the graph, by calculating the a posteriori probabilities of the hidden variables, according to the values of the observed variables, and by obtaining the value of the utility for each possible decision.

3 Road Transport Risks Assessment

Logistic performance is a multi-criteria variable, the most traditional being: cost, quality, reliability, deadlines, and information. The new features of the logistic information system should not overshadow the traditional functionalities that provide support for activity forecasting, logistic devices design, and operations planning and conduct. Transport is one of the logistic activities that requires an information system as modern as possible. The transport profession includes land transport (roads and rails), waterway-transport (maritime and rivers), and air transport.

Road transport of merchandise is one of the main families of transport professions. The volumes of road transport of merchandise have increased significantly in recent years. This evolution is accompanied by an increase change in risks. What is needed to control the risks associated with this mode of transport is to define a risk assessment indicator, with an evaluation mechanism to measure the overall safety of a road transport mission. Good indicators can provide better information on the level of risk. Use a BN to analyze and diagnose the various risks associated with road transport in order to identify and quantify the most important indicators, and to determine the relationships between these indicators (causes-consequences). Data obtained from survey and experts evaluation represent the a priori probabilities for the BN nodes.

3.1 Problem formulation

3.1.1 Establishment of the Bayesian network structure

The action plan to prevent road risk in the company is based on the highlighted risks. A part of the actions to be implemented concerns the management and organization of human and material resources. Preventing road risks consists to act on [13]:

- Displacements and infrastructure
- Vehicles and maintenance
- Communications during displacements
- Skills and training

Several risks are associated with road transport. If personal injury risks are the most significant, others, such as those related to driving activity, are to be considered. The variables listed above are the input variables in the Bayesian model. By projection of intuitive reasoning and using graph theory, it is possible to model the causality between risks and the variable „merchandise supply" taken as a variable representing the logistic function. The Bayesian network, given in Fig. 2, facilitates the dialogue between modeler (expert) and user (Manager / Business Leader).

3.1.2 Example modeling

In this paper, we deal with the case of a supply of goods between two harbor companies in Algeria, the harbor company of Algiers, EPA (capital of Algeria), and the harbor company of Jijel, EPJ (located in eastern Algeria). Through an evaluation of the risks related to the transport of goods, we evaluate the influence of these risks on the supply activity.

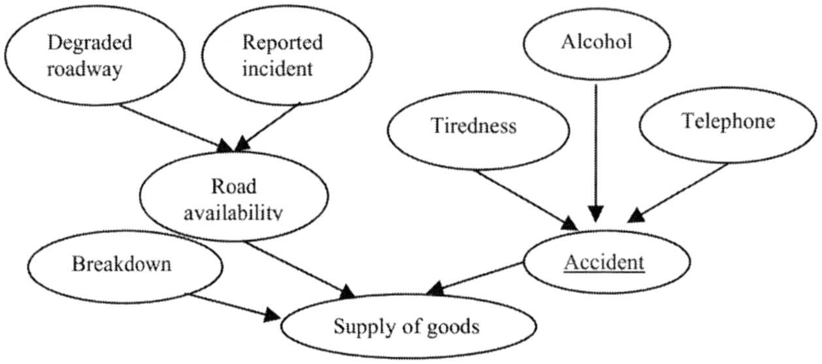

Fig. 2. Causal network modeling the influence of different risks on supply.

A. Definition of the Bayesian network parameters

Network parameters are given by the a priori probabilities of each input variable or node (see Table1.). The data processed in this case study are binary true or false.

Goods are transported by semi-trailer traveling east-west between EPA and EPJ over a distance of 241.91km, of which 34.11km represents a degraded segment.

Table 1. Input probabilities for the Bayesian model

Variable	A priori Probability
Telephone	Yes at 95% - No at 5%
Alcohol	True at 2% - False at 98%
Tiredness	True at 15% - False at 85%
Reported incident	yes at 50% - No at 50%
Degraded roadway	Good at 85.9% - Bad at 14.1%
Breakdown	True at 22% - False at 78%

3.2 A posteriori probabilities calculation

Graph and probability theories are at the origin of Bayesian networks. In this context, the construction of the Bayesian network, in Fig. 3, is essentially based on the definition of its graphical structure and probabilistic space. Mainly the a priori probabilities are the inputs of the model. The calculation phase gives rise to the definition of the a posteriori probabilities. This phase is called inference in the Bayesian approach; it consists of calculating conditional probabilities of events connected to each other by causal relationships. The inference, in a network of causalities, consists of propagating one or more information within this network, to deduce how the beliefs concerning the other nodes are modified. Fig. 3 shows the Bayesian network that models the influence of road risks on the merchandise supply. This network is based on the Netica software. In case there is no prior information on the variable, the value of 0.5 will be given, by the software, to its probability.

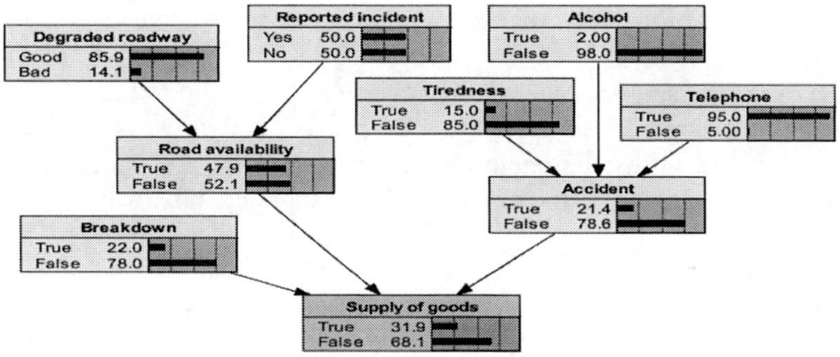

Fig. 3. BN modeling the influence of different risks on supply.

4 Inference Results Analysis

The independence between the variables of the network (Fig. 3) is very clear. Therefore, the inference becomes simple and easy. Also the a posteriori information, on the supply, will be available after inference in this network. A probability that the supply will be insured by 21.4% is not acceptable by the majority of companies active in logistics field. But if we return back to the causes of this high probability and by analysis of the network of Fig. 3, we notice that the availability of the road and the occurrence of an accident are the two variables that most influence the supply variable. Now by reasoning and reverse reading in the network of Fig. 3, it is possible to deduce that telephone communications are the main cause of road accidents. Strong beliefs in one cause at the expense of another change the results of the inference. For the case study presented in this article, the a priori probabilities of the declared incident variable are all equal, since there is no a priori information on the events that occurred on the road, but the strong belief in roadway degradation has not strongly influenced the probability of road availability.

Once the a posteriori probabilities are calculated, the corrective actions can be defined. In a certain reasoning environment and based on the results found, the variable "telephone" (95% true) must be treated as a priority. Awareness campaigns can minimize the impact of this variable. Another action concerns the variable availability of the road (50% true). Itinerary planning is an example of possible solution. Now once the actions are defined, it remains to anticipate and predict the impact of these actions on the supply function. The main question that arises is why giving preference to this or that action? In order to anticipate decisions and move from an uncertain decision-making environment to a more secure one, influence diagrams are used.

5 Solution Modeling With Influence Diagram

The variable that characterizes the logistics, in the Bayesian model, is the merchandise supply. Any decision made must be evaluated in terms of utility. In the logic of influence diagrams, the utility function models the impact of consequences. For our example, this utility function represents the cost of a supply delay, an accident risk, and a break-down risk. For decision-making purpose, utility will be calibrated with respect of the preferences of the manager, or the planner. The circumstances and consequences of the risks vary. They show that risk has two aspects: probability and consequences. For consequences, these are characterized by the supply: availability of merchandise at the customer level. For a general case, it can be the protection of persons, the environment, but also protection of the transport tool or the merchandise itself.

There are two ways to reduce risks:

- decrease of the probability of occurrence of the undesirable event (non-supply of merchandise),
- attenuation of the consequences of the adverse event,

The influence diagram is an extension of Bayesian networks. An influence diagram, which calculates the utility function measured by a discrete variable, which is the satisfaction of the customer or the supplier of the supplier, plus a decision variable are illustrated by Fig. 4.

Two things are considered here: the decision node "course alteration" to another road like the national road, and the utility node that measures satisfaction. This network also allows the re-evaluation of the supply function with uncertainty measurement. On the influence diagram in Fig. 4, our decision-making is certain, it is 45%, and the probability that the supply of the goods is insured increases to 50%.

There are links from Course alteration and supply of goods to Satisfaction, the ideal is to ensure supply without changing the road (utility = 100), Then when the supply will be sure even with change of road (utility = 90). Finally it is more penalizing if the goods do not arrive in time (utility = 0).

The decision-making studies intend essentially to assess the probability of the occurrence of various risks by taking into account, from the definition of logistic processes and strategies, all the initiating factors:

- technical factors: equipment and transported products (including ergonomics, dangerousness, manufacturing, and quality assurance issues),
- human factors: quality of training especially in safety, ergonomics, qualification,
- environmental factors: natural risks, ambient environments (dust, gas, noise, ...).

Anticipating the impact of decisions has become possible with influence diagrams. In general, if the Bayesian network shown in Fig. 3 models the influence of road risks on merchandise supply, the influence diagram illustrated in Fig. 4 models the decision risk upon the utility function.

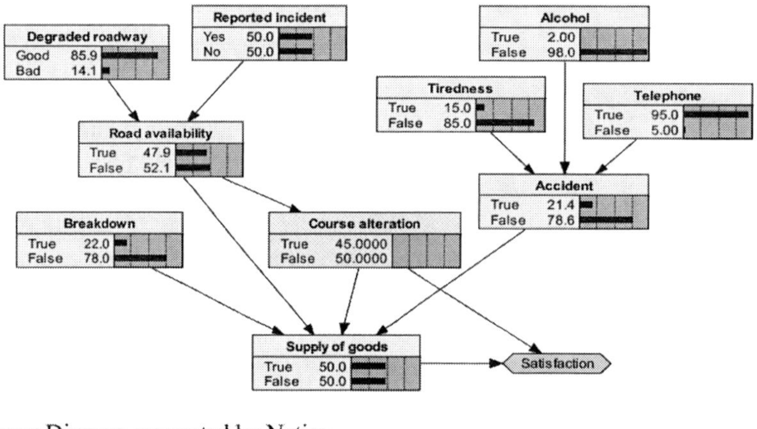

Fig. 4. Influence Diagram generated by Netica.

6 Conclusion

In this study, an inventory of the main risks related to the logistic function illustrates the very wide variety of variables for the Bayesian model. This makes the determination of the network structure a step that deserves special attention. A confrontation of experts opinions can figure out this problem, but forgetting a variable will have a serious impact on the results. Logistic activities differ for the majority of them. The simple model, developed in the example of this study, shows the impact of the road unavailability on the supply of the merchandise.

Through this paper, it is possible to conclude that through surveys, reliable statistics, and experience feedback, it will be possible for the decision-makers and managers, in charge of the logistics function, to make certain decisions. Also, it is possible to select the causes of the most unfavorable risks by characterizing parameters, such as probability of occurrence and severity. The advantage of the model presented in this article is its simple structure, which could be understood by even a non specialist. Another parameter is the intuitive progress (path, evolution, movement) of information (cause - effect). Furthermore, it is extremely important to mention that some (a part) of the studied risks are part of a process leading initially to evaluate the probability of ensuring the supply of merchandise. The prediction tools proposed in this article do not allow, except in the case of redesign of strategy and logistic processes, modifying the incidence of a risk. The presented Bayesian network and influence diagram allow only reducing the severity of the effects induced by a risk on the logistic function. All the risks presented in this article allow carrying out a complete study aiming to improve the logistic processes and to reduce the criticality of risks consequences to an acceptable level. Other parameters, mainly concerning the nature of the merchandise and the behaviour of the driver, can be taken into consideration in the Bayesian model.

References

1. H.O. Nguyen, J. Tongzon, Transport Policy, **17**, 135-146 (2010)
2. W. Hachicha, M. Elmsalmi, Journal of Risk Research, **17**, 1301-1324 (2014)
3. A. Toubol, *Transport intermodal. Technique de l'ingénieur* (AG 8 160v2, 2007).
4. Y. Wu, H. Nakamura, M. Asano, J. Mod. Transport **21**, 224-235 (2013)
5. M. Hassan, H. Abdel-Ata, M. Ahmed, A.S. Al-Ghamdi, Transportation Research Part C, **24**, 288-298 (2012)
6. B.Uffe, A. Kjærulff, L. Madsen, *Bayesian Networks and Influence Diagrams, A Guide to Construction and Analysis* (Springer, LLC, 2008)
7. T. Bedford, R. M. Cooke, *Probabilistic Risk Analysis: Foundations and Methods* (Cambridge University Press, New York, 2001)
8. M.E. Paté-Cornell, Reliab. Eng. Syst. Safe **54**, 95-111 (1996)
9. V.M. Bier, L.A. Cox, *Advances in Decision Analysis-From Foundations to Applications, Probabilistic risk analysis for engineered systems* (Cambridge, 2007).
10. Norsys – Netica Application – Norsys Software Corp. http://www.norsys.com/netica .html
11. P. Damien, *Bayesian theory and applications* (Oxford University Press, 2015)
12. P. Naim, P.H. Wuillemin, P. Leray, O. Pourret, A. Becker, *Réseaux bayésiens. 2 ed.* (Eyrolles, France, 2004).
13. A.S. Valladeau, B. Andéol-Aussage, *Transport routier de marchandises. Guide pour l'évaluation des risques professionnels* (INRS, ED 6095, 2011)

Analysis of problems during implementation of Lean Manufacturing elements

Manuela Ingaldi[1], *Szymon T.* Dziuba[2,*], and *Anna* Cierniak-Emerych[2]

[1]Czestochowa University of Technology, Faculty of Management, Poland
[2]Wroclaw University of Economics, Poland

Abstract. Changes are inherent in activity of any enterprise. Several methods or concepts can be employed to introduce changes. One of the concept which has become popular recently and improves enterprise's functioning is Lean Manufacturing. There are multiple techniques and tools which function within this method. Unfortunately, in the case of any changes, also in the case of implementation of Lean Manufacturing components, one can meet the resistance of employers and a number of technical and organizational problems. The aim of the paper is to collect and analyse information concerning the problems that occur during implementation of the Lean Manufacturing in Polish enterprises. The data were collected by means of a direct questionnaire survey. The results can be a suggestion for the enterprises which are planning to implement the Lean Manufacturing and indicate the problems they should be prepared for in the process.

1 Introduction

With changing environment and customer requirements, enterprises have to introduce changes also in their structure. The changes can concern many aspects of enterprise activity, concerning both production and non-production aspects.

One of the concepts which is often used in such situations is the Lean Manufacturing (LM). It is not only a slow but continuous process of rationalization of the entire organization but also its relations with the environment. This means „introduction of substantial changes in terms of activity, property structure and organizational and management methods" [1].

The LM concept allows enterprises to determine value, organize activities that add value with their optimal order, performing the activities without disturbances when the need arises and they are used more and more effectively. Womack and Jones found that LM methodologies are lean because they enable a business to do more with less (less human effort, less equipment, less time, and less space) while always coming closer to meeting customers' exact needs [2-3].

The LM is based on the use of various management methods and techniques, with the most popular including: 5S, Just in Time, Kanban, Kaizen, SMED (Single Minute

* Corresponding author: szymon.dziuba@ue.wroc.pl

Exchange of Die), TPM (Total Productive Maintenance), Just in Time, One Piece Flow, work standardization, TQM (Total Quality Management), Poka-Yoke, Hoshin Kanri, VSM (Value Stream Mapping), visual management. This list is not ended [4-5].

Unfortunately, as it is the case with any change, techniques and tools used in the LM are not easy to be implemented. Introduction of nearly every change in enterprises is challenged with employee resistance. Employees can react to changes in two manners: through attitudes, which include emotional internal human inclination to changes and behaviours that manifest themselves in specific activities [6-9].

However, apart from the resistance of the employees, who are often afraid of new duties, increased responsibility for the assignments, introduction of changes also faces several technical and organizational problems connected with the lack of various resources.

Much attention in the literature concerning implementation of the Lean Manufacturing and its elements has been devoted to the outcomes of changes and methods to implement and plan changes. However, the related problems are often neglected [10-16].

The aim of this paper is to indicate problems during and after implementation of the elements of the Lean Manufacturing in Polish enterprises. The examinations were of character of direct questionnaire, with the results based on the answers of the employees from Wrocław and its surroundings in which the LM was implemented. The paper presents only selected results of the analysis.

2 Metodology

The examinations concerning the analysis of the problems of implementation of the Lean Manufactuing in Polish enterprises were based on an oral (direct) questionnaire survey. The employees in the enterprises where the mean method was implemented answered to the questions according to the previously prepared form. The interviews with the employees were made in the enterprises they work with but in the absence of direct supervisors. The answers were usually answered by several employees in order to prevent from a direct reactions of the managers to the results. The questionnaire data were collected from both supervisors (one from each enterprise) and low-rank employees.

The results of the research presented in the article are preliminary research, carried out in the period January-March 2018. The research will be continued until December 2018. Firstly the authors sent questions regarding the implementation of Lean Manufacturing and the willingness to participate in the research. The questions were sent to most of the small and medium-sized enterprises from Wrocław and its surroundings.

The survey was performed in 29 small and medium-sized enterprises, with the answers provided by 248 people in total. These were enterprises from the food, electronic and electrical, metal, clothes and chemistry sectors, mostly with Polish capital. These are the results of surveys conducted in enterprises that have implemented selected elements of the Lean Manufacturing and that have agreed to conduct the research as the first, what's more the research was carried out by the end of March 2018. Respondents, who participated in the research, also took part in the implementation of the LM and its elements in the research enterprises. Further research in other enterprises that have agreed to participate in the research should allow to identify the problems of the LM implementation in particular industries.

Implementation of the LM or its elements had started at least 7 years before in 11 enterprises, 2 to 6 years before in 16 enterprises and less than 2 years before in 2 enterprises.

3 Results

The results concerning the major causes of implementation of the LM in the enterprises were first presented (Fig. 1). The employees were allowed to choose more than one answer. None of the respondents checked the „Other" answer.

According to the employees, the most frequent cause of implementation of the LM was to reduce waste in the enterprise (72%) and improve production efficiency (69%). The third indication was popularity of the LM (54%). These are the major assumptions of the LM.

The requirements of the customers did not exceed 40%, whereas market requirements reached 27%. This means that the decision on the implementation of the LM was made internally in the enterprise and was not forced by the environment.

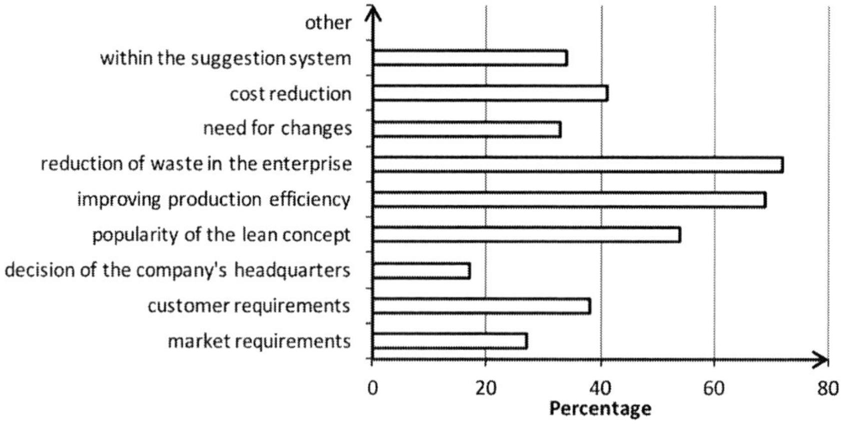

Fig. 1. The major causes of implementation of the Lean Manufacturing or Lean elements in an enterprise [own study].

As with any change, various reactions were observed during the implementation of lean Manufacturing (Fig. 2). In most cases (31%), the most frequent reaction of employees was resistance to the new. Further reasons was feeling of threat (25%) and criticism (17%). This is the most standard approach of employees to implementation of changes. They are accustomed to what happens in the enterprise and know what they are responsible for and what their assignments are.

Fig. 2. Main reactions of employees to implementation of the Lean Manufacturing [own study].

Changes often involve new duties, the need for modification of previous behaviours and habits and learning. Therefore, it is important that implementation of any Lean elements in the enterprise should be preceded by talking to employees and training which will make them more open to the new and what can happen.

During implementation of the Lean Manufacturing, employees in the enterprises encountered several problems (Tab. 1). Also in this case, the employees could indicate several answers.

The most frequent cause of problems during implementation of the LM was improper communication in the enterprise (65%). This is the serious problem since adequate communication between the employees determines the success of the enterprise. Unfortunately, problems with communication are a typical Polish problem in enterprises. Despite the specific organizational structure, the employees do not know who they should turn to with problems and who they should listen to. Information flow is often difficult.

Another important cause of the problems is excessive pressure of the owners (61%) with the lack of support of the top managers (57%). Owners often require impossible things from their employees and give instructions, expect results while they are not willing to help. However, the success of the Lean Manufacturing requires involvement of all employees and the managers should be the role models for other employees.

A substantial problem is reluctance of employees to give up previous privileges (47%). This is connected with previous components i.e. communication in the enterprise and attitudes of managers.

43% of people indicated insufficient amount of training on the Lean Manufacturing. Individual techniques and tools of the LM are not difficult to be used. However, it should be remembered that all employees have higher education level and do not have the most basic knowledge of production engineering. Therefore, employees should be instructed how each instrument works, provided with practical classes and should see the results obtained by other enterprises. This can suggest them how to use these instruments and reduce the resistance or reluctance of employees to changes. Training should be the second stage of the implementation of the LM, after presentation of the causes and expectations of managers concerning implementation of the concept and showing the support for the employees.

Table 1. The main causes of problems during implementation of Lean Manufacturing [own study].

The main causes of problems during implementation of Lean Manufacturing	**Percentage**
high pressure of the owners	61
hurry	53
no general indications for implementation of Lean Manufacturing	19
no support of top managers	57
no financial funds	29
results only from the standpoint of reduction of organizational costs	37
inefficient communication in the enterprise	65
thinking that resources are insufficient	24
Individual large projects are implemented, without the long-term plan to implement Lean Manufacturing	39
insufficient amount of training on Lean Manufacturing	43
employee's reluctance to give up previous privileges	47
other	1

The respondents were also asked about major causes of insufficient effects of implementation of the LM elements in the enterprise studied (Fig. 3). Respondents were allowed to indicate several answers.

It turned out that this was related mainly to the lack of employee's commitment (73%), lack of self-discipline (69%) and return to old habits (67%). This is connected with answers to the previous question i.e. the lack of adequate communication in the enterprise, the requirement and the lack of support of the managers and insufficient amount of training concerning the Lean Manufacturing. Showing the need for changes and commitment of managers would impact on reduction of the percentage of these causes.

The problem can be connected with another cause i.e. huge rotation of employees (62%). This is the problem typical of many Polish enterprises, not only those which implement the Lean Manufacturing. New employees should be trained on e.g. LM and learn about challenges the enterprises have to face and how the Lean enterprises function. This involves additional time and expenses. New employees need some time to become part of team in the enterprise and even more in order to identify with changes that occur in the enterprise.

This is related with poorly developed LM culture, which was directly indicated only by 48% of the respondents. Building such a culture in the enterprise and, consequently, feeling of association with the enterprise and opportunities for suggesting improvement is the basis for a success of the Lean Manufacturing.

Procedures created within the Lean Manufacturing are too complex and difficult to employees (61% of the indications). This translates into the lack of commitment and return to old habits. If one does not know how to do something, they do as they can and as they did before. The procedures should be created with the employees the procedure concern. They can help create procedures and give clues about the language that should be used in order for the procedure to be understood by others.

17% of the respondents indicated the answer „Other". However, no additional information was given despite such an opportunity.

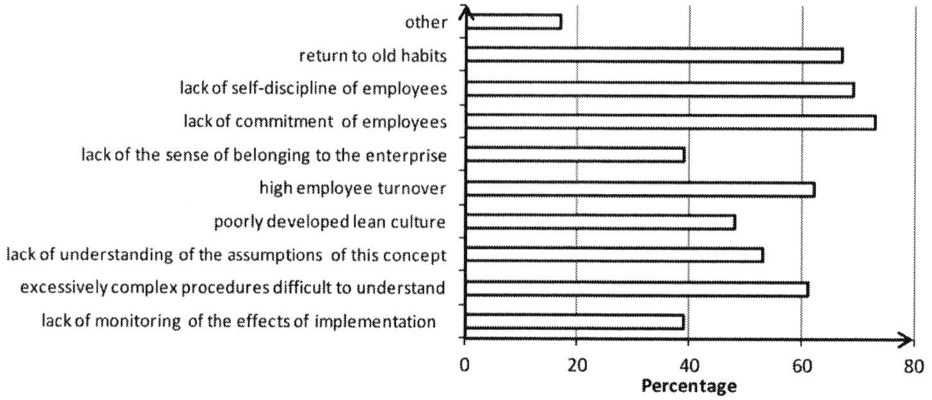

Fig. 3. Main causes of insufficient effects of implementation of Lean Manufacturing [own study].

4 Summary and conclusion

The paper presents the results of a direct questionnaire survey which was conducted in Polish enterprises and which concerned the problems encountered during implementation of the Lean Manufacturing and its instruments. The employees indicated several problems with implementation, with the most important being insufficient communication in the enterprise, excessive pressure of the owners and lack of support of top managers. The most important causes of insufficient effects of the implementation of the Lean Manufacturing is the lack of commitment of employees, lack of self-discipline and tendencies for returning to old habits.

Everybody is afraid of changes, which is a natural process. It is important that to familiarize with these changes, their course and causes, which will help accept the situation and understand the need for implementation.

Each change can be connected with several problems which are encountered at various stages. The problems should be analysed and their causes should be identified in order to cope with them. It is critical not to give up and to implement changes as they are an important stage of each enterprise.

Reference

1. J. Lipecki, Ekonomika i Organanizacja Przedssiębiorstw **8**, 12-15 (1998)
2. J.P. Womack, D.T. Jones, *Odchudzanie firm. Eliminacja marnotrawstwa – kluczem do sukcesu* (Centrum Informacji Menedżera, Warszawa, 2001)
3. J.P. Womack, D.T. Jones, D. Roos, *Machine that changed the world* (Simon and Schuster, New York, 1990)
4. J. Czerska, *Doskonalenie strumienia wartości* (Difin, Warszawa, 2009)
5. M. Podobiński, Nauki o Zarządzaniu - Management Science, **3**(24), 112-122 (2015)
6. M. Nowicka Skowron, R. Ulewicz, *25th International Conference on Metallurgy and Materials* (Brno, Czech Rep., 1962-1967, 2016)
7. D. Klimecka-Tatar, Production Engineering Archives, **17**, 40-44 (2017)
8. R. Ulewicz, D. Jelonek, M. Mazur, Management And Production Engineering Review, **7**, 89-94 (2016)
9. A. Pacana, R. Ulewicz, Polish Journal of Management Studies, **16**, 165-174 (2017)
10. S. Gobinath, D. Elangovan, S. Dharmalingam, International Journal of ChemTech Research **8**, 44-51 (2015)
11. A.P. Puvanasvaran, M.H.M.A. Megat, S.H. Tang, M.R. Muhamad, A.M.S. Hamouda, American Journal of Applied Sciences, **5**, 504-511, (2008)
12. P. Jedynak, Jagiellonian Journal of Management, **1**, 51-64 (2015)
13. G.A. Marodin, T.A. Aurin, Journal of Manufacturing Technology Management, **26**, 57-79 (2015)
14. R. Ulewicz, J. Selejdak, S. Borkowski, M. Jagusiak-Kocik, *22nd International Conference on Metallurgy and Materials* (Brno, Czech Rep., 1926-1931, 2013)
15. M. Nowicka Skowron, R. Ulewicz, *24th International Conference on Metallurgy and Materials* (Brno, Czech Rep., 1707-1712, 2015).
16. K. Knop, M. Ingaldi, M. Smilek-Starczynowska, *Advances in Manufacturing. (red.) Hamrol A., Ciszak O., Legutko S., Jurczyk M., Reduction of Errors of the Conformity Assessment During the Visual Inspection of Electrical Devices* (Springer International Publishing, Cham, 2018)

Safety in ensuring the quality of production – the role and tasks of standards requirements

Adam Górny*

Poznan University of Technology, Faculty of Management Engineering, Poznan, Poland

Abstract. During the performance of manufacturing tasks, safety – which is a primary responsibility of employers - has a major impact on production quality. To achieve such safety, employers must ensure working conditions that reflect worker needs and meet technical and organizational requirements. Safety requirements are designed to foster threat- and strain-free work in the workplace. By striving to comply with such requirements, organizations improve the conditions in which their production is performed. In their effort to incorporate essential principles into their systems, employers can benefit from applying the ISO 45001 standard, which defines the systemic requirements that must be embraced in their business in the same manner as the requirements adopted in any other field of activity, such as production quality. The article refers to standards derived from ISO 45001 that facilitate manufacturing improvements. It outlines the impact of systemic enhancements on workers' ability to perform work safely.

1 Introduction

Measures aimed at ensuring safe working conditions are commonly adopted to make certain that production proceeds in conditions that are compatible with the needs of persons who remain on the production floor [1-3]. Once working conditions meet the expectations of such persons, direct and indirect benefits are achieved, among them greater efficiency, lower absenteeism and reduced strenuousness. On the other hand, unsafe working conditions disturb production processes [1, 4].

Risk factors, unsafe working conditions and work-related injuries reduce the efficiency of operations and impair the functioning of processes. Occupational accidents, diseases and incidents negatively affect performance, as revealed by relevant indicators [5, 6]. Such events impact on workers' ability to perform production tasks efficiently and effectively and the company's capacity to make improvements defined as the ability to meet the expectations of the concerned parties [7].

Organizations are responsible for identifying and either eliminating or mitigating threat impacts. To that end, they should select areas for monitoring so as to ensure that, among others, their effect on manufacturing is continuous and effective [1]. To make certain that the selected measures are indeed effective, it is essential to embrace occupational safety

* Corresponding author: adam.gorny@put.poznan.pl

standards. One such standard is ISO 45001, which defines principles for systemic occupational health and safety management. The article outlines some of the key effects and applications of the above standard for shaping the manufacturing environment.

2 Standards requirements in the systemic management of production safety

The aim of managing an organization's activities is to achieve specific benefits. This principle can also be applied to production management.

A management system should include areas of importance for securing the effective implementation of specific activities. These include [8, 9-11]: an organization structure that facilitates task implementation, people taking ownership of the decisions they make and the tasks they perform, and proper planning of actions, goals and resources. It is therefore critical in the technical, organizational and human aspects to define rules and procedures for the normal course of work as well as emergencies, keep a record of any measures and decisions taken, ensure efficient and effective information flows along vertical and horizontal lines in the organization's management structure, and ensure proper purchasing and subcontracting [12]. Such activities are identified in the most common management standards [13, 14]. Equally important is the extension of the requirements to human and cultural aspects of the organization's functioning as well as its relations with any concerned parties [7, 15].

An organization whose management is process-centered should place a great deal of emphasis on any processes that enable it to meet customer needs and expectations. Of essential importance in this approach are links across processes. The organizations that take proper account of such links will benefit from optimal interactions between processes, fulfilling needs of the concerned parties and gaining the ability to bring their management to the desired level of maturity [8]. To improve the environment in which production is performed, organizations should ensure they are able to either eliminate or reduce the impacts of the causes of irregularities on workers. To that end, they should focus on the threats and nuisances that impair their ability to carry out production tasks [16, 17].

For a number of years now, safety requirements have featured prominently in the overall standards landscape. Safety standards are defined as guidelines for managing the work environment and ensuring that production proceeds efficiently and effectively [8]. It is equally common to incorporate the principles of systemic safety management into TQM requirements, often in the form of Total Safety Management [3, 18-20]. TQM allows organizations to optimize the way problems are solved with proper account taken of improvements in efficiency and stakeholder satisfaction, which in turn create an environment that fosters better work performance.

To create a friendly working environment, organizations commonly employ standards governing the management of working condition and work safety, risk assessment and management, and threat prevention and mitigation. They accomplish these aims by implementing corrective and preventive measures. At the heart of work safety management lies risk assessment. Such assessment should apply to both worker performance and their ability to carry out tasks.

To help meet the above goals, a new occupational health and safety management standard named ISO 45001 has been developed. Under its premises, improvements relying on the application of ISO 45001 are to help standardize and improve systemic activities, primarily in organizations that have previously implemented such activities in other areas of management [3, 7]. Improvement measures can be performed in the same manner as those carried out in other areas of an organization's operations.

3 Occupational safety management standards based on ISO 45001 requirements

3.1. Objectives and applications of ISO 45001 standard

ISO 45001 should be seen as support in integrating an operational control system [17]. The operational parameters of processes, including those of the environment in which the processes are performed, require adequate monitoring [1, 13]. Such monitoring is crucial for averting the risks that arise in the course of process implementation affecting the achievement of intended and desired effects.

The structure of the standard mimics that of the management guidelines adopted in ISO 14001 (EMS) and ISO 9001 (QMS). Its approach to systemic management relies on the continual PDCA improvement loop, in which a special role is ascribed to leadership and the management's responsibility for the measures taken [21]. The standard is also intended to enable the organization to deliver on its commitments [22]. Compliance with systemic rules requires a broad view on any issues. Risk assessment and operational control over activities alone are therefore considered to be insufficient [12].

An organization that establishes an ISO 45001 compliant management system should be able to:

- Adopt effective work safety management principles tailored to the specific nature of the organization's operations, in keeping with the PDCA model, consistently with the improvement principles applied in other areas of the company's management,
- Define any processes that play a role in reducing work strain, taking into account their context and operating risks as well as any applicable legal requirements,
- Identify any threats and assess any risks that compromise the organization's ability to perform tasks,
- Raise worker awareness of hazards and improve the organization's ability to describe hazards based on risk assessment,
- Assess the effectiveness of the measures taken to improve occupational safety, prevent injuries and protect people's health,
- Develop an occupational safety and health protection culture across the organization,
- Promote health education, with training recognized as essential for continuous improvement and for the effectiveness of improvement measures,
- Provide employees with opportunities to actively contribute to the promotion of safety at work.

The outcomes of the measures taken should be viewed as worker health improvements achieved by way of systemic occupational health and safety management [3, 14]. To enjoy such benefits, organizations need to ensure that both their workers and management engage in promoting occupational safety [7, 19, 20].

3.2. ISO 45001 requirements

The ISO 45001 standard offers manufacturing safety guidelines that recognize the special role played by human operators. Once implemented, the standard can be said to help enhance processes and specifically improve:

- The effects of activities and how they interrelate with each other,
- Organizational structures and the performance of tasks within organizations,
- Human and cultural aspects of organizations' functioning,
- Infrastructure, the work environment and technologies,
- Relations and interdependencies with all concerned parties.

At its core, systemic management focuses on [7, 9, 16, 17, 21]:

• The context of an organization's functioning

Requirements applicable to operations should be associated with management objectives highlighting the organization's duty to ensure occupational safety and comply with the laws that the organization has committed to observe. To that aim, the environment in which the company is operating needs to be recognized in its management system. Hence, an analysis of the organization's environmental impacts will be required [4].

• Assessment of the effects of measures

To comply with requirements, organizations need to rely on indicators that describe their activities. The standard relies on a dual assessment of the results, focused on efficiency and system performance. In the realm of occupational safety, this is achieved by preventing worker injuries and any adverse health impacts.

• Scope of management system

The system must encompass an occupational safety policy. Its scope extends to any company activities, products and services that in any way affect occupational safety. Both existing and potential threats should be taken into account. Such factors need to be assessed and monitored. The verification should include any outsourced work and any external processes that affect the organization's functioning.

• Leadership

The management is obliged to strive to improve the effectiveness of the work safety management system and support any people responsible for such effectiveness. It should therefore fully integrate system requirements with ongoing production processes. Specifically, organizations should include management effectiveness issues in their strategic planning.

• Concerned parties

Organizations have a duty to name the parties that influence the functioning of their management system and define the rules that such parties should follow.

• Risk assessment and management

Risk assessment should be viewed as an instrument for verifying the conditions in which humans operate in the work environment. In addition to its traditional safety applications, risk assessment provides a way to gauge the impact of uncertainty on the effects of an organization's operations and the ability to achieve intended results that are consistent with the level of management maturity achieved by the organization.

4 Production safety requirements

ISO 45001 relates to managing occupational safety. The improvements sought through compliance with systemic requirements allow companies to take effective measures to meet the needs of all stakeholders, including those in their external environment. This contributes to making the measures more effective, mainly as a result of the skills acquired and the ability to anticipate changes, prepare the organization for their occurrence and ensure it responds appropriately. This allows organizations to maintain growth and improve their capacity to carry out production tasks.

Table 1 presents selected impacts of the normative requirements specified in ISO 45001 on the ability to safely implement production processes. Such safety is achieved chiefly by striving to improve the ability to perform work and augment the environment in which production tasks are performed.

Table 1. The potential for safe performance of work achieved through compliance with ISO 45001

Area of ISO 45001 requirements	Measures taken to ensure safe performance of work
Context of organization's functioning	- Measures that enable organizations to understand the need and see the potential for performing tasks that contribute to safety improvements, - A better understanding of the need to strive to enhance occupational safety management systems with a view to improving occupational safety, - Ensuring an effective identification of threats and opportunities for their prevention, - Clearly defining organizations' responsibility for identifying occupational safety objectives, - Extending supervision over the obligations and tasks that contribute to occupational safety improvements, in keeping with applicable national and international legislation.
Scope of OHS management system	- Incorporating all actions, products and services whose performance and provision affects worker safety, - Ongoing oversight over all efforts made with proper account taken of occupational safety issues, - Oversight and monitoring of any outsourced work, - Oversight and monitoring over internal processes that affect the organization's operation, - Assessment of identified impacts against acceptable reference values adopted by the organization, - Use of the findings of a review of the causes of errors and complaints as well as of customer expectations and satisfaction to improve occupational safety management system.
Leadership	- Ensuring full integration of business processes with systemic requirements, - Accounting - in the adopted approach - for the commitments and rules derived from the principles of systemic occupational safety management, - Demonstrating commitment to identify and take measures aimed at increasing safety levels, - Incorporating applicable laws and other obligations adopted by the organization into regular practice.
Concerned parties	- Identifying all parties whose actions may affect the functioning of the occupational safety management system, - Using third party commitments to improve system functioning and occupational safety, - Identifying (and establishing the priority of) the requirements that need to be observed by the concerned parties to help improve occupational safety, - Ongoing checks of the compliance of all concerned parties with applicable laws adopted and recognized as significant for ensuring the proper operation of the organization, - Prioritizing improvements on the basis of input data received from the concerned parties.

The measures specified in Table 1 allow companies to benefit in many areas of their operation. These include [3, 12, 14, 21]:

- Increased ability to achieve compliance with new mandatory legal requirements,
- Reduced costs of accidents and strain caused by failures to comply with requirements,
- Reduced costs of production downtime and disruptions,
- Lower insurance premiums as a result of decreased accident rates,

- Reduced absenteeism affecting the timeliness of task performance.

However, to mitigate strains on work safety, organizations commonly need to modify their management practices and principles [2].

5 Conclusions

The adoption of the international ISO 45001 standard may contribute to improvements in the way working conditions are managed with positive effects on the quality of production processes. Normative requirements are of great use in the implementation of improvement activities, as they are crucial for organizations' ability to perform production tasks. Success in achieving the intended effects depends vitally on ensuring the active engagement of companies' workers and representatives.

References

1. E. Bottani, L. Monica, G. Vignali, Safety Science, **47**(2), 155-162 (2009)
2. T. Karkoszka, Procedia Manufacturing, **13**, 1463-1469 (2017)
3. A. Górny, Management Systems in Production Engineering, **3**(14), 106-111 (2014)
4. R.L. Granerud, R.S. Rocha, Safety Science, **49**(7), 1030-1039 (2011)
5. B. Fernández-Muñiz, J.M. Montes-Peón, C.J. Vázquez-Ordás, Safety Science, **70**, 295-307 (2014)
6. A. Górny, MATEC Web of Conferences, **94**, 04011 (2017)
7. A. Górny, Modern Management Review, **XX**(22), 73-88 (2015)
8. ISO 9001:2015, *Quality management systems. Requirements, International Organization for Standardization* (International Organization for Standardization, Geneva, 2015)
9. ISO 45001:2018, *Occupational health and safety management systems. Requirements with guidance for use* (International Organization for Standardization, Geneva, 2018)
10. A. Górny, *Proceedings of International Conference on Innovative Technologies, The work environment in the structure of management system* (Faculty of Engineering University of Rijeka, Rijeka, 2013)
11. J. Tabor, Polish Journal of Management Studies, **7**, 188-200 (2013)
12. S. Nenonen, The Service Industries Journal, **33**(1), 99-114 (2013)
13. E. Lafuente, J. Abad, Safety Science, **103**, 12-22 (2018)
14. K. Frick, Safety Science, **49**(7), 974-987 (2011)
15. P. Babović, Acta Medica Medianae, **48**(4), 22-26 (2009)
16. ISO/CD 45001, *Occupational health and safety management systems. Requirements with guidance for use* (BSi, London, 2014)
17. ISO 45001, New International Standard for Occupational Health & Safety Management Systems. http://www.bsigroup.com/en-GB/ohsas-18001-occupational-health-and-safety/ISO-45001 (date of access: September, 15, 2017)
18. J. Álvarez-Santos, J.-Á. Miguel-Dávila, L. Herrera, M. Nieto, Safety Science, **101**, 135-143 (2018)
19. A. Górny, Communications in Computer and Information Science, **529**, 136-142 (2015)
20. A. Górny, *Occupational Safety and Hygiene, Work environment in quality assurance*, (Portuguese Society of Occupational Safety and Hygiene, Guimarães, 2016)
21. A. Górny, MATEC Web of Conferences, **94** (2017)
22. Fernández-Muñiz, J.M. Montes-Peón, C.J. Vázquez-Ordás, J. of Loss Prevention in the Process Industries, **50**(B), 403-415 (2017)

Healthcare products and food supplements in Poland – a comparison

Katarzyna Hys[1,*]

[1]Opole University of Technology, 45-758 Opole, Poland

Abstract. The analysis of the level of consumption of food supplements by Poles raises increasing controversy. Many communities pay attention to the alarming situation, including, among others, Central Apothecary Council, National Institute of Medicine, Office of Competition and Consumer Protection, National Sanitary Inspectorate or the Supreme Chamber of Control. For the purpose of this paper, the studies were conducted with the use of the desk research technique. There are many factors which affect the increase in food supplement sales in Poland. One of them is the lack of basic client knowledge concerning the differences between healthcare products and food supplements. Clients identify supplements with medicine, which is wrong. Due to the above, the purpose of the paper is to demonstrate the differences between healthcare products and food supplements. As result, a set of criteria will be established, which will be used to conduct a comparative analysis for a healthcare product and food supplement.

1 Introduction

In sales, both traditional and online, there are many pharmaceuticals with various applications and effects. This paper will focus on the market of food supplements and healthcare products.

In general, as part of the studied product group, the following medications are distinguished: prescription products, OTC medicine, food supplements and cosmetics. These medications constitute separate product groups. However, clients often view them as generic products with the same effect and impact on human heath, which is actually not the case. Unfortunately, clients demonstrate a lack of knowledge on food supplements and healthcare products [1], [2]. They do not see any differences between them and in the case of many product they cannot specify their correct application. Clients wrongly think that the effects of products from the specified groups have medicinal consequences, aimed at improving health or healing of a (self-)diagnosed medical condition [3-7].

Due to the above, this paper will include a comparative analysis, the purpose of which is to clearly determine the differences between food supplements and healthcare products. Cosmetics will be purposefully omitted in this comparison, because they do not constitute the subject of this paper.

* Corresponding author: k.hys@po.opole.pl

2 Methodology

As part of the defined goal, the paper includes studies conducted with the use of the literature analysis method. The studies concerned the nomenclature used in the scope of the studied product group and features used to describe them.

Then, criteria were established based on literature studies and an in-depth subject literature analysis using the desk research technique. In the latter part of the elaboration, the criteria were used to compare food supplements with healthcare products. In consequence, a catalogue of features was developed along with a detailed description of the studied medications in terms of the differences existing between them.

The studies were conducted in two stages. Each of the stages constituted an execution of a defined research purpose. The research purpose was to demonstrate the differences between food supplements and healthcare products. The purpose will be achieved in several stages: identification of the evaluation criteria of the studied medications, benchmarking of the studied products in terms of the defined criteria as well as elaboration and description of the differences.

3 Results

The activities aimed at achieving the research purpose included studies conducted in two stages. The first stage consisted of the identification of the evaluation criteria of food supplements and healthcare products. Then, an analysis was conducted, the results of which were used to evaluate and compare these products based on the defined criteria. These activities were executed as part of the second stage of the research.

1.1 Definition of evaluation criteria

The literature analysis in the scope of food supplements and healthcare products allowed for establishing a set of criteria. In order to clearly interpret particular evaluation criteria, they were described in details (Table 1).

Table 1. Identified evaluation criteria.

Criteria	Description
Basic criteria	
product group	food supplement or healthcare products (including prescription and OTC)
product type	product type
intended purpose	practical purpose of the given product
properties	set of characteristic features for the given group
indications	recommendations or instructions for use
effect	consequence, result of use

Table 1. Identified evaluation criteria cont.

Criteria	Description
Safety criteria	
safety of use - supervision over production and use	set of regulations governing the safe conditions of producing the product, ensuring safety of its consumption
dosage	standardised and strictly measured portion/dose
information leaflets	contents and substantive content
Criteria concerning product marketing	
legal regulations/concerning product registration on the market	legal regulations concerning the requirements/obligations imposed on producers to register the product for market sales
method of marketing (sales)	current regulations on and procedures of product marketing
documentation required for product marketing	type of documentation required during product marketing
designation of the number of permit for marketing	individual number constituting a permit for marketing
packaging designation	designation of the product group used on the packaging
Market appeal criteria	
availability	product market availability, sales points
packaging	originality, imitation

Sixteen criteria were identified based on the literature studies. These criteria were classified in four thematic groups. The following criteria were distinguished: basic, concerning consumption safety, product marketing and market appeal.
In the latter part of the paper, the defined criteria will be used to develop a profile of the studied product groups. Then, the profile will be used to provide a description of the differences between food supplements and healthcare products.

1.2 Food supplements vs. healthcare products - a benchmark

The characteristic features of food supplements and healthcare products were defined for the designated evaluation criteria (Table 2).

Table 2. Food supplements vs. healthcare products [8].

Criterion	Product Group	
	Food supplement	**Healthcare products** (including prescription and OTC)
Basic criteria		
product type	food product	medical product

intended purpose	diet supplementation	diagnosis or restoration, improvement or modification of the organism's physiological functions
properties	source of vitamins or mineral nutrients	properties of prevention or treatment of diseases
indications	no indications	specified in the medicinal product description, detailed in the leaflet
effect	nutritious or other physiological effect	pharmacological, immunologic or metabolic
Safety criteria		
safety of use - supervision over production and use	no statutory requirement of continuous monitoring of safety of use	continuous supervision and monitoring of the quality by the pharmaceutical inspectorate
dosage	administration in small measured unit quantities (universal doses)	strictly according to physician's recommendations
information leaflets	not subject to approval	subject to approval by the URPLWMiPB
Criteria concerning product marketing		
legal regulations/concerning product registration on the market	does not require registration	requires registration
method of marketing (sales)	notification of the Chief Sanitary Inspectorate and submission of packaging design	based on an application submitted to URPLWMiPB or EC
documentation required for product marketing	no specific documentation	specific documentation
designation of the number of permit for marketing	no number	number of permit for marketing on the packaging
packaging designation	phrase: "food supplement" always aside the trade name	phrase: "non-prescribed drug" or "medicinal product issued without physician's recommendations - OTC"
Market appeal criteria		
availability	unlimited	limited
packaging	mainly copy of the medicine's packaging	prototype

4 Discussion

Studies were conducted for two product groups: food supplements and healthcare products. The analysis of the feature descriptions of food supplements and healthcare products allowed for defining the differences between them.

The key difference between a food supplement and a healthcare product is the product type. In the case of the supplement, it is a food product, i.e. the same as bakery products, vegetables and fruits, meat, etc. Healthcare products, i.e. prescription medicine or OTC constitute medicinal products. This feature is reflected in the description of each next parameter.

The purpose of consuming food supplements is only (!) to supplement normal diet. In the case of healthcare products, they are used at the time of occurrence of a medical condition, in order to make a diagnosis or restoration, improvement or modification of the organism's physiological functions. Supplements constitute a source of vitamins or mineral nutrients and the purpose of their consumption is to supplement their quantities in the organism. On the other hand, healthcare products have properties that prevent or treat diseases. Therefore, the purpose of use of the two is completely different. This can also be seen when analysing the indications of the studied medications. In the case of supplements, there are no indications for use, whereas in the case of healthcare products indications are precisely specified in the product characterisation included in the information leaflet. The differences are also apparent in the effects or impact of a human organism. In the case of food supplements, only a nutritious or other physiological effect is achieved. The effects of consuming healthcare products are pharmacological, immunological or metabolic. The described differences between the studied products were classified according to the basic criteria.

As part of the criteria groups concerning safety, the safety of use - supervision over production and use was defined first. In the case of food supplements, there is no statutory requirement of continuous monitoring of the safety of preparation production and use. On the other hand, all healthcare products are subject to rigorous procedures of supervision monitored by the pharmaceutical inspectorate. It is interesting that in the case of food supplements, dosage is universal. Clients consume a constant dose regardless of the organism's demand for given vitamins or mineral nutrients, because no tests are conducted which would determine the organism's demand for them. In the case of healthcare products, doses are determined by the treating physician. It is important to be aware that the contents of supplements' information leaflets are not subject to approval. The contents of are only subject to approval by the URPLWMiPB in the case of healthcare products.

The marketing of food supplements does not require product registration. It is only necessary to notify the Chief Sanitary Inspectorate and deliver a packaging design. No specific documentation is provided and the supplement does not have a number of permit for marketing. On the other hand, in the case of healthcare products, it is necessary to register and approve the products for marketing. These are based on an application submitted to the Office for Registration of Medicinal Products, Medical Products and Biocidal Products (URPLWMiPB) or the European Commission, after presenting specific documentation. On this basis, a healthcare product is assigned a number of permit for marketing. The supplement packaging must always have the phrase "food supplement" aside the trade name. On the other hand, healthcare products must have the phrase "non-prescribed drug" or "medicinal product issued without physician's recommendations - OTC" on their packaging.

In the case of food supplements, a client has almost an unlimited access to them. It is possible to purchase them in traditional or online sales points, such as pharmacies, small and medium food stores, supermarkets, petrol stations. The sale of healthcare products is limited mainly to pharmacies and pharmacy points. A food supplement packaging imitative - it is very similar to medicine packaging. The packaging of healthcare products is a prototype.

4 Conclusion

The conducted analysis served as a basis for developing a catalogue of features along with a detailed description of the studied medications in terms of the differences existing between them. The research purpose was therefore achieved. The analysis of the studied notions allowed for defining the following conclusions.

Food supplements and healthcare products constitute various product groups. They differ in terms of product type, intended use, properties, indications for use, effects. Completely different regulations and principles of qualitative safety and marketing also apply to both groups. In the case of food supplements, it is possible to state that the existing entries and requirements are dangerous from the point of view of the client's health and life. They do not take into account the basic standards and principles of safety applicable to healthcare products. Widespread availability and omnipresent promotional activities have blurred the boundaries between supplements and healthcare products in the clients' awareness. In effect, clients stock up on supplements on a mass scale by introducing a so-called self-diagnosis and self-treatment. The carefree behaviour of clients who uncritically purchase and consume mass quantities of supplements affects their state of health. In a long period of time it can also result in a catastrophe related to the loss or permanent damage to health.

Due to the above, it is absolutely necessary to conduct educational activities aimed at informing the clients about the differences between food supplements and healthcare products.

References

1. H. Bojarowicz, P. Dźwigulska, Hygeia Public Health, **47**(4), 427-432 (2012)
2. H. Bojarowicz, P. Dźwigulska, Suplementy diety. Część II. Wybrane składniki suplementów diety oraz ich przeznaczenie, Hygeia Public Health, 47(4), 433-441 (2012)
3. A. Brzozowska, Żywność. Nauka. Technologia. Jakość, **4**(29), 16-28 (2001)
4. M. Kozłowska-Wojciechowska, *Żyjmy w zdrowiu czyli nowa piramida żywienia* (Wydawnictwo Prószyński i S-ka, 2004)
5. M. Jarosz, *Suplementy diety a zdrowie* (PZWL, Warszawa, 2008)
6. G. Krasnowska, T. Sikora, Żywność. Nauka. Technologia. Jakość, **4**(77), 5-23 (2011)
7. K. Hys, *Managing business growth in a volatile environment, (ed.) I.V. Raguz, Mechanisms stimulating actions customer - decision maker on the market: marketing mix instruments* (Dubrovnik International Economic Meeting, Dubrovnik, 2017)
8. K. Hys, *Zarządzanie. Teoria i Praktyka Management, Impact of advertising on the sale of medicinal products and food supplements in Poland* (Theory&Practice, 2017)

MATEC Web of Conferences **183**, 01007 (2018) https://doi.org/10.1051/matecconf/201818301007
QPI 2018

Monitoring of information security system elements in the metallurgical enterprises

Justyna Żywiołek[1,*]

[1]Częstochowa University of Technology, Faculty of Management, Poland

Abstract. The article concerns the monitoring of elements of the information security system in an enterprise. The purpose of the research was to determine the reasons for monitoring information flows in the surveyed enterprise. The identification of information flows facilitates information management, empowering individuals to process information and preventing information security incidents. The implementation of information management methods facilitates monitoring the information security status.

1 Introduction

The company should maintain its assumed level of security, which results from the current state of the elements of the security system, that is, the state of resources, vulnerabilities, threats and safeguards. This state is described in the model used for risk analysis, based on the resource model. In practice, it turns out that these elements are subject to changes, which should be constantly monitored, because they affect the overall level of information security and services in the institution. The subject of monitoring activities are changes taking place in the elements of the security system (risk factors), that is in the state of resources, vulnerabilities, threats and safeguards.

2 Changes cause the system's monitoring

Regardless of the reasons for changing the status of elements of the security system, their impact on the assumed level of security for the institution is subject to a thorough analysis, conducted as part of the monitoring process, which defines the types of impact on safety [1]: indirect - caused by significant changes in the systems, detected by the change management process, which take place in the context of their impact on the elements of the security system, direct - caused by changes occurring spontaneously in elements of the security system.

Figure 1 presents typical relationships expressing the impact of various factors on the monitored elements of the security system. Resources are monitored not only in terms of changing their value, but also in terms of changing the security objectives of teleinformation systems (the so-called security purpose may change).

* Corresponding author: justyna.zywiolek@wz.pcz.pl

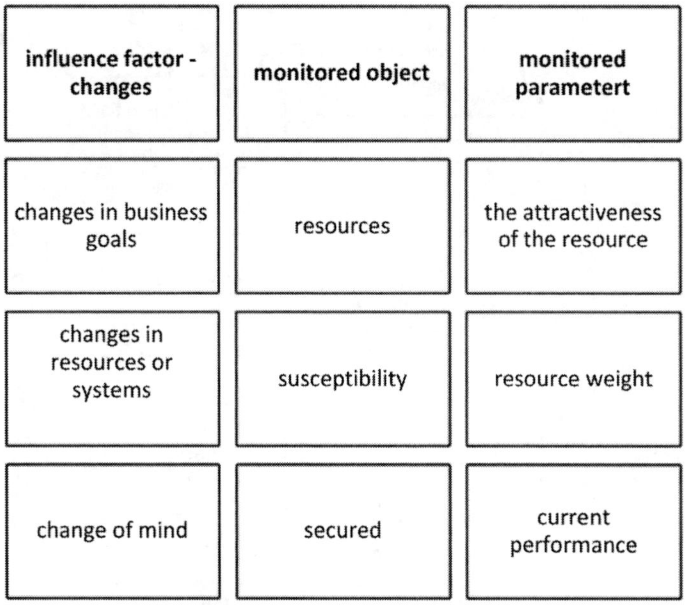

Fig. 1. The impact of changes in systems and their environment on the need to monitor the security system.

Threats and vulnerabilities are monitored in terms of changes in their significance, resulting from changes in systems and their environment, as well as in resources [2]. It should be noted that in the case of resources, the decisive factor may be a change in their attractiveness.

In order to maintain compliance of systems with the security policy, it is recommended to constantly monitor [4]: existing security, new systems or services introduced, planned changes to systems and services.

The security attribute is associated with monitoring, that is, accountability and event logs connected to it. Many security measures generate logs that can be subject to statistical analysis, enabling early detection of anomalies, changes in trends and repetitive incidents [5-7]. The institution should appoint people responsible for constantly performing such activities. Event logs are a quite universal mechanism, also used by other post-implementation processes [4, 9, 14].

3 Control of security status through safe data deletion, perimeter control of the facility, document workflow

In many companies, information protection does not work at all, or, as in the surveyed enterprises, it is limited to the use of security in the form of agreements on the confidentiality of information with employees and physical protection for personal and car traffic [3]. Each employee who has access to sensitive information is required to sign an appropriate statement. There are severe penalties for disclosing information that is the subject of the contract. Control of the security system allows to learn about the reasons for the controls [11]. The main reasons for the control, including the people responsible for the controls, are shown in Figure 2.

Almost 70% of documents that are currently created in companies or are created by users are digital [10]. Almost 90% of them, however, will never be printed and will remain only in the memory of computer disks, memory cards and other media, the newer versions of which appear each year on the market [12].

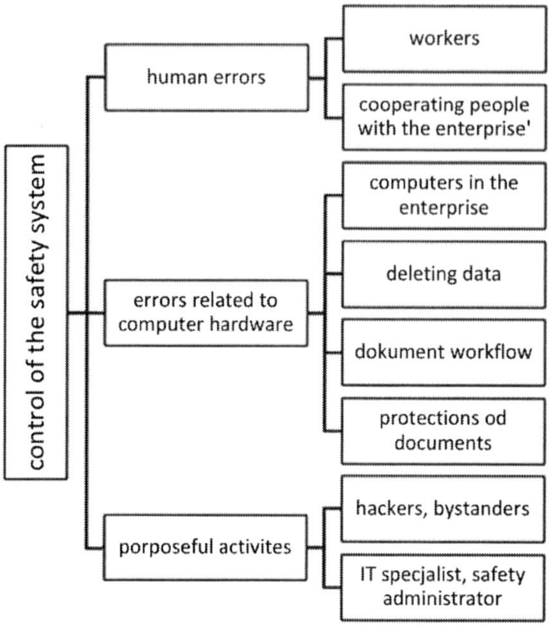

Fig. 2. Reasons for control - diagram Why? Why?

4 Possible control activities of the information security system

In the surveyed enterprises, after the research audit, it was recognized that the greatest threat to the company is the ignorance of employees. For this reason, in addition to the planned audits and trainings, the company decided to be interested in possible measures to control employees' security activities [11, 13].

Conducting research required the use of selected methods of information management. For the design of the information system, one of the methods of information management has been used, counted among the classic techniques of registration of information processes. The Clark chart is a graphical technique of recording and examining the course of the information process, allowing to capture the involvement and activity of the company's organizational units. In addition, it enables detection of unnecessary activities and determination of time losses. The Clark column chart is built from a table of several or a dozen or so columns depending on the number of organizational units. All of the surveyed enterprises have in their structure the following departments: production, logistics, trade, research as well as administration and management for which Clark charts were developed (Figure 3-5).

Functions	The entities						
	Inside					Outside	
	A secretary's employee	Security administrat	Employee	IT specialist	Accountan t	Supplier / seller	Security auditor
Shipping / delivery of a fixed asset						1	
Adoption of the measure	2						

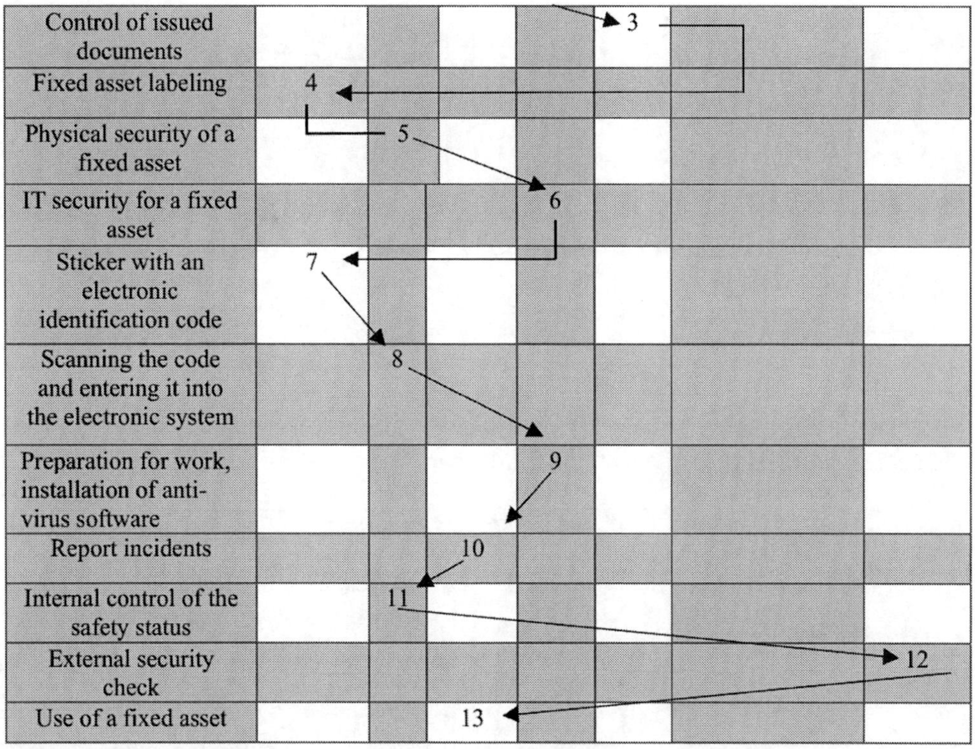

Fig. 3. Clark chart for the procedure of taking fixed assets on the example of a laptop

The Clark chart describing the procedure for accepting a fixed asset is used to design the information process from its delivery to the company until acceptance and use. The individual functions present the stages of the information process, while the entities, divided into external and internal ones, indicate the persons or institutions to whom the information is received, are subjected to processing. Information processing includes the acceptance of information, operations related to its processing, as well as security, if it is required by security rules and sending information for further activities. By dividing the information process into individual stages, the procedures for the information security of the company can be accurately designed. Clark charts also allow you to avoid situations in which elements of the information process reach people who are not well-intentioned or unscheduled in a given information process.

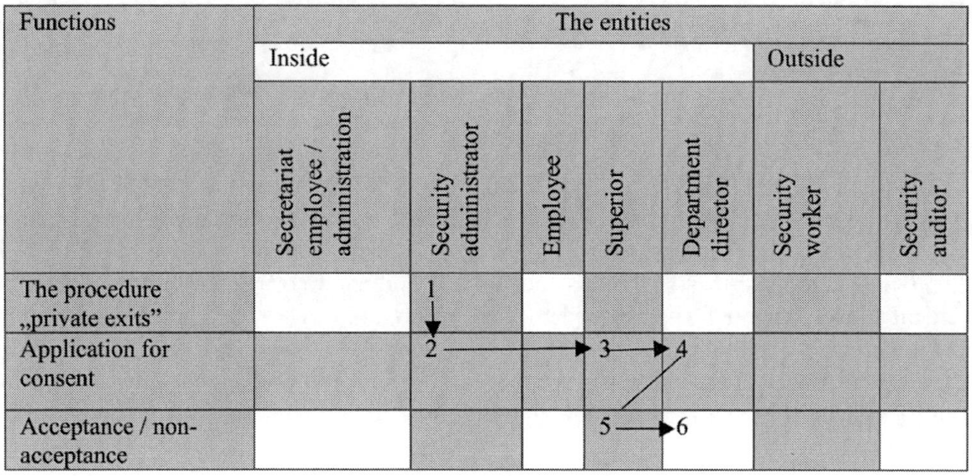

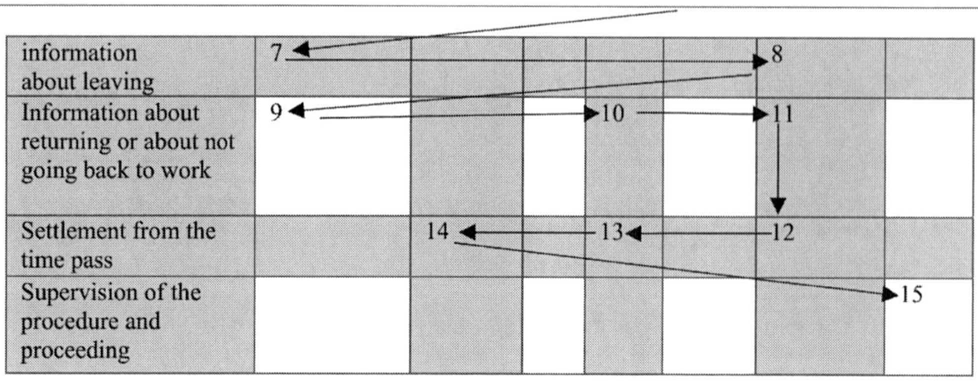

Fig. 4. Clark chart for the employee's output during work

Another element that should be examined in detail to ensure security is incident management. For this type of event, the Clark chart was also developed (Fig. 5).

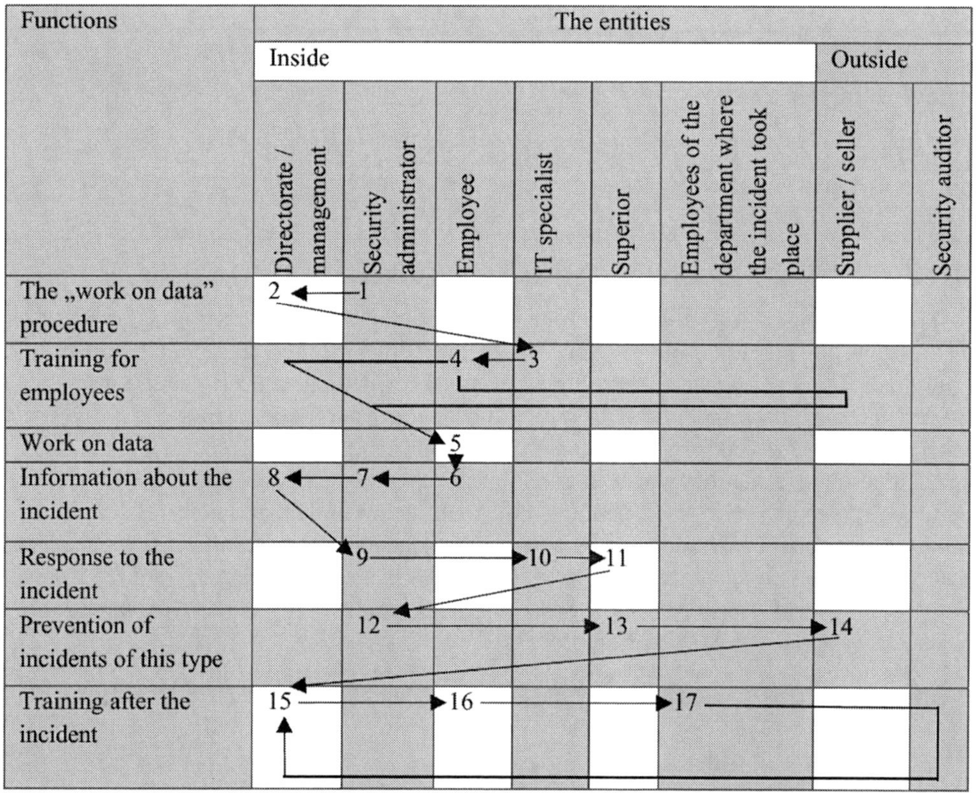

Fig. 5. Clark chart for the possibility of an incident

The occurrence of incidents has a particular impact on reducing the level of information security in the enterprise. Avoiding their occurrence is an important element for security. The developed Clark chart assumes a way of working with data, provides for the possibility of an incident and instructs how to proceed after it. Indication of people informed about the incident allows to avoid further or wider loss of information. The Clark chart also illustrates how important the training activities for employees are during the incidents.

Clark charts prepared and presented are only examples. While designing the information security system, it is necessary to learn about information processes affecting safety.

Summary

It is necessary to take preventive measures in order to be „smarter before harm" and effectively take care of information security within the organization. As in the case of computer viruses, there is unfortunately no recipe for 100% protection against unauthorized information from employees. However, it is necessary to implement certain safeguards leading to minimizing the danger of this type of threats, because they most often result from the intentional action of the employee, not from his ignorance or accident. Appropriate care for information security should be preserved throughout the life of information: from its inception to its permanent destruction.

The information security policy is the key to explaining the actual intentions of the company and people managing information security. This document should be formal and subject to constant supervision, the more so as it is an element of the company's strategy. The assumption of the strategy is the recognition of information processes, their classification and examination of information flows. Implementation of the information security strategy is possible thanks to the use of information management methods. Enterprises have implemented cryptographic keys, a program for sending e-mails to many recipients, and also use electronic signatures. Facilitation of securing results from the use of a clean desk policy, monitoring access to rooms. The solution that brought the greatest effect is the electronic circulation of documents.

References

1. A. Białas, *Bezpieczeństwo informacji i usług w nowoczesnej firmie* (WNT, Warszawa, 2007)
2. P. Zou, P. Lun, D. Cipolla, S Mohamed, Safety Science, **98** (2017)
3. A. Lemańska-Majdzik, K. Smoląg, *Role and Importance of Fanpage in Promotion of Products and Services, The Economies of Balkan and Eastern Europe Countries in the Changed World* (EBEEC, Split, Chorwacja, 2016)
4. M. Chałon, *Ochrona i bezpieczeństwo danych oraz tendencje rozwojowe baz danych* (Oficyna Wydawnicza Politechniki Wrocławskiej, Wrocław, 2007)
5. M. Nowicka-Skowron, R. Ulewicz, *METAL 2015* (24th International Conference on Metallurgy and Materials, 1707-1712, 2015)
6. M. Kowalewski, A Ołtarzewska, *Telekomunikacja i Techniki Informacyjne*, **3-4** (2007)
7. K. Liderman, *Bezpieczeństwo informacyjne* (PWN, Warszawa, 2012)
8. A. Brzozowska, D. Bubel, A. Pabian, Procedia - Social and Behavioral Sciences, **213**,992-999 (2015)
9. G. Stewart, Information Security Technical Report, **14** (2009)
10. J. Żywiołek, E. Staniewska, Logistyka, **6**, 382-385 (2012)
11. J. Żywiołek, Logistyka, **6**, 653-658 (2012)
12. J. Żywiołek, Production Engineering Archives, **2**, 36-39 (2016)
13. J. Pietraszek, A. Gądek-Moszczak, T. Torunski, Adv. Mat. Res.-Switz., **874**, 139 (2014)
14. R. Ulewicz, M. Nowicka-Skowron, *26th International Conference on Metallurgy And Materials* (2338-2343 (2017).

Interpretive Structural Modeling of identified Barriers to Lean Implementation in SMEs

Arvind Kumar Shrimali[1,*], *Vimlesh Kumar* Soni[2], and *Shashank Singh* Pawar[1]

[1]Department of Mechanical Engineering, Chameli Devi Group of Institutions, Indore, M. P., India
[2]Department of Mechanical Engineering, Maulana Azad National Institute of Technology, Bhopal, M. P., India

Abstract. Lean practices are implemented in manufacturing companies and services to find hidden waste and attain continuous improvement. Various enterprises have experienced difficulties in the Lean implementation. Following the use of appropriate lean instruments and techniques, there are many other factors that affect success lean implementation process. Researchers have identified a huge number of barriers to implementation of Lean. Understanding the barriers and the interactions between them can be crucial to the success of lean implementation.
Interpretive Structural Modeling (ISM) is one of the established methodologies to bring forward the interrelationships among parameters of an issue or a problem.
Purpose- The purpose of this paper is to create the hierarchy of the various barriers to Lean Implementation according to their importance using the approach of ISM to facilitate Small and Medium enterprises (SMEs) across India. The study is specific to the Small and Medium Enterprises (SMEs) from India.
Keywords: Lean implementation, Small and Medium enterprises (SMEs), Interpretive Structural Modeling (ISM)

1. Introduction

Lean implementation is fast becoming popular in India and the world to eliminate the wastes within an organization and thus improve the performance. Lean strives to create an efficient operation and by bringing together the best practices and concepts of productivity and problem-solving.

It is difficult to move from a traditional production system to a Lean production system. This change requires attention for both processes and people. The availability of extensive documentation on the benefits of the lean manufacturing implementation could not extend it to a large number of companies.

The present study focuses on the barriers to the implementation of the Lean concept in Small and Medium Enterprises (SMEs) in India and to create the hierarchy of the various

* Corresponding author: arvindshrimali@gmail.com

barriers to Lean Implementation according to their importance using the approach of ISM to facilitate SMEs across India. ISM methodology has been used in this study to understand and analyse contextual relations and hierarchical levels of barriers in the implementation of Lean.

The body of this paper is organized as follows. After Introduction in Section 1, section 2 presents the literature study on barriers to implementation of Lean. Section 3 explains the ISM methodology. Section 4 presents the details of the ISM approach to the barriers to implementation of Lean model. The discussion on the outcome of the study is presented in Section 5. And Section 6 presents the Conclusions.

2 Literature review

2.1 Background of Lean Implementation in SMEs

The Lean manufacturing is a manufacturing strategy that eliminates waste and increases the value of production activities to achieve a smooth flow of production. Without a Lean manufacturing strategy, an organization cannot succeed in today's global competition, where higher quality, faster delivery, and lower costs are needed [1].

Dora et al. [2] under a study examined the circumstances or key factors and their impact on limited production in small and medium enterprises. Its own characteristics, as well as the extremely volatile demand and supply, constitute barriers to the introduction of Lean processing. Unlike large companies, the environment of SMEs has a great resistance to adapt to change.

This study focuses on an analysis of the barriers to the implementation of the Lean approach in SMEs in India.

2.2 Barriers for adopting Lean in SMEs

The Literature survey has thrown light on various kinds of barriers that hinder an organization from going Lean. The critical obstacles to the implementation of the Lean have been examined by several researchers. Z. Radnor et al. [3] identified three problems that organizations face: the issue of people, the issue of the process and the issue of sustainability. In a similar type of classification by M. F. Bollbach [4], he explained the social and technical barriers that are necessary for the implementation of Lean. The obstacles in the Lean building were studied by H. M. Alinaitwe [5]. The obstacles to the introduction of lean health have been reviewed by L. Brandão de Souza and M. Pidd [6].

The findings derived from a research study by Jadhav et al. [7] led to the identification of 24 Lean barriers. The authors conclude that the success of the Lean implementation will not be completely based on the application of appropriate tools and techniques, but also on top management participation and leadership, attitude of employees, resources and organization culture.

According to a study [8] the small size of the enterprises, the traditional design and the rigid design make it difficult to implement the lean manufacturing in small and medium enterprises. The study also helps experts anticipate possible barriers and take the appropriate measures to address them in a lean implementation.

The most frequent problems enlisted by the researchers Mariusz Bednarek and Luis Fernando Niño Luna [9] include:

- Overcome the rigid corporate culture;
- Ignorance of the tools and methods of implementation;

- Efforts are focused on individual tasks instead of complex objectives or development of the business as a whole;
- Resistance to work;
- The use of unidentified models according to the characteristics of each company;
- Lack of an adequate implementation strategy;
- Lack of training and required skills;
- Lack of resources needed for lean implementation;
- Insufficient support for the Board of Directors and mid-level officials when projects are implemented.

Shrimali and Soni [10] have studied and ranked the barriers in the implementation of Lean Manufacturing and improve understanding with the Lean philosophy in small and medium enterprises in India. The researchers have presented the results of a survey conducted on the Indian SMEs with the main purpose of identify the barriers during the implementation of lean manufacturing practices and comparing these barriers. (Table 1).

Table 1. Results on Lean Implementation Barriers (Shrimali & Soni, 2017) [10].

S. No.	Factor	Mean	Rank
1	Resistance to Change the Middle Management	4.62	1
2	Lack of Flexible Working Arrangement	3.36	2
3	Absence of Lean Implementation Team	3.26	3
4	Lack of Reward System	3.17	4
5	Little Support from Top Management	2.37	5
6	Poor Lean Training	2.21	6
7	High Cost/Investment	2.08	7
8	Absence of Consultant	1.76	8
9	Other	1.69	9

The study shows that the biggest barriers in implementing lean manufacturing practices is Resistance to change by middle management and Absence of a lean implementation team or Lack of skilled people and followed by Lack of flexible working arrangements and Little Support from the top management.

3 Problem description

The study has taken the research of Shrimali and Soni [10] on the barriers to the implementation of Lean in SMEs as the base paper for this study. The total feedbacks analysed by the researchers are 149. The study mentions that out of the 149 responses analysed, 16 are from Medium sized, 127 are Small sized and 2 are Micro Enterprises belonging to different industry sectors.

This current paper aims to study, understand and analyze the contextual relations and hierarchical levels of barriers in implementation of Lean. The authors have used the interpretive structure modeling (ISM), an established method of identifying relationships between different parameters of the system.

The purpose of this paper is:

1. To identify the most critical barriers associated with the implementation of Lean in SMEs,
2. To determine the relationship between these identified barriers to implementation of Lean by means of interpretive structural modeling,
3. To suggest a structural model for the barriers to Lean Implementation, and
4. To organize the identified barriers in different categories using MICMAC analysis.

4 Solution methodology

4.1 Interpretive Structural Modeling (ISM)

The Interpretive structural modeling (ISM) is a process of interactive learning. Presence of indirect or direct-connected elements complicates the structure of any system. It is difficult to handle such a system, whose structure is not clearly defined. Therefore, a methodology is needed to identify a structure. Such a method is- Interpretive structural modelling. The various components of a process are analyzed in a comprehensive system model. There are many examples of the use of ISM in the research literature. The two basic concepts to understand the ISM system are- reachability and transitivity.

ISM's basic idea is to use the experts' practical experience and expertise to develop a complex system (element) into subsystems (elements) and create a multi-level structural model. The ISM method is the rule and guidance on the complexity of the relationship between the elements of the system elements [11, 12].

The various steps in the ISM methodology [13] are the following:

Step 1. The variables (criteria) are considered for the system under consideration.

Step 2. Based on the variables identified in step 1, a context relationship between the variables is established to determine which variables are to be examined.

Step 3. The structure of the personalized interaction matrix (SSIM) is developed for the variables, which indicates the pairing relationships between the variables of the evaluated system.

Step 4. The access matrix is developed from SSIM and the matrix transition is verified. The transition of a contextual relationship is a fundamental premise in ISM. It establishes that if variable A is related to B and B is related to C, then A is necessarily related to C.

Step 5. The scope matrix obtained in step 4 is divided into different levels.

Step 6. Based on the previous relationships in the availability matrix, an objective graphic is drawn and the transitive links are eliminated.

Step 7. The resulting digraph is converted to ISM by replacing the report node variables.

Step 8. The ISM model developed in step 7 is reviewed to verify the conceptual inconsistency and make the necessary adjustments. The previous steps are illustrated in Figure 2.

5 Application of ISM

ISM has been used by some researchers across the globe to develop a relationship among issues/enablers/barriers in different fields.

5.1 Data collection

The identified barriers taken under the study are from the base paper by Shrimali & Soni [10]. These barriers are-

1. Little Support from Top Management
2. Resistance to Change the Middle Management
3. Poor Lean Training
4. Absence of Lean Implementation Team
5. Lack of Flexible Working Arrangement
6. Absence of Consultant
7. Lack of Reward System
8. High Cost/Investment

5.2 Structural self-interaction matrix (SSIM)

In the first step towards generating a matrix, considering the contextual relationship for each variable, the existence of a relation between any two barriers (i and j) and the associated direction of the relation is questioned. Following four symbols are used to depict the direction of a relationship between the barriers (i and j):

V : Barrier i will promote barrier j;
A : Barrier j will promote barrier i;
X : Barrier i and j will promote each other; and
O : Barriers i and j are unrelated.

The SSIM for the barriers in the implementation of green supply chain is given in Table 2. The following section explains the use of the symbols V, A, X, and O in the SSIM.

Table 2. Structural Self Interaction Matrix (SSIM).

	i j	1	2	3	4	5	6	7	8
1	Less Support from Top Management		X	V	V	O	V	V	X
2	Resistance to Change the Middle Management			V	X	V	V	V	X
3	Poor Lean Training				A	O	O	O	X
4	Absence of Lean Implementation Team					V	O	O	A
5	Lack of Flexible Working Arrangement						O	O	A
6	Absence of Consultant							O	A
7	Lack of Reward System								O
8	High Cost/Investment								

5.3 Initial Reachability Matrix

The SSIM has been converted into a binary matrix, called the initial reachability matrix by substituting V, A, X and O by 1 and 0 as per the case. The substitution of 1s and 0s are as per the following rules:

i. If the (i, j) entry in the SSIM is V, the (i, j) entry in the reachability matrix becomes 1 and the (j, i) entry becomes 0.
ii. If the (i, j) entry in the SSIM is A, the (i, j) entry in the reachability matrix becomes 0 and the (j, i) entry becomes 1.
iii. If the (i, j) entry in the SSIM is X, the (i, j) entry in the reachability matrix becomes 1 and the (j, i) entry also becomes 1.
iv. If the (i, j) entry in the SSIM is O, the (i, j) entry in the reachability matrix becomes 0 and the (j, i) entry also becomes 0.

Table 3. Initial Reachability Matrix.

S.	i j	1	2	3	4	5	6	7	8
1	Less Support from Top Management	1	1	1	1	0	1	1	1
2	Resistance to Change the Middle Management	1	1	1	1	1	1	1	1
3	Poor Lean Training	0	0	1	0	0	0	0	1
4	Absence of Lean Implementation Team	0	1	1	1	1	0	0	0
5	Lack of Flexible Working Arrangement	0	0	0	0	1	0	0	0
6	Absence of Consultant	0	0	0	0	0	1	0	0
7	Lack of Reward System	0	0	0	0	0	0	1	0
8	High Cost/Investment	1	1	1	1	1	1	0	1

5.4 Final Reachability Matrix

After incorporating the transitivity as described in Step 4 of the ISM methodology, the final reachability matrix is shown in Table 4, the driving power and dependence of each variable is also shown. Driving power for each variable is the total number of variables (including itself), which it may help to achieve. On the other hand, dependence is the total number of variables (including itself), which may help in achieving it. These driving power and dependencies will be later used in the classification of variables into the four groups of autonomous, dependent, linkage and drivers (independent).

Table 4. Final Reachability Matrix.

S.	i j	1	2	3	4	5	6	7	8	Driving Power
1	Less Support from Top Management	1	1	1	1	0	1	1	1	7
2	Resistance to Change the Middle	1	1	1	1	1	1	1	1	8
3	Poor Lean Training	0	0	1	0	0	0	0	1	2
4	Absence of Lean Implementation Team	0	1	1	1	1	0	0	0	4
5	Lack of Flexible Working Arrangement	0	0	0	0	1	0	0	0	1
6	Absence of Consultant	0	0	0	0	0	1	0	0	1
7	Lack of Reward System	0	0	0	0	0	0	1	0	1
8	High Cost/Investment	1	1	1	1	1	1	0	1	7
	Dependence Power	3	4	5	4	4	4	3	4	**31**

5.5. Level partitions

The reachability and antecedent [14] for each barrier is obtained from the available final matrix. The reachability set for a given variable consists of the variable itself and other variables that it can help achieve. A set of antecedents consists of the variable itself and other variables that can help achieve them. Subsequently, the intersection of these sets is derived for all the variables. The variable for which the reachability and intersection sets are the same is given by the top level variable in the ISM hierarchy that would not help to achieve any other variable above its own level. After identifying the top-level element, the remaining variables are discarded. In this study, 8 barriers are presented together with their reachability set, background set, antecedent set and levels, as shown in Table 5. The process of identifying these barriers is completed in eight iterations.

This iteration continues until the levels of each variable are reached. The specified levels help build the digraph and the final ISM model (Fig 1).

Table 5. Iteration table for level partition.

S. No.	Reachability Set	Antecedent Set	Intersection Set	Level
1	1,2,3,4,6,7,8	1,2,8	1,2,8	II
2	1,2,3,4,5,6,7,8	1,2,4,8	1,2,4,8	I
3	3,8	1,2,3,4,8	3,8	III
4	2,3,4,5	1,2,4,8	2,4	III
5	5	2,4,5,8	5	IV
6	6	1,2,6,8	6	IV
7	7	1,2,7	7	IV
8	1,2,3,4,5,6,8	1,2,3,8	1,2,3,8	I

Table 6. Driving power and dominance diagram for lean barriers.

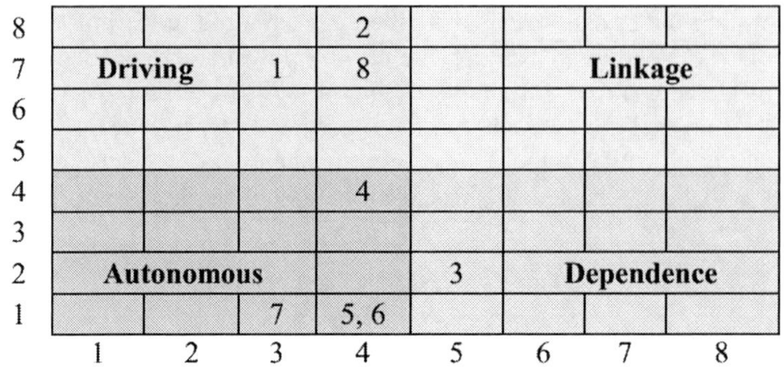

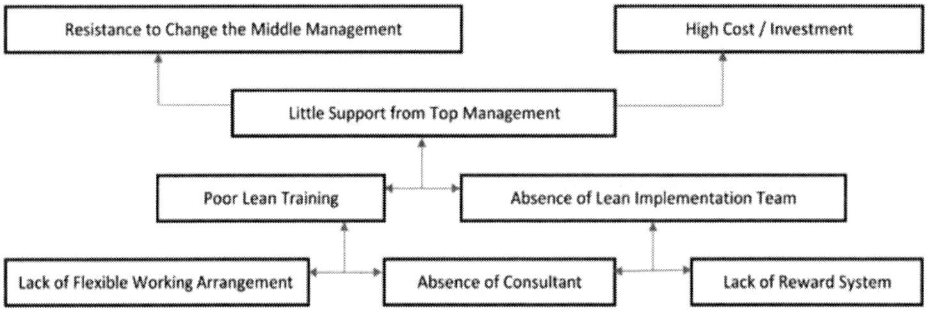

Fig. 1. ISM based model for barriers to lean implementation.

5.7 MICMAC analysis

Matriced' Impacts croises-multipication applique' and classment (cross-impact matrix multiplication applied to classification) is abbreviated as MICMAC. The MICMAC principle is based on multiplication properties of matrices [15]. The purpose of MICMAC analysis is to analyze the drive power and dependence power of enablers. This is done to identify the key enablers that drive the system in various categories. Based on their drive power and dependence power, the enablers, in the present case, have been classified into four categories as follows:

1. Autonomous enablers: These enablers have weak driving power and weak dependence. They are relatively disconnected from the system, with which they have few links, which may be very strong. These enablers are represented in Quadrant I.
2. Dependent enablers: This category includes those enablers which have weak drive power but strong dependence power and placed in Quadrant II.
3. Linkage enablers: These have strong driving power as well as strong dependence and are placed in Quadrant III. They are also unstable and so any action on them will have an effect on others and also a feedback effect on themselves.
4. Independent enablers: These have strong driving power but weak dependence power. These are represented in Quadrant IV.

6 Results and discussion

Autonomous variables generally appear as weak driver as well as weak dependent and are relatively disconnected from the system. These variables do not have much influence on the other variables of the system. Absence of lean implementation team, lack of flexible working arrangement, absence of consultant and lack of reward system are falling under autonomous variable category and these are linked with each other in the structure. Thus, these are stable factors. While Little support of Top Management, Resistance to change from the middle management and Higher investment cost are driving the barriers. Emphasis is needed to resolve these barriers to ensure successful lean implementation.

7 Conclusions

To be more competitive a lean implementation should be well coordinated and responsive.

This paper developed a model to understand the barriers to Lean Implementation in SMEs. Although, ample literature is available on lean manufacturing and various issues related to it. The relationship between barriers to lean implementation in SMEs has not been modelled for manufacturing organizations. The present model will help managers and lean practitioners to understand the relationship in detail. This research assumes significant contribution in this regards.

This study has considered eight barriers. The interpretive structural modeling (ISM) approach has been employed to develop the structural relationship among these barriers. ISM approach helped in determining driving and dependence power of all variables. It is observed that 'resistance to change from the middle-level management', and 'high cost of investment' are major drivers for of lean barriers. Attention must be paid on these barriers to have proper lean practice.

The present study was aimed to bring forward the contextual relations and hierarchical levels of barriers in the implementation of Lean. This has brought forward the parameters responsible as drive factors among the barriers to the implementation of Lean in SMEs. Linear structural relationship can be developed between these lean barriers so as to test the validity of the barriers of such hypothetical models for future study.

References

1. N. Nordin, B. M. Deros, D.A. Wahab, IJIMT, **1**(4), 374-380 (2010)
2. M. Dora, M. Kumar, X. Gellynck, Production Planning & Control, **27,** 1-23 (2016)
3. Z. Radnor, P. Walley, Public Finance (28 July–3 August), 16-19 (2006)
4. M.F. Bollbach, http://hdl.handle.net/2134/10094 [2013 01 20].
5. H.M. Alinaitwe, Journal of Construction in Developing Countries, **14**, 1 (2009)
6. L. Brandão de Souza, M. Pidd, Public Money & Management, **31**, 59-66 (2011)
7. J.R. Jadhav, S.S. Mantha, S.B. Rane, International Journal of Lean Six Sigma, **5**, 122-148 (2014)
8. A.K. Shrimali, V.K. Soni, IJMPERD, **7**, 283-300 (2017)
9. M. Bednarek, L.F.N. Luna, IFIP, **257**, 239-247 (2008)
10. A.K. Shrimali, V.K. Soni, IJMET **8**, 1-9 (2017)
11. J.W. Warfield, IEEE Transactions on Systems, Men and Cybernetics, **4**, 51-81 (1974)
12. A.P. Sage, *Interpretive Structural Modeling: Methodology for Large-Scale Systems.* (NY: McGraw-Hill, New York, 1977)
13. G. Kannan, S. Pokharel, P.S. Kumar, Resources, conservation and recycling **54**, 28-36 (2009)
14. J.W. Warfield, *Societal Systems: Planning, Policy and Complexity* (John Wiley & Sons, Inc., New York, 1976)
15. H.D. Sharma, A.D. Gupta, Sushil, Technological Forecasting and Social Change, **48,** 285-309 (1995)

Accreditation of research in the thermo-climatic chamber – various aspects

Artur Gawlik[1], *Piotr* Kucybała[1] , *Danuta* Owczarek[1], and *Janusz* Pobędza[1,*]

[1]Cracow University of Technology, Mechanical Department , Al. Jana Pawła II 37, 31-864 Kraków, Poland

Abstract. One of the important aspects in the implementation of new products for production in the open field is the assessment of the impact of environmental conditions on their operation. Typically thermo-climatic research is carried out for such project. The laboratory of Techno-climatic Research and Heavy Duty Machines of Cracow University of Technology, equipped with a large-sized thermo-climatic chamber conducts this type of research. Bearing in mind the assurance of high quality of the conducted research, the quality management system (QMS) was developed and implemented. The article describes the requirements, scope and process of obtaining a Polish Centre for Accreditation (PCA) certificate.

1 Introduction

1.1 Motivation for conducting techno-climatic research

The constantly increasing range of requirements as to the quality and reliability of manufactured machines and devices force not only the implementation of production quality management systems, but also the development of standards and requirements that the machines and devices must fulfill. It is also necessary to develop experimental test methods and procedures that allow checking and verifying whether the requirements are met. Such procedures are developed, amongst others, in accredited laboratories and institutions that rely on many years of research and design experience, and guarantee a very high quality of conducted research.

Currently, in the phase of machines and devices designing process, advanced IT tools are used as well as the constructors' shared experiences and knowledge exploited. This makes it possible, to a large extent, to meet the functional requirements, strength, durability and reliability conditions. However, despite the use of more and more perfect systems for computer modeling and simulation, there is still a need to conduct experimental research that will verify the effects of design works [1].

Vehicles, mobile machines and engineering devices intended for open field operation, must be adapted to various and changeable climatic conditions, and thus it is necessary to conduct research determining the impact of various climatic exposures on the functionality,

* Corresponding author: janusz.pobedza@mech.pk.edu.pl

reliability and durability of the constructed and manufactured equipment. In addition, ensuring the comfort of machine operators, vehicle drivers and passengers is also an important issue.

1.2 Experience in techno-climatic studies

Research on the impact of environmental exposures on machines and engineering objects has been carried out at the Cracow University of Technology since 1973. For this purpose, a large-sized thermo-climatic chamber has been used, with dimensions: 20m x 6m x 6m and with temperature range down to -40°C. Various objects have been subjected to research: vehicles, earth moving machines, military vehicles, construction and road machines, postal and vending machines, and even a helicopter and a railway wagon [1-5].

As part of the research, particular attention was paid to the operation of diesel engine start-up systems. Much attention was also devoted to determining the characteristics of hydraulic drive and control systems and assessing their reliability under conditions of low temperature [6, 7]. The evaluation of the heating and air-conditioning systems performance was also an important area of research [8].

Currently, techno-climatic research is carried out in a new facility, characterized by high cooling capacity and the largest, in this part of Europe, dimensions of the test space. The examples of objects tested in a thermo-climatic chamber are presented in Fig. 1.

Fig. 1. The examples of objects tested in a thermo-climatic chamber: military vehicle, fire truck.

The main area of the thermo-climatic chamber is the test space with the following dimensions: length 19.4 m, width 7.7 m and height 8 m. An additional space, which can be used during tests, is the , 3.7 m long entry gate (Fig. 2). Large dimensions of the test space ensure sufficient distance of the tested object from the walls of the chamber, and thus allow to minimize the impact of possible heat sources on measuring transducers.

The equipment of the thermo-climatic chamber consists of a refrigerating and heating systems, supervised and monitored by an integrated control unit, whereas the main temperature registration system is composed of thirty Pt100 temperature sensors (class A) connected in a three-wire system. The location of the sensors is determined each time, individually for the tested object, in accordance with the requirements of the relevant standard and/or as agreed with the client.

To increase the veracity of the obtained tests results, the research quality assurance system has been elaborated and introduced in the Laboratory of Techno-climatic Research and Heavy Duty Machines, which is confirmed by the certificate of Polish Centre for Accreditation (PCA).

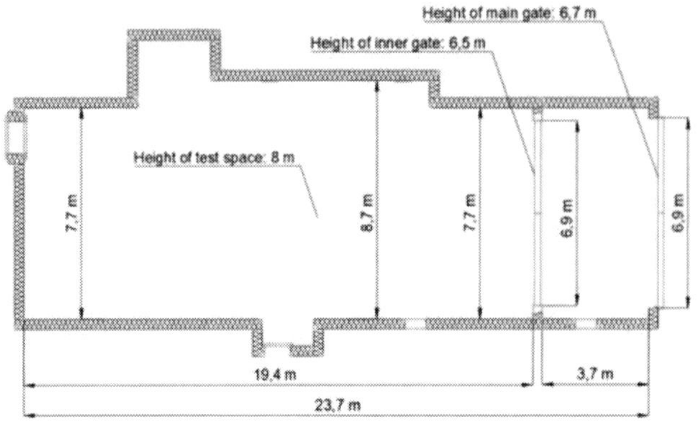

Fig. 2. The view of the thermo-climatic chamber testing space.

2 Accreditation procedure

2.1 PCA accreditation requirements

Accreditation should be understood as a formal recognition by the authorized accreditation body of competences of organizations operating in the field of conformity assessment, that is, certification and inspection organizations or technical laboratories. The rules of accreditation are included in international standards and guidelines, in which, amongst others, are defined requirements for accredited organizations. Getting the accreditation means that accredited body has been positive assessed according to these standards and guidelines. In Poland in this field operates the Polish Centre for Accreditation (PCA) which is the national accreditation body authorized to accreditation of conformity assessments organizations [9].

At present, PCA conducts accreditation and supervises processes, included but not limited to testing, calibration and processes of medical laboratories. The Laboratory of Techno-climatic Research and Heavy Duty Machines is understood here as a testing laboratory. The PCA application rules are presented in DA-01 document [9] and more detailed requirements for testing laboratories are presented in DAB-07 [9] document. Both can be downloaded from PCA website. The laboratory can gain the accreditation after fulfilling all accreditation requirements specified in DAB-07 document while performing tests using methods applied for accreditation. The accreditation requirements for testing laboratories are briefly as follows:

- PN-EN ISO/IEC 17025: General requirements for the competence of testing and calibration laboratories [10],
- DA-05 policy for participation in proficiency testing [9],
- DA-06 policy for ensuring traceability [9],
- documents describing test or calculation methods used in a laboratory, normative documents or own laboratory procedures,
- additional documents used for laboratory accreditation.

In the process of gaining the accreditation the Laboratory of Techno-climatic Research and Heavy Duty Machines had to fulfil all accreditation requirements, which was done at the turn of the year 2017/2018 and finally it was granted accreditation on 12 April, 2018 – AB 1678.

2.2 Preparation to accreditation – QMS of Laboratory of Techno-climatic Research and Heavy Duty Machines

The accreditation applying process of Laboratory of Techno-climatic Research and Heavy Duty Machines started from completing and checking all necessary elements, such as competent staff, machines/equipment, finances, consent of the authorities and good quality management system (QMS) composed of various procedures which also included technical procedure.

The most time-consuming stage was the creation of QMS because the idea was to have the system which would be well-suited to the Laboratory specifics and which would be used for many different types of research. This system is composed of five sets of documents, as described below.

I. The Quality Manual – which, according to PN-EN ISO/IEC 17025 [10], includes or makes reference to the supporting procedures, also the technical ones. The goal of the quality manual is to outline the structure of documentation of the quality system. It is composed of seven chapters: (1) Presentation of the Laboratory of Techno-climatic Research and Heavy Duty Machines, (2) The Quality Policy Statement, (3) Supervision over the Quality Manual, (4) Management, (5) Technical requirements, (6) Annexes and (7) Related documents. One of the most important information of the Quality Manual is the Table 1, which presents the scope of research.

Table 1. Scope of research.

Subject of research / product	Type of activity / researched features / method	Related documents	No of research done
Cars and vehicles. Mobile machines. Objects and engineering devices. Objects with dimensions up to: L 22 m x W 7.7 m x H 8 m	Resistance to temperature. Temperature distribution in the thermo-climatic chamber. Temperature distribution in the tested objects. Scope: from -50 ° C to + 70 ° C Direct measurement method.	PB01 issue 4 of 19/02/2018	M
Special vehicles, including military, police and rescue applications. Objects with dimensions up to: L 22 m x W 7.7 m x H 8 m	Total resistance to elevated ambient temperature. Scope: to +70 °C Direct measurement method.	NO-06-A107:2005 p.4.2 PB01 issue 4 of 19/02/2018	M
	Total resistance to reduced ambient temperature. Scope: to -50 °C Direct measurement method.	NO-06-A107:2005 p.4.3 PB01 issue 4 of 19/02/2018	

II. The Quality Policy Statement – including, but not limited to: the Laboratory management's commitment to good professional practice and to the quality of testing in servicing of Laboratory customers, the management's statement of the Laboratory's standard of the service, the purpose of the management system related to quality and all other declarations required by PN-EN ISO/IEC 17025 [10].

III. Eight Quality Management System procedures:

P1 "Supervision of documents and records",
P2 "Overview of inquiries, offers and contracts, customer service, complaints",
P3 "Shopping",
P4 "Deviations, supervision of non-compliant tests, corrective and preventive actions",

P5 "Internal audit and management review",
P6 "Staff",
P7 "Accommodation and environmental conditions, equipment, measurement consistency",
P8 "Ensuring the quality of results",

IV. One technical procedure – PB01 "Temperature resistance tests and temperature distribution tests in the thermo-climatic chamber and in the objects tested in it" together with one related instruction: I01/PB01 „Operation of refrigeration aggregates and heaters in a thermo-climatic chamber".

V. External documents - such as: norms for example NO-06-A107:2005 [11], PCA/EA documents, legal documents such as: Rector's orders, health and safety documents.

2.3 Accreditation process

The general accreditation process is described in DA-01 [9] document. It is always composed of documents review and on-site assessments. First, the organization has to submit an application, then the application is checked by the PCA staff and the assessment process is organized together with the appointment of the evaluation team. Next, the submitted documentation is checked by the team of auditors and an initial visit of auditors is optionally arranged. Then, an on-site assessment is carried out and after that auditors check the implementation of corrective actions. Finally, PCA makes a summary of the assessment results and the decision about accreditation and its scope is taken. If the decision is positive, the Laboratory is awarded the accreditation and is subjected to an annual supervision process. According to the act, the accreditation process cannot last longer than 12 months.

3 Guidelines and conclusions

The development of documentation as well as accreditation obtained by the Laboratory of Techno-climatic Research and Heavy Duty Machines contributed to the organization and improvement of the techno-climatic studies carried out at the Cracow University of Technology. The increasing range of requirements related to the resistance of machines and devices to extreme climatic conditions, prompts the Laboratory staff to develop and implement further accredited procedures: the ones related to the determination of the impact of humidity and/or light radiation. Another issue is to obtain the Accreditation in Defence and Security, which will confirm the competence of the Laboratory to perform activities related to the assessment of compliance of products for the needs of national defence and security in terms of features related to their resistance to different climatic conditions.

Regardless of researches focused on external commercial client, the chamber may be used to provide the extreme environments (temperature, humidity, light) in scientific investigations involving biotechnology processes [12, 13] or material properties [14-19] in real scale. The chamber has been already used to test human resistance to extreme conditions, what may be useful in construction models of human-company relationships [20-23]. Such experiments lead to huge datasets which typically require analysis [24, 25] deeper than typical statistical postprocessing. Certainly, the climatic chamber will be intensively used.

References

1. W. Cichocki, S. Michałowski, J. Pobędza, *Examination of working machines and engineering objects under environmental exposure conditions* (Wyd. PK, Kraków, 2015)
2. W. Cichocki, A. Garbacik, K. Golec, K. Szewczyk, *A new concept of numerical modeling of heating and start-up systems* (Wyd. PK, Kraków, 1996)
3. J. Pobędza, A. Sobczyk, Adv. Mat. Res.-Switz., **849,** 100 (2014)
4. A. Sobczyk, *Improvement of Hydraulic System Efficiency by Means of Energy Recuperation* (Wyd. PK, Kraków, 2011)
5. K. Szewczyk, E. Lisowski, J. Pobędza, *Technical report of the R&D project CPBP 02.05* (Kraków, 1990)
6. K. Szewczyk, J. Pobędza J., *Proc. of 6th Sci. Conf. on Control, Drive, Fatigue Strength and Designing of Construction Machines* (Rynia, 1994).
7. K. Szewczyk, *The influence of reduced temperature on the characteristics of hydraulic systems of heavy machinery* (Wyd. PK, Kraków, 1991)
8. A. Gawlik, P. Kucybała, J. of KONES **24,** 35 (2017)
9. *Description of the accreditation system DA-01 9th Edition.* (Polish Centre for Accreditation. http://www.pca.gov.pl, acc. date 2018-04-26)
10. Technical standard PN-EN ISO/IEC 17025.
11. Technical standard NO-06-A107:2005.
12. E. Skrzypczak-Pietraszek, J. Pietraszek, Acta Biol. Crac. Ser. Bot., **51**, 62 (2009)
13. E. Skrzypczak-Pietraszek, I. Kwiecien, A. Goldyn, J. Pietraszek, Phytochem. Lett., **20**, 443 (2017)
14. R. Ulewicz, J. Selejdak, S. Borkowski, M. Jagusiak-Kocik, *Proc. 22rd International Conference on Metallurgy and Materials* (Brno, Czech Republic, 1926, 2013)
15. R. Ulewicz, P. Szataniak, F. Novy, *Proc. 23rd International Conference on Metallurgy and Materials* (Brno, Czech Republic, 784, 2014)
16. P. Szataniak, F. Novy, R. Ulewicz, *Proc. 23rd International Conference on Metallurgy and Materials* (Brno, Czech Republic, 778, 2014)
17. E. Augustyn, M.S. Kozien, *Acta Phys. Pol.* **125**, 164 (2014)
18. A. Dudek, A. Wronska, L. Adamczyk, *J. Solid State Electrochem.* **18**, 2973 (2014)
19. E. Lisowski, G. Filo, *Flow Meas. Instrum.* **53**, 269 (2017)
20. Kozien, E., *Proc. 20th Int. Sci. Conf. on Economic and Social Dev. (ESD)* (Prague, Czech Republic, 538, 2017)
21. A. Pacana, R. Ulewicz, *Polish J. Environ. Studies* **16**, 165 (2017)
22. Kozien, E., Kozien, M.S., *Proc. 26th Int. Sci. Conf. on Economic and Social Dev. – Building Resilient Society (ESD)* (Zagreb, Croatia, 210, 2017)
23. Kozien, E., Kozien, A., *Proc. 26th Int. Sci. Conf. on Economic and Social Dev. – Building Resilient Society (ESD)* (Zagreb, Croatia, 326, 2017)
24. J. Pietraszek, A. Gądek-Moszczak, T. Torunski, *Adv. Mat. Res.-Switz.,* **874**, 139 (2014)
25. J. Pietraszek, A. Goroshko, *Adv. Mat. Res.-Switz.,* **874**, 145 (2014)

ICT development on the example of selected software applied in manufacturing area

Agnieszka Szczotok[1,*], and *Renata* Dwornicka[2]
[1]Silesian University of Technology, Faculty of Materials Engineering and Metallurgy, Z. Krasinskiego Str. 8, 10-019 Katowice, Poland
[2]Cracow University of Technology, Institute of Applied Informatics, Al. Jana Pawła II 37, 31-864 Cracow, Poland

Abstract. This paper provides an insight into a few selected aspects of computer-aided software usage in the manufacturing processes. This research aims to point out a diversity of software used in manufacturing area which is available on the market and to summarize most of the functionalities of three selected types of software for production zone backing.

1 Introduction

Nowadays, we can observe a geographic increase in the number of individuals, producers and companies, as well as growing availability of distribution channels for products (e.g. online shopping, web enterprise management, etc.). It is related to the evolution of ICT (Information and Communication Technologies) and its increasing effect in the business world. ICT generally includes any communication device or application, including but not limited to radio, television, cellular phones, computer and network, hardware and software, satellite systems, and etc, as well as various services and applications associated with them, such as videoconferencing and distance learning. ICT has changed the concept of market place, buying and selling habits of consumers, advertising media, business operations and many more. ICT is most often used in mass production, as computer control makes it possible to produce many identical items very quickly. But ICT is also useful in small batch and even one-off work, because computer control enables complicated shapes to be produced more accurately than made by hand. ICT is used in manufacturing high quality products and it reduces the cost of production by minimizing the use of human labour in many industries and businesses. The importance of the ICT sector within OECD economies has been growing over the 1990s.

Digital technologies and access to high-speed broadband positively impacts the operations and profitability of all businesses across just about every industry and business type. Since high quality products are manufactured at low cost, people can buy good quality products at a relatively low price.

Quality control systems utilize production line information. They need visualization methods, data mining technologies, and statistical process control to analyze and improve

* Corresponding author: agnieszka.szczotok@polsl.pl

quality of products. The background for these functions is provided by ICT according to Fig. 1.

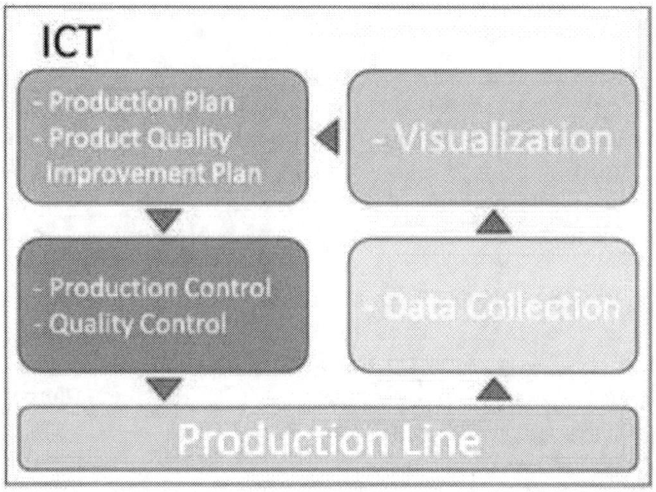

Fig. 1. Integration of quality management in the production zone with ICT.

An ICT system is a set-up consisting of hardware, software, data and the people who use them. It commonly includes communications technology, such as the Internet.

This paper presents the application of ICT in the range of software area to improve productivity and product quality through the development of production management systems is presented. The selection of examples of software in manufacturing application presented in the article was not random. The authors' intention was to point out a diversity of software used in manufacturing area that is available on the market and to summarize most of the functionalities of three selected types of software for production zone backing.

2 Overview of selected software

2.1 Impuls EVO

The first example of the presented software is Impuls EVO created by BPSC company [1]. This software is an example of the integrated MRPII/ERP class system. In the production area the software provides two functionalities: Discrete Manufacturing and Process Production.

The first one mentioned here enables: production planning for different periods of time, reflection of the multi-level structure of product, calculating technical production cost margin for different product configurations according to the technology and different calculation algorithms, real time management of the whole production process, as well as registration of tasks according to work cards, touch screens and the MES system.

The second functionality - Process Production - consists of the following possibilities: representation of material flow in the full structure of departments and production warehouses, freely chosen number of alternative recipes for every product, tree-type structure of recipes, registration and a precise analysis of the product path (traceability), planning tools that allow creation of tasks for job positions in production routes, estimation of cost of production and reacting to changes in prices of resources or production technologies, as well as supporting the fulfillment of HACCP system's requirements.

Undoubtedly an advantage of Impuls EVO software is a solution which makes it possible to connect majority of the external solutions functioning on the market. Usually they are tools for a limited, narrow field and clients are used to them. Among such solutions there are for example billing programs, B2B or B2C. Thanks to the integration of such solutions with the ERP system the management of all processes in the company will be conducted more comprehensively.

2.2 MONACO®

MONACO® is created by the company GT Systems 2 company [2]. It is a specialized software tool for record-keeping, archiving and creation of technical documentation for manufacturing enterprises. The concept technical documentation is not understood only as a set of computer data needed for correct production planning and management in ERP systems, but it covers a much greater scope, e.g. Items, Drawings, Bills of materials, Technological procedures, Inputs - outputs (tools, additional material...), Technological regulations (testing, welding...), Measuring and control regulations (CAQ,...), CAD Models, Technical regulations (constructions), Technical specifications - requirements of customers, Technical specifications - catalogues, Service regulations, Projects (works schedules), Calculations (reviews of expenses).

In the production zone MONACO® is a tool used especially for effective: offer proceeding (creation of preliminary technical designs, estimates of expenses and terms, preliminary specification of required inputs - material and production needs); production planning (creation of items, bills of materials, technological procedures, requirements regarding capacity); production (creation of production documentation - drawings, operations sketches, regulations, tools equipment, regulations for quality management (CAQ), etc.).

2.3 Tecnomatix

Tecnomatix is an example of digital manufacturing solutions that help to digitalize manufacturing and the process of transforming innovative ideas and raw materials into real products [3]. With Tecnomatix software, it is possible to design and configure 3D factory layouts for efficient factory operations using an integrated line planning environment, synchronize process design to deliver better plans, conduct assembly simulation for virtual verification of all the process operation steps and their details defined in manufacturing plans, optimize production logistics and material flow, integrate electronic work instructions with manufacturing planning, and even program and simulate robots. All these functionalities help to maximize production efficiency.

3 Discussion

When selecting the computer-aided quality control system, clients first focus on the functionality and capabilities of the production part. This is of key importance to them. A specification of clients' requirements emphasizes this part of the system. The company which plans to implement ICT, especially in the production zone, should most importantly, check how their business processes are conducted in this area and have a closer look at them.

ERP Impuls EVO supports scheduling and management of process production comprehensively. Thanks to the connection between a logistic chain, production processes, and a financial flow it is possible to improve the level of customer service and to rationalize of ordering and performance of contracts together with production optimization.

MONACO® technical information system gives possibilities for integration with ERP and CAD systems as well as commercial policy of the company. The system offers tools for keeping basic requirements for technical documentation: precision – quality (the more precise and "truer" the documentation is, the smoother and more floating the production is); documentation elaboration speed (its creation should take as little time and means as possible; at the same time the technical documentation department should provide information at the right time); generality (the system enables creation of technical documentation with an aim to minimize changes of "non-technical" nature, e.g. materials changes, swaps of departments, swaps of capacities - machines, etc.); portability (production shifting between respective plants or corporations should not have any influence, or only minor one, on technical documentation).

Tecnomatix connects the disciplines of planning and engineering with automation and control to create a smart factory model with all of the production process details, including the automation hardware and software.

On the other hand, it should be noted that such systems have numerous limitations in interfaces between them and other management systems utilized in industry e.g. environmental management [8]. The support for R&D is also limited only to the record-keeping without built-in analytical tools e.g. very interesting interdisciplinary investigations involving academia and industry (e.g. interviewed authors of the long-term research from biotechnological branch [9-13], materials science [14-19]) only hardly may be supported by these tools. In a consensus opinion, the interviewed authors believed that more sophisticated analytical tools must be provided from the outside and individually adapted to the nature of the their work, regardless of whether they are directly related to the optimization of production [20, 21] or research [22-25].

The functionalities of the analyzed software pointed in the work indicate many possible benefits for a manufacturer from having ICT tools available on the market. The results suggest both efficiency and economical gains through an increased usage of ICT tools.

4 Conclusion

Thanks to digital manufacturing solutions it is possible to improve management on all decision-making levels. Such solutions support effectiveness and efficiency, lower the costs of activity and increase its profitability.

Having analyzed the presented types of software and their functionalities for manufacturing, it is therefore considered that the knowledge of possibilities of the various commercially available types of software is necessary to decide which functionalities could be the best options for a particular manufacturing process. The ideal software application could require a connection of various functionalities from different types of software to meet all wants and needs.

Among the advantages of ICT the producers frequently point e.g. [4-7]: time and labor costs reduction and production increase, greater consistency and accuracy in production, systematic growth of capacity, safety of the workers involved, quality of the final product, improvement of customer service, possibility of making complex predictions quickly, and ability to communicate and work globally. It seems to be possible to achieve these goals in the case of taking into account the specificity of the particular industry and enterprise during selection and implementation of the manufacturing software.

References

1. BPSC homepage. http://www.bpsc.com.pl/
2. GT Systems 2, s.r.o. homepage. http://www.gtsystems2.sk

3. Siemens Product Lifecycle Management Software Inc. homepage. http://www.plm.automation.siemens.com
4. K. Czech-Dudek, Mechanik, **7**, 149 (2015)
5. M. Miecielica, W. Wiśniewski, *Computer-aided design of technological processes* (PWN Publisher, Warszawa, 2005)
6. R. Sirirattrakul, *Capacity Building Workshop on Information Society Statistics: Infrastructure, Household and other Indicators* (Bangkok, 2007)
7. Thailand National Statistical Office, *Measuring the impact of ICT use in business: the case of manufacturing in Thailand* (UNCTAD, New York and Geneva, 2008)
8. A. Pacana, R. Ulewicz, Polish J. Environ. Studies, **16**, 165 (2017)
9. L. Skrzypczak, E. Skrzypczak-Pietraszek, E. Lamer-Zarawska, B. Hojden, Acta Soc. Bot. Pol., **63**, 173 (1994)
10. E. Skrzypczak-Pietraszek, J. Pietraszek, Acta Biol. Crac. Ser. Bot., **51**, 62 (2009)
11. E. Skrzypczak-Pietraszek, J. Pietraszek, Chem. Biodivers., **11**, 562 (2014)
12. E. Skrzypczak-Pietraszek, New Biotech., **33**, 155 (2016)
13. E. Skrzypczak-Pietraszek, I. Kwiecien, A. Goldyn, J. Pietraszek, Phytochem. Lett., **20**, 443 (2017)
14. T. Pieczonka, J. Kazior, A. Tiziani, A. Molinari, J. Mater. Process. Technol., **64**, 327 (1997)
15. N. Radek, K. Bartkowiak, *Proc. Sixth Int. Wlt Conf. on Lasers in Manufacturing, , Physics Procedia Vol. 12* (Munich, Germany, 499, 2011)
16. T. Pieczonka, J. Kazior, A. Szewczyk-Nykiel, M. Hebda, M. Nykiel, Powder Metall., **55**, 354 (2012)
17. R. Ulewicz, P. Szataniak, F. Novy, *Proc. 23rd International Conference on Metallurgy and Materials* (Brno, Czech Republic, 784, 2014)
18. P. Szataniak, F. Novy, R. Ulewicz, *Proc. 23rd International Conference on Metallurgy and Materials* (Brno, Czech Republic, 778, 2014)
19. N. Radek, A. Sladek, J. Broncek, I. Bilska, A. Szczotok, Adv. Mat. Res.-Switz., **874**, 29 (2014)
20. J. Pietraszek, *Proc. 11th Int. Conf. on Artificial Intell. and Soft Comp. (ICAISC), Lect. Notes. Artif. Int. 7267* (Zakopane, Poland, 310, 2012)
21. J. Pietraszek, E. Skrzypczak-Pietraszek, Adv. Mat. Res.-Switz., **874**, 151 (2014)
22. T. Styrylska, J. Pietraszek, ZAMM-Z. Angew. Math. Mech., **72**, T537 (1992)
23. J. Pietraszek, A. Gądek-Moszczak, T. Torunski, Adv. Mat. Res.-Switz., **874**, 139 (2014)
24. N. Radek, J. Pietraszek, B. Antoszewski, Adv. Mat. Res.-Switz., **874**, 29 (2014)
25. A. Gadek-Moszczak, J. Pietraszek, B. Jasiewicz, S. Sikorska, L. Wojnar, *Proc. 6th Int. Conf. on Comp. Collective Intell. (ICCCI)* (Seoul, South Korea, 127-136 2014)

Problems in waste management in the aspect of the secondary use of plastics from WEEE

Katarzyna Kapustka[1,*], *Gerhard* Ziegmann[1], and *Dorota* Klimecka-Tatar[2]

[1]University of Technology, Clausthal Centre of Material Technology, Agricolastraße 2, 38678 Clausthal-Zellerfeld, Germany, EU
[2]Czestochowa University of Technology, Department of Production Engineering and Safety, ul. Armii Krajowej 19 B 42-200 Czestochowa, Poland, EU

Abstract. Waste of electrical and electronic equipment is one the fastest growing waste streams in the EU, with some 9 million tonnes generated in 2005 and expected to grow to more than 12 million tonnes by 2020. Electrical and electronic products contain substances, which are valuable as well as often also critical. The main aim of the paper is presentation the methodological approach to identification of bromine or chlorine presence in components (in WEEE). This followed by assessment of strengths and weaknesses of the most popular methods. The main analysis for identification of bromine and chlorine in plastics have been presented.

1 Introduction

The production of electric and electronic equipment (EEE) is one of the fastest growing areas, it has resulted in an increase waste of electrical and electronic equipment (WEEE) amount. In view of the ecological problems involved in the management of WEEE, European countries and organizations have drafted national legalisation to improve the reuse, recycling and other forms the point of waste treatment but also from recovery of valuable materials. Electrical and electronic products contain substances, which are valuable as well as often also critical (e.g. mass metals: copper, aluminium etc., precious metal: gold, silver etc., critical metals: indium, gallium etc.) and pose risks to the environment and human health (e.g. heavy metals: mercury, cadmium etc., flame retardants: pentabromophenol etc.) [1,2]. On the other hand, waste management, in WEEE area, gives the opportunity to re-use elements, the occurrence of which is quite limited, and the technology of production is expensive and complicated. Such an example may be the reprocessing of magnetic materials based on rare earth elements from electric motors and hard disk [3]. Therefore, the management of electro-intangible has become one of the important cognitive aspects, aims to combine the experience of engineering sciences with collecting and separation in the field of electronic waste management [4]. In Europe, exist high political interest for converting waste into a resource and a proper management of this waste flow. While secondary plastics increase at the process are largely redirected to the manufacture due to their high fraction purity, secondary plastic waste accruing after product

* Corresponding author: katarzyna.kapustka@tu-clausthal.de

use is recovered on a significantly smaller scale. The energetic recovery of plastic waste is still dominant in European country.

2 Global production of plastics and WEEE generation

This is in clear contradiction with the emerging circular economy policy framework where the value of products, materials and resources is maintained in the economy for as long as possible [5, 6]. In 2014, the world production of plastics was 270 million tonnes (Mt), with an EU production and demand of 50 Mt (18.5%) and 49 Mt, respectively and 2.8 Mt (5.8%) of this EU production used for electric and electronic equipment. The waste of electric and electronic separately collected in 2013 reached 3.5 Mt (6.5 kg/inhabitant), of which 2.7 Mt were recovered. The plastic fraction in European WEEE depends on categories and products and was estimated to be 2.3 Mt or about 20.6% of the weight of EEE in 2008 [7]. The mass flow of recycled plastics from WEEE separately sorted in the EU can be estimated to 0.48 Mt (0 2.4 Mt WEEE recovered/ 20% plastics in WEEE). It is worth noting that in WEEE the flame retardant (FR) additives are also found - chlorinated and brominated organic compounds or polymers are mixed with the plastics used in electric and electronic equipment to prevent fire damage. On average, 30% of the plastics in EEE contain FRs and halogenated compounds [8-10]. The life cycle environmental impacts of post-consumer plastics production from mixed plastic-rich WEEE treatment residues from perspective of the customers delivering the residues and the customers buying the obtained post-consumer recycled plastic is clearly superior to the alternatives (i.e. municipal solid waste incineration (MSWI) and virgin plastics production) [6].

Directive 2012/19/EE of European Parliament and of the Council on waste electrical and electronic equipment (WEEE Directive) aims to address this issue by implementing inter alia the principle of the extended producer responsibility and collection, recovery as well as joint recycling/preparing for reuse targets (Art. 11, Directive 2012/19/EU). The Commission implemented Regulation 2017/699 validating a regular methodology for: the calculation of the weight of EEE placed on the national market. In 2017, the Commission adopted a legislate proposal to introduce the use of substances in electrical and electronic equipment as hazardous. The activities provides for the creation of collection schemes where consumers return their used waste EEE free charge. The objective of these schemes is to increase the recycling and re-use of such products (Fig.1).

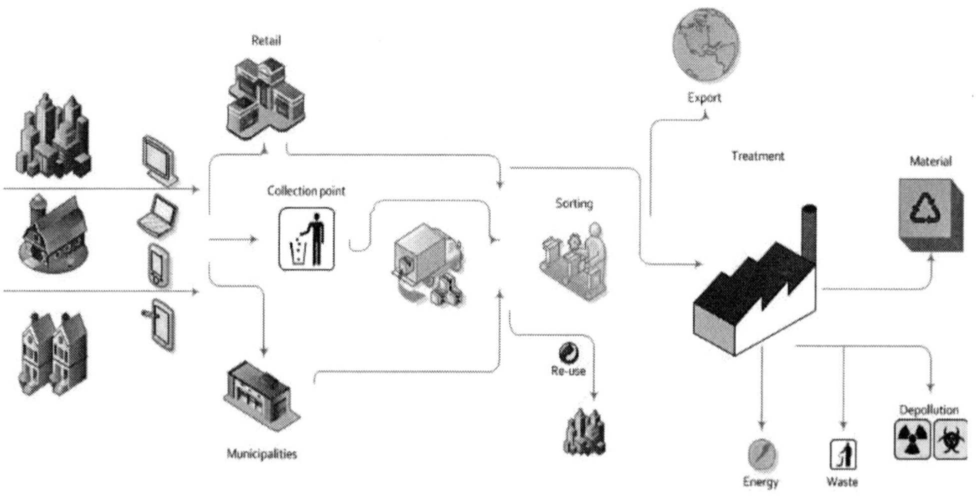

Fig. 1. Circuit diagram, from electric and electronic equipment (EEE) through waste of electrical and electronic equipment (WEEE) to recycled material.

3 Management of materials obtained from WEEE

The WEEE is called the mixture of materials and components due their hazardous content, and if not properly utilize, can cause major environmental and health problems. The role of WEEE utilization management is improvement of resource by increased collection, transformation and recycling of electronics at the end of their life is essential. Secondary market operations for EEE, which involve repair, replacement of spare parts, refurbishment and reuse, and retrofitting, should be facilitated to promote a circular economy in the Union [11]. Besides the problem of low collection rates, the recycling of plastics are very low [12 14]. Plastics are a considerable fraction of WEEE and contributes to the total generation of post-consumer plastic waste in the EU-27, Norway and Switzerland in 2008 at 5% [15], but the presence of brominated flame retardants (BFR) as well as various plastic types and missing incentives (e.g. economic benefits of energy recovery) hampers the recycling of plastic obtained from WEEE [7,13]. The high halogen content in theses plastics with respect to their end of life management are the reasons the popular environmental and health issue problem. In fact, halogenated plastics under heating treatment can release irritant and corrosive gases, such as HCl or HBr. Also, in the presence of oxygen, metal oxides in the ashes, and temperatures between 200 and 500°C, HCl and HBr can be possible precursor of toxic substances such as polyhalogenated dioxins and furans and various other organic halogen compounds [16]. Equal if they undergo pyrolysis, a thermal treatment at low temperatures under inert atmosphere, dehalogenation of pyrolysis oil and gas is essential to make them commercially sufficient. For these reasons, quality parameters of WEE plastics is depend on the concentrations of bromine and chloride.

4 Analysis of bromine and chlorine presence in WEEE plastics

Sample pre-treatment followed by gas chromatography-mass spectrometry (GC-MS) is instrumental analysis technique used to distinguish restricted from non-restricted BFRs and for their density [17, 18]. Pre-treatment let on a complete dissolution of the polymer in an relevant solvent, normally at room temperature, in sequence to avoid the thermal degradation of BFR [19, 20]. Alternatively, a solid-liquid extraction is often adopted along with a series of extraction procedures, such as those performed by shaking, ultrasonication, microwave extraction, pressurized solvent, and supercritical CO_2 [20-25]. In addition to GC-MS, other analytical techniques used after the sample pretreatment and high performance liquid chromatograpy coupled with MS or UV detector, infrared analysis, and GC with an electron capture detector [21, 26-28]. When the BFRs are copolymerized into the backbone of the polymer, their separation and quantification by the methods can be very complicated to achieve [29, 30]. Unfortunately, as the sample amount to be subjected to pyrolysis is very small (less than 1 mg), this technique is not appropriate for heterogeneous plastic waste [31]. Furthermore, some pyrolysis products may negatively affect the results of the analysis and identification of all the compounds.

The alternative destructive but quite simple analysis that could be identically used for plastic materials is the elemental analysis of chlorine and bromine by ion chromatography (IC) or titration. The analysis consists of high-pressure oxygen bomb combustion of the sample followed by halogens recovery in a basic suspension [32, 33]. Methods based upon non-destructive technics bring a useful perspective to check the eventual presence of bromine and chlorine. They involve infrared equipped with an attenuated total reflectance accessory, energy dispersive X-ray fluorescence spectroscopy (ED-XRF) and Raman spectroscopy [18, 21, 34, 35]. ED-XRF presents elementary results for bromine and chlorine, while Raman spectroscopy or infrared spectroscopy can be helpful in determining specific halide compounds. Kikuchi et al. in their research [35] indicated that BFRs in acrylonitrile butadiene styrene (ABS) that Raman spectroscopy is less sensitive to matrix interference, the results are greatly affected by this problem, which may mask the results, in particular for small compound contents [18]. Morf et al. published a quantitative XRF analysis [36], has been implemented on grained samples mixed with a matrix modifier (silicates and/or lime), for calibration used a similar matrix (stand on a calibration for geological samples with standard reference material). Although Aldrian et al. assessments a handled XRF [37], provided with the manufacturer's software, for a quantitative analysis on WEEE. Many authors observed the importance of routine calibration of the XRF analyser [33], as in the case of manufacture installed elemental rates are appreciably different from calibration standards. To conclude, the lack of standards with matrices as identical as possible to those of the samples can lead to less exact results. Consequently, it might be profitable to investigate the results secured by using more authoritative laboratory methods, and regard XRF as a very usable checking test for determining elemental concentration [33,38]. In table 1 the concentration of bromine and chlorine based on results reported in literature are presented.

Table 1. Review of Cl and Br analysis of WEEE plastics by different techniques [38-41].

Type of obtained plastic material	Br, %	Cl, %	Possible methods of analysis
TV and monitors	0.01-1.3	0.01-10.6	Semiquantitative XRF
small household appliances - small WEEE	4.5	1.2	Not mentioned HXRF XRF ICP-OES
WEEE		0.02-3.9	Total reflection XRF HXRF
Cooling devices - fridge freezers	1.3	-	IC

According to literature data in the plastic obtained from the small pieces of household equipment (small WEEE) identified even 4.4% of Br and 1.2 of Cl, while in waste from TV and PC even 10.6 % Cl [38-41]. Such high contents bromine and chlorine in plastic material obteined from differend kind of WEEE make these materials particularly demanding in terms of processing and disposal.

5 Summary

Literature review and familiarization with the basic legal acts force in Europe, allows to draw the conclusion, that waste management is a strategic point in the development of materials management. It can be noticed that the significant development of civilization contributes to the increase in the number of manufactured goods (in particular electrical and electronic equipment), which service is limited and, as a consequence, increases the number of waste. Difficulties arising from the recovery of WEEE materials arise from the complexity of this waste. WEEE is called the mixture of materials and components due their hazardous content, and if not properly utilize, can cause major environmental and health problems. The most difficult material to recover are polymeric materials, and in particular those that contain bromine and chlorine in chemical composition. The high halogen content in WEEE plastics with respect to their end of life management are the reasons of the popular health problem, cause the product of precursor of toxic substances such as polyhalogenated dioxins and furans and various other organic halogen compounds.

Therefore, methods are sought for easy identification of chemical components in plastic waste, which will allow quick reactions to WEEE utilization.

References

1. B.H. Robinson, Sci. Total. Environ., **408**(2), 183 (2009)
2. O. Tsydenova, M. Bengtsson, Waste Manag., **31**(1), 45 (2011)
3. K. Kapustka, G.Ziegmann, D.Klimecka-Tatar, Production Engineering Archives **17**, 36 (2017)
4. M. Ingaldi, S. Jursovà, *METAL 2013: 22nd International Conference On Metallurgy And Materials* (TANGER, Ostrava, 1779, 2003)
5. European Commission. Communication from the Commission to the European Parliament, the Council, the European Economic and Social Committee and the Committee of the Regions - Closing the loop - An EU action plan for the Circular Economy [Internet]. Brussels, Belgium, (2015)
6. P. Wäger, R. Hischier, Sci. Total. Environ., **529,** 158 (2015)
7. Plastic Europe, *Plastics the facts 2016* (2017)
8. BMUB. Elekro - und Elektronikgeräte in Deutschland: Daten 2009 und 2010 zur Erfasung, Behandlung und Wiederverwendung [Internet] (2013)
9. H. Wilts, N. von Gries, I. Dehne, R. Oetjen-Dehne, N. Buschow, J. Sanden. *Entwicklung von Instrumenten und Maßnahmen zur Steigerung des Einsatzes von Sekundärrohstoffen - mit Schwerpunkt Sekundärkunststoffe* (Report No., FKZ 3712 933 340)
10. M. Alaee, P. Arias, A. Sjödin, A. Bergman. Environ. Int., **29**, 683 (2003)
11. M. Bientinesi, L. Petarca. Waste Manag., **29**, 1095 (2009)
12. R. Widmer, H. Oswald-Krapf H. D. Sinha-Khetriwal, M. Schnellmann, H.Böni. Environ. Impact Assess. Rev., **25**, 436 (2005)
13. Directive of the European Parliament and the Concil amending Directive 2011/65/EU on the restriction of the use of certain hazardous substances in electrical and electronic equipment (2017)
14. J. Baxter, M. Wahlstrom, M. Zu Castell-Rüdenhausen , A. Frane, M.Stare, S.Lokke, et al. *Plastic value chains: Case: WEEE (Waste Electric and electronic equipment) in the Nordic region* (Nordic Council of Ministers 2016)
15. J. Huisman, P. Leroy, F. Tertre, M. Ljunggren, et al., *Prospecting Secondary Raw Materials in the Urban Mine and mining waste (ProSUM)- Final Report*, (Brussels, Belgium, 2017)
16. H. Friege, Waste Manag. Res., **30**, 3 (2012)
17. B. Beccagutti, L. Cafiero, M. Pietrantonio, S. Pucciarmati, Sustainability-Basel, **8**, 1101 (2016)
18. M. Brebu, T. Bhaskar, A. Muto, Y. Sakata, Chemosphere, **64**, 1020-1025 (2009)
19. E. Plinke, N. Wenk, G. Wolff, D. Castiglione, M. Palmark, *Mechanical Recycling of PVC Waste, Study for DGXI of the European Commission (B4-3040/98/000821/MAR/E3)* (European Commission, Brussels, Belgium, 2001)
20. S. Kemmlein, D. Herzkeb, R.J. Law, J. Chromatogr, **1216**, 320 (2009)
21. S. Hamm, M. Strikkeling, P. Ranken, K. Rothenbacher, Chemosphere, **44**, 1353 (2001)

22. V. Zitko, Chemosphere, **38**, 629 (1999)
23. Altwaiq, M. Wolf, R. van Eldik, Anal Chim, Acta, **491**, 111 (2003)
24. M. Schlummer, L. Gruber, A. Maurer, G. Wolz, R. van Eldik, Chemosphere, **67**, 1866 (2007)
25. T. Imai, S. Hamm, K. Rothenbacher, Environ. Sci. Technol, **37**, 652 (2003)
26. G.T. Tomy, T. Halldorson R. Danell, R.K. Law, et.al, Mass Spectrom., **19**, 2819 (2005)
27. B. Binici, M. Bilsel, M. Karakas, I. Koyuncu, A.C. Goren, Talanta., **116**, 417 (2013)
28. M. Schlummer, F. Brandl, A. Mäurer, R. van Eldik, Chromatogr., **1065**, 39 (2005)
29. Ballesteros-Gomez, S.H. Bdandsma J. de Boer, P.E.G Leonards, Chemosphere, **116**, 10 (2014)
30. M.P. Luda, A.I. Balabanovich, M. Zanetti, D. Guaratto, Poly. Degrad., **92**, 1088 (2007)
31. F.C-Y. Wang. J. Chromatogr., **866**, 225 (2000)
32. J. Aguado, D.P. Serrano, G. San Migueal, J. Polym. Environ., **15**, 107 (2007)
33. J.A. Onwudili, P.T. Williams, Chemosphere, **74**, 787 (2009)
34. E. Dimitrakakis, A. Janz, B. Bilitewski, E. Gidarakos, Waste Manag., **29**, 2700 (2009)
35. M. Riess, T. Ernst, R. Popp, B. Müller, H. Thoma, O. Vierle, M. Wolf, R. van Eldik, Chromosphere, **40**, 937 (2000)
36. S. Kikuchi, K. Kawauchi, S. Ooki, M. Kurosawa, H. Honjho, T. Yagishita, Anal. Sci., **20**, 1111 (2004)
37. L.S. Morf, J. Tremp, R. Gloor, F. Schupisser, M. Stengele, R. Taverna, Waste Manag., **27**, 1306 (2009)
38. Aldrian, A. Ledersteger, R. Pomberger, Waste Manag., **38**, 297 (2017)
39. R. Taurino, P. Pozzi, T. Zanasi, J. Anal.Chem., **368**, 235 (2000)
40. B.M. Caballerom I. de Marco, A. Adrados, et. al., Waste Manag., **57**, 226-234 (2016)
41. O. Sindiku, J. Babayemi, M. Schlummer, M- Schleup, A. Watson, R. Weber, Environ. Sci. Pollut. Res., **22**, 14489 (2015)

Health problems among employees in small enterprises as a result of improper OHS management

Marta Niciejewska[1,*], and *Dorota* Klimecka-Tatar[1]

[1]Czestochowa University of Technology, Department of Production Engineering and Safety, ul. Armii Krajowej 19 B, 42-200 Czestochowa, Poland, EU

Abstract. Occupational health problems are rarely underestimated both by employers and employees themselves. Lack of awareness about the impact of the environment on the health of the human body results in more frequent accidents, the sources of which are seen in the health problems of employees. Lifestyle, as well as the management of occupational health and safety, affect work quality and safety. It this paper ailments that have a significant impact on the health of the employee, which are the result of improper management of occupational safety and health in small enterprises, have been presented. The obtained results also indicate factors that cause various health problems among employees, with particular emphasis on stress symptoms.

1 Introduction

Health problems resulting from work may have a different origin. The literature on the subject indicates a whole range of threats that negatively affect the health and even the life of the employee - in the physical, biological, chemical and psychosocial areas [1-3]. Taking into account the level of occupational safety as well as health care in enterprises, it can be stated that all health-related problems are due to the employee's insufficient working conditions, but also to the lack of awareness of safe work [4-6]. To the most frequently indicated health problems associated with professional work include musculoskeletal disorders [5,7-9]. According to numerous reports and publications musculoskeletal disorders are the most widespread health-related problems. Almost 24% of employees from EU countries complain about back pain, and 22% - for muscle pain. These diseases are more common in the new Member States, with 39 and 36% respectively. Musculoskeletal disorders do not only cause physical suffering and loss of income in the individual dimension, but they are also costly for businesses and the economy in general. Although they can occur with every employee, they can be effectively prevented by assessing the professional activities performed and by introducing appropriate preventive measures and controlling their effectiveness [10]. In Poland, the Central Statistical Office also indicates musculoskeletal disorders as those that are the most common among employees of the

* Corresponding author: marta.niciejewska@wz.pcz.pl

following three sectors: health care, transport, trade as well as professional builders. The reasons are not seen only in the hazards of the working environment, but also in the phenomenon of increased work time. Work often dominates in human life, and the lack of balance between work and private life results in health problems [1,11]. Time pressure causes, that employees forget about individual health prophylaxis, e.g. basic blood control tests, cholesterol levels, a balanced diet, outdoor activities. In addition, the sedentary nature of the work, as well as the non-ergonomically designed work place result in the occurrence of musculoskeletal disorders, exposure to noise generates irritability and headaches, and the disorder of interpersonal relationships in turn creates a sense of burnout and stress [13-15]. Consequently, these factors can contribute to the creation of dangerous occupational diseases.

The information from European Agency for Safety and Health at Work, EU-OSHA reports, that more often "hidden" threats, which include psychosocial risks and stress, have a decisive impact on human health. Psychosocial hazards arise from poor work planning, poor work organization and poor management, and from an unfavorable social work environment. They can lead to negative psychological, physical and social effects such as occupational stress, burnout or depression [14,16]. As with other issues related to mental health, the problem of stress is often misunderstood or stigmatized. The effect of psychosocial risks is reaching for stimulants - cigarettes, alcohol, antidepressants. They, in turn, generate further health problems among employees. However, it is difficult to determine which of these stimulators are only a reflection of the working conditions. There is a constant need, for employers to implement and for workers to compliance, preventive actions that will increase the level of safe working conditions [1].

2 Methodology of research

In this work, to achieve cognitive goals, the original questionnaire regarding assessment of factors that affect job security in small enterprises has been used. Respondents, in the number equal to 1006 came from small businesses - manufacturing, service, retail and mixed, survey questionnaires were filled out anonymously by employees (one questionnaire = a single economic entity). For the presentation and careful analysis of the received data the statistical methods have been used. Data are presented in nominal and ordinal scale. There were used mainly structural indicators, correlation analysis using the coefficient tau-Kendall, $\chi 2$ test of independence and φ- Yule correlation coefficient. In the field of mathematical statistics, the Mann-Whitney test was used as an alternative to the medium test.

3 Results and discussion

During the survey, the respondents had to indicate in the questionnaire which of the mentioned ailments, in their opinion, are related to the work performed on a daily basis. The obtained results are presented in Tables 1-6 (according to the average). In Table 1 a distribution of indicated factors is presented which, in the opinion of respondents, most often occur in work conditions - according to the sector of activity. Respondents did not indicate high significance of the arduous working conditions effects. The maximum average rating was 3.145 for musculoskeletal disorders. In each case, the assessment for the manufacturing enterprises sector was higher than the ratings of the service sector. Significant differences were noted for assessments of musculoskeletal disorders ($Z = 2.252$, $p = 0.024$), symptoms of chronic fatigue ($Z = 7.358$, $p < 0.001$) and burnout ($Z = 5.103$, $p < 0.001$). In this aspect, physical fatigue associated with the production sector generates higher ratings. In the other hand monotonous work inhibits creativity and in the long run causes burnout.

Table 1. Distribution of indications of the effects of arduous work conditions occurring most often in the opinion of respondents and the results of the the Mann-Whitney test - significance of differences between the production and service sector ($\alpha = 0.05$).

Consequences of arduous work conditions	Average rating in enterprises					Median	Rezults of M-W test	
	Production	Service	Trade	Mixed	Total		Z	P
Musculoskeletal disorders	3.207	3.132	3.000	3.129	3.145	3.000	2.252	0.024
Chronic fatigue	2.751	2.368	2.778	2.621	2.510	2.000	7.358	0.000
Burnout	2.390	2.178	2.500	2.293	2.256	2.000	5.103	0.000
Cardiovascular disease	2.212	2.156	2.250	2.250	2.186	2.000	1.506	0.132

The distribution of the most frequent indications according to musculoskeletal disorders is presented in Table 1.2. Respondents did not indicate very important reasons for musculoskeletal disorders. Repeated activities ($Z = 3.621$, $p < 0.001$), non-ergonomic work positions ($Z = 3.115$, $p < 0.001$) and forced work position ($Z = 1.977$, $p = 0.048$) were clearly indicated as important in production enterprises. Manual transport of materials was rated as higher by the employees of service industry service ($Z = -4,336$, $p < 0.001$). In the service sector, which is more often associated with white-collar workers, manual material transport seems to be too heavy work, which cannot be said about the production sector.

Table 2. Distribution of indications of the couses of musculoskeletal disorders occurring most often in the opinion of respondents and the results of the the Mann-Whitney test - significance of differences between the production and service sector ($\alpha = 0.05$)

Couses for musculoskeletal disorders	Average rating in enterprises					Median	Rezults of M-W test	
	Production	Service	Trade	Mixed	Total		Z	P
Repeated activities	3.241	3.100	2.750	3.093	3.120	3.000	3.621	0.000
Incorrect work position	3.029	2.963	2.667	2.736	2.936	3.000	1.406	0.160
Forced position at work	3.062	2.975	2.917	2.936	2.988	3.000	1.977	0.048
Manual transport of materials	2.656	2.980	2.750	2.550	2.834	3.000	4.336	0.000
Non-ergonomic work station	2.572	2.358	2.389	2.143	2.370	2.000	3.115	0.002

In table 3 a distribution of indications of the effects of arduous work conditions was presented, which occur most often in the opinion of respondents according to their age. Between the impact assessments arduous working conditions and the age of the respondents there are weak, statistically significant positive correlations. They concern the assessment of the significance of musculoskeletal disorders ($r\tau = 0.105$, $p < 0.001$), symptoms of chronic fatigue ($r\tau = 0.140$, $p < 0.001$), symptoms of burnout ($r\tau = 0.073$, $p = 0.001$) and cardiovascular disease ($r\tau = 0.141$, $p < 0.001$). With age, the assessment of the significance of these ailments increases.

Table 3. Distribution of indications of the effects of arduous work conditions occurring most often in the opinion of respondents and the value of the tau-Kendall correlation coefficient between the age of musculoskeletal disorders (α = 0.05)

Consequences of arduous work conditions	Age						Total	Correlation	
	18-24	25-34	35-44	45-54	55-64	65 and more		Tau	P
Musculoskeletal disorders	2.841	2.722	3.168	3.218	3.273	3.250	3.145	0.105	0.000
Chronic fatigue	2.638	2.051	2.469	2.621	2.909	3.250	2.510	0.140	0.000
Burnout	2.174	1.871	2.346	2.242	2.333	3.250	2.256	0.073	0.001
Cardiovascular disease	2.087	1.847	2.139	2.321	2.303	2.750	2.186	0.141	0.000

Table 3 presents the results concerning the distribution of indications of the causes of musculoskeletal disorders, which occur most often, in the respondents' opinion, relatively to their age. There are weak, statistically significant positive links between the assessment of the causes of musculoskeletal disorders and the respondents' age. They concern the assessment of the causes of musculoskeletal disorders such as: frequent repetitive activities ($r\tau = 0.108$, $p < 0.001$), incorrect position at work ($r\tau = 0.078$, $p < 0.001$), forced position at work ($r\tau = 0.102$, $p < 0.001$), manual material transport ($r\tau = 0.100$, $p < 0.001$) and non-ergonomically designed and equipped work stations ($r\tau = 0.077$, $p < 0.001$). With age, the assessment of the importance of these reasons increases.

Table 4. Distribution of indications of the couses of musculoskeletal disorders occurring most often in the opinion of respondents and the value of the tau-Kendall correlation coefficient between the age of musculoskeletal disorders (α = 0.05)

Consequences of arduous work conditions	Age						Total	Correlation	
	18-24	25-34	35-44	45-54	55-64	65 and more		Tau	P
Repeated activities	2.870	2.733	3.076	3.222	3.364	3.500	3.120	0.108	0.000
Incorrect work position	2.725	2.522	2.969	2.966	3.242	3.250	2.936	0.078	0.000
Forced position at work	2.725	2.561	3.024	3.024	3.273	4.000	2.988	0.102	0.000
Manual transport of materials	2.391	2.443	2.890	2.925	2.758	2.750	2.834	0.100	0.000
Non-ergonomic work station	2.246	2.063	2.353	2.406	2.636	3.000	2.370	0.077	0.000

In the study, respondents made indications of stress symptoms that occur most frequently in their lives and are equated with work. Table 5 presents results on the distribution of stress symptoms, which most often appear, in the opinion of respondents, relatively to their age. There is a positive weak correlation between age and stress symptoms such as: increased blood pressure ($r\tau = 0.050$, $p = 0.018$), increased cholesterol ($r\tau = 0.146$, $p < 0.001$), anger ($r\tau = 0.077$, $p < 0.001$) and smoking ($r\tau = 0.120$, $p < 0.001$). With age, the assessment of the significance of these symptoms of stress increases. There was also a negative statistically significant correlation between age and the assessment of stress symptoms such as: depression ($r\tau = -0{,}050$, $p = 0.017$), absenteeism at work ($r\tau = -0.0545$, $p = 0.011$) and desire to change job ($r\tau = -0.190$, $p < 0.001$). Older employees have a decidedly less willingness to change their job and, partly because of the fear, they try to not abuse absenteeism.

Table 5. Distribution of indications of stress symptoms occurring most often in the opinion of respondents according to the age of respondents, and the value of the correlation coefficient of tau-Kendall between age and stress symptoms (α = 0.05)

Consequences of arduous work conditions	Age						Total	Correlation	
	18-24	25-34	35-44	45-54	55-64	65 and more		Tau	P
Increased blood pressure	2.377	2.153	2.380	2.474	2.576	3.000	2.429	0.050	0.018
Acceleration of heart rate	2.406	2.231	2.432	2.491	2.636	3.000	2.478	0.024	0.248
Increased cholesterol	1.899	1.890	2.207	2.317	2.212	2.500	2.205	0.146	0.000
Anger	2.551	2.071	2.437	2.502	2.697	3.250	2.455	0.077	0.000
Anxiety	2.145	1.827	2.089	2.061	2.152	2.500	2.084	0.012	0.559
Depression	1.971	1.851	2.037	1.997	1.939	2.000	2.031	-0.050	0.017
Smoking	2.362	2.349	2.647	2.802	2.879	3.500	2.687	0.120	0.000
Alcohol abuse	2.043	1.867	2.215	2.048	2.152	2.750	2.132	-0.015	0.464
Insomnia	2.188	2.051	2.241	2.263	2.424	2.750	2.270	0.028	0.187
Absenteeism at work	2.043	1.831	2.073	2.020	1.970	2.000	2.053	-0.054	0.011
Tendency to make mistakes	2.275	1.882	2.113	2.205	2.121	2.500	2.157	0.024	0.259
The desire to change jobs	2.507	2.235	2.387	2.191	2.242	2.500	2.367	-0.190	0.000

Table 6. Distribution of indications of stress symptoms occurring most frequently in the opinion of respondents by the sector of activity, as well as results of the test of significance of differences between the indications for the production and service sector (α = 0.05)

Couses for musculoskeletal disorders	Average rating in enterprises					Median	Rezults of M-W test	
	Production	Service	Trade	Mixed	Total		Z	P
Increased blood pressure	2.510	2.365	2.444	2.557	2.429	2.000	2.457	0.014
Acceleration of heart rate	2.556	2.390	2.611	2.679	2.478	2.000	3.382	0.001
Increased cholesterol	2.137	2.248	2.222	2.136	2.205	2.000	-3.634	0.000
Anger	2.444	2.394	2.778	2.650	2.445	2.000	0.299	0.765
Anxiety	2.066	2.061	2.417	2.129	2.084	2.000	-0.597	0.551
Depression	1.975	2.037	2.194	2.057	2.031	2.000	-2.197	0.028
Smoking	2.614	2.756	2.194	2.650	2.687	3.000	-2.374	0.018
Alcohol abuse	2.041	2.188	2.000	2.086	2.132	2.000	-2.859	0.004
Insomnia	2.187	2.324	2.389	2.157	2.270	2.000	-3.160	0.002
Absenteeism at work	2.062	2.051	2.083	2.036	2.053	2.000	0.478	0.633
Tendency to make mistakes	2.170	2.129	2.528	2.157	2.157	2.000	1.141	0.254
The desire to change jobs	2.386	2.312	2.889	2.429	2.367	2.000	0.417	0.677
Increased blood pressure	2.510	2.365	2.444	2.557	2.429	2.000	2.457	0.014

Table 6 presents results on the distribution of stress symptoms, which occur, in the opinion of respondents, according to the indications of the sector of activity. The surveyed employees perceive possible symptoms of stress only to a limited extent - as significant or moderately significant. Nevertheless, there are significant differences in the assessment of these symptoms from the employees of the production and service sectors. In the service sector, the role of increased cholesterol is significantly higher ($Z = -3.634$, $p < 0.001$), insomnia ($Z = -3.160$, $p = 0.002$), alcohol abuse ($Z = -2.859$, $p = 0.004$), smoking ($Z = -2.3374$, $p = 0.018$) and depression ($Z = -2.197$, $p = 0.028$). On the other hand, the production sector more often sees symptoms of stress such as acceleration of heart rate ($Z = 3.382$, $p = 0.001$) and increased blood pressure ($Z = 2.457$, $p = 0.014$).

4 Summary

Health prevention activities in the workplace, elaborated among others by EU-OSHA are available for every enterprise. Unfortunately, in the smallest business entities this subject is downplayed and often overlooked. Based on the presented results it can be concluded that musculoskeletal disorders result mainly from the activities of a frequently repeated, non-ergonomic workplace or forced body position during the performance of tasks; chronic fatigue and burnout are also a problem of a modern employee who works too hard and too hard in relation to the time spent on strength regeneration and rest; workload intensify the use of stimulants, which seemingly support professional performance and eliminate fatigue.

References

1. N. Ulutasdemir, M. Kilic, Ö. Zeki, F. Begendi, Ann. Glob. Health., **81**, 503 (2015)
2. L.L. Schmitz, J. Health Econ., **50**, 183 (2016)
3. J. Mullen, E.K. Kelloway, M. Teed, Safety Sci., **91**, 405 (2017)
4. D. Klimecka-Tatar, M. Niciejewska, Production Engineering Archives, **13**, 49 (2016)
5. P. Arezes, F. Serranheira, Int. J. Ind. Ergonom., **60**, 1 (2017)
6. Y. Kim, J. Park, M. Park, Saf .Health Work, **7**, 89 (2016)
7. https://oshwiki.eu/wiki/Risk_factors_for_musculoskeletal_disorders_development:_hand-arm_tasks, EUOSHA.
8. L.E. Charles, C. C. Ma, C. M. Burchfiel, R. G. Dong, Saf .Health Work (to be published)
9. Z.G. d. Santos, L. Vieira, G. Balbinotti, Procedia Manufacturing, **3**, 5947 (2015)
10. J. Costa-Font, M. Ljunge, Econ. Hum. Biol., **28**, 119 (2018)
11. M.E. Major, N. Vézina, Int. J. Ind. Ergonom., **48**, 149 (2015)
12. D.J. Weber, W. A. Rutala, Infect. Dis. Clin. N. Am., **30**, 729 (2016)
13. V. Schiele, H. Schmitz, J. Health. Econ., **49**, 59 (2016)
14. T. Rosenthal, A. Alter, J. Am. Soc. Hypertens., **6**, 2 (2012)
15. K. Hirokawa, T. Ohira, M. Nagayoshi, M. Kajiura, H. Imano, A. Kitamura, M. Kiyama, T. Okada, H. Iso, Preventive medicine reports, **4**, 61 (2016)
16. M. Gavrilă-Ardelean, O. D. Moldovan, Procedia - Social and Behavioral Sciences, **159**, 589 (2014)
17. Pacana, R. Ulewicz, Polish Journal of Management Studies, **16**, 165-174 (2017)

MATEC Web of Conferences **183**, 01013 (2018) https://doi.org/10.1051/matecconf/201818301013
QPI 2018

Level of compliance with the requirements of corporate social responsibility in Czech enterprises

Zdenka Hofbruckerová[1,*], *Petra* Halfarová[1], and *Andrea* Uherková[1]

[1]VŠB – Technical University of Ostrava, Department of Quality Management, 17. listopadu 15/2172, 708 33 Ostrava – Poruba, Czech Republic

Abstract. Corporate Social Responsibility (CSR) concept is focused on understanding and managing the impacts a company has on the world around it, and doing so as a part of core business activities. CSR contains three parts: Profit (economical part), People (social aspects – employees and stakeholders) and Planet (environment). The papers deal with basic features, objectives and outputs of CSR concept in practical life of Czech enterprises. The main aim of the research was to define if organizations referring themselves as socially responsible do implement activities of all the three pillars equally and integrate them into their day-to-day activities.

1 Introduction

The most companies nowadays practice some form or part of Corporate Social Responsibility concept (CSR). The goal of the paper is to analyse if organizations participating on Governor Award for Social Responsibility and Social Responsibility Poll in Moravian – Silesian Region know and are able to interpret required aspects in all three pillars of CSR management system according to ČSN 01 0391 standard in parallel.

1.1 Corporate Social Responsibility

Corporate social responsibility (CSR) is voluntary organization's initiative to assess and take responsibility for the company's effects on environmental management, economical growth and social progress. The term generally applies to efforts that go beyond law requirements with respect to communication with stakeholders. The concept of social responsibility of organizations is built on three pillars, which are collectively referred to as the triple bottom line (TBL) or 3P strategy. They are: Profit / economic, People / social and Planet / environmental area. The areas represent a threefold overall impact of business on society. [1] That shows that it is possible to achieve positive financial results, which lead to a reduced impact on the environment and society without dealing additional damage [2].
This attitude are accepted all around the world for example in India [3], Greece [4] or in Italy [5].

* Corresponding author: zdenka.hofbruckerova@vsb.cz

1.2 ČSN 01 0391:2013

ČSN 01 0391 is a national Czech standard issued in 2013 which specifies requirements for the management system of social responsibility. The standard tries to offer the way how to integrate CSR management system requirements for the individual pillars (economic, environmental and social) and requirements for improvement (in accordance with the PDCA cycle). Standard can be used for certification on the national level. This standard is most suitable for organizations that has already implemented any management system (according to ISO 9001, 14001, OHSAS), as corresponds to the different chapters of these international standards. Appropriate chapters and the similar structure can help organizations to transform requirements of CSR according to ČSN 01 0391 to existing integrated management system easier [6].

1.3 Award of Governor for Social Responsibility and Social Responsibility Poll

Governor of Moravian – Silesian Region together with Quality Council of Czech Republic announces Award for Social Responsibility since 2009. The aim of the competition is to appreciate the companies, public sector organizations and municipalities activities in the Moravian-Silesian Region in the field of corporate social responsibility. The competition is divided into categories:

- business entities (businesses) up to 50 employees, from 51 to 250 employees, over 250 employees,
- public sector organizations: up to 50 employees, over 50 employees,
- municipality: municipalities with extended competences, other municipalities.

Social Responsibility Poll participants were nominated by the public and employees as the social responsible organizations and therefore it is likely that the CSR concept is part of their strategic focus.

2 Backgrounds

The issue of corporate social responsibility is now becoming an increasingly important part of strategic planning and business development. The cause is in particularly increasing consumer pressure to take responsibility for the level of social, environmental and economic impacts of corporate activities and expectations towards producers, service providers as well as public administration organizations or educational institutions.

The concept of corporate social responsibility can be an opportunity to further develop the organization after the introduction of a quality management system or an integrated management system and thereby become another element of the organization's competitiveness [7].

However, it is a fact that organizations are not making full use of their potential in the field of CSR implementation. Organizations often do not know the essence of the concept of CSR and for that reason it is not included in their strategic development plans. Also there are many aspects that management perceives as significant barriers to introducing CSR in a complex (three-pillar) form. Organizations independent of size indicate as ones of the main barriers the fact, that CSR activities do not affect the customers preferences and do not differ organization from competition and on the other side cause the limitation of profit maximalization and higher costs of realization of CSR activities [8]. However, these organizations have not mapped the possibilities of overcoming these barriers. According to many researches, it is possible to say that customers prefer organizations that declare and can prove their social responsibility. [9] Equally, it is possible to obtain funding for the implementation of CSR programs from EU funds. [10]

The main aim of the research is to define if organizations referring themselves as socially responsible do implement activities of all the three pillars equally.

The respondents were chosen from all organizations that participated in Social Responsibility Poll (promulgated by Institute for Social Responsibility) or applied in the evaluation for Award for CSR in Moravian-Silesian Region from 2010 to 2014. This target group was chosen intentionally because participated organizations are likely that the CSR concept is part of their strategic focus.

3 Survey

On the basis of literature studies, informal interviews with managers of different types of organizations, and experience with organizational assessment in terms of meeting CSR requirements, the following hypothesis has been specified:

Companies participated in Award of Governor for Social Responsibility and Social Responsibility Poll meet the requirements of all three pillars of CSR in a mutual balance (from more than 70% in each researched pillar).

Data for verifying were gained in an empirical survey by the questioning. Questionnaires was designed in google application and only prolink to it was sent to the email addresses of organization heads, top manager, PR managers and other managers from September to December 2016. 114 respondents from 336 companies answered in the survey, which is a return of 34%. The respondent group consisted of 21,05% small (less than 25 employees), 36,84% middle and 42,11% big (more than 250 employee) companies. Service enterprises had the biggest share (61,54%), production companies built more than one fifth (20,51%) and organizations of public sector were represented by 17,94%.

The questionnaire was designed according to requirements of ČSN 01 0391 standard to the three pillars of CSR (social, environmental and economic pillar). Individual questions and positive responses are shown in Table 1.

Table 1. Formatting sections, subsections and subsubsections.

Pillar	Question	Positive responses
Social pillar	Do you verify compliance of CSR rules in your suppliers?	23,08%
	Have you created a pay system for work, a system of benefits and extraordinary rewards?	**97,44%**
	Do you take into account in some way the work-life balance?	64,10%
	Do you support a local community?	**71,79%**
	Do you ensure consumer protection by providing complete and comprehensible product information?	**91,89%**
	Do you provide consumer protection by providing complete information on warranty and dealing with complaints and disputes?	**85,29%**
	Have you set up a system for protecting consumer privacy?	**81,08%**
	Can you prove that the organization does not practice unfair commercial practices?	**100,00%**
Environmental pillar	Do you identify the environmental aspects (those that have or may have an impact on the environment) for existing activities and products?	**71,43%**
	Do you identify environmental aspects for planned, new and changed activities and products?	57,14%
	Have you determined the significance of the identified	51,52%

	environmental aspects?	
	Are you trying to minimize emissions in product management?	**85,29%**
	Are you trying to minimize the potential for interference in product management?	32,35%
	Are you trying to minimize the consumption of feedstock and energy in product management?	**73,53%**
	Are you trying to minimize waste generation in product management?	55,88%
Economical pillar	Do you evaluate the suppliers also on the basis of his / her eligibility to comply with the principles of corporate social responsibility?	41,03%
	Do you have a procedure for identifying stakeholders?	45,49%
	Have you set ethical principles towards stakeholders?	59,46%
	Do you have a procedure for recording the findings of non-compliance with ethical principles?	30,56%
	Are you planning investments into the community?	41,03%
	Do you maintain your current community status listings?	30,77%
	Do you employ workers from the surrounding community?	43,59%
	Do you select suppliers mainly from the surrounding community?	33,33%
	Do you report the fees and penalties applied to your organization?	**76,92%**

Each pillar was evaluated separately, respectively answers to its questions were evaluated separately. As a question that meets the majority requirement in a given pillar, a question has been included in which the percentage of positive responses exceeds 70%. If these questions (with a positive response rate of more than 70%) in a particular pillar were more than 70%, it was possible to argue that the pillar was fulfilled with the majority of requirements. If all three pillars of CSR have been fulfilled, then it was possible to argue that companies meet most (more than 70%) the requirements of all three pillars of CSR.

4 Results Analysis

Survey in the area of the social pillar was based on the eight requirements according to ČSN 01 0391 standard. Six of them crossed the positive response threshold above 70%. The average percentage of positive responses for these requests is relatively high, namely 87.3%. In percentage terms, this means that the social pillar was filled by a total of 75%. Therefore, it can be argued that companies fulfill the majority requirements of the social pillar.

The area of environmental pillar research consisted of seven requirements, but only three of them have exceeded the 70% threshold. They were focused on the identification of environmental aspects for existing activities and products, as well as efforts to minimize the impact from the point of view of emissions and, finally, efforts to minimize the consumption of feedstock and energy. The average percentage for these three positive responses was 76.8%. Since the set seventy percent threshold exceeded only three of the seven requirements, this pillar was only filled by 42.9%. It is necessary to say that this pillar, from the point of view of majority fulfillment, has not been fulfilled.

The economic pillar requirements were formulated in nine questions, but only one requirement exceeded 70% in the percentage of positive responses. This requirement concerned the recording of fees and sanctions applied to the organization. Unfortunately, the remaining eight requirements were below 60 percent of the positive response, even seven requests were only under the 50 percent threshold. Therefore, it should be noted that this pillar has not been met from the point of view of the majority fulfillment of requirements. The results of meeting the requirements of all three pillars are shown in the Table 2.

Table 2. Comparison of the fulfillment of the requirements of the individual CSR pillars.

Pillar	Fulfillment
Social	75.00 %
Environmental	42.90 %
Economic	27.27 %

Therefore, the hypothesis was not fulfilled. Based on the questionnaire survey, it is not possible to claim that companies fulfill most (more than 70%) the requirements of all three pillars in the field of CSR in mutual balance. It can be argued that companies fill the most of requirements only in social pillar. The mutual balance in all three pillars is met only in 27.27% of organizations.

5 Conclusion

In general, based on the findings, it is possible to state that organizations claiming the concept of corporate social responsibility do not fulfill the prerequisite for a comprehensive fulfillment of the requirements of all three pillars in mutual balance. The least attention is paid to the economic pillar. Most of the organizations surveyed (75 %) perform the social pillars in particular in relation to employees. This fact follows from the historical development of the concept of CSR in our environment, when a certain form of employee care was implemented even during the socialist regime. On the contrary, the requirements that the other pillars include in the current concept of CSR have not been taken into account at all. However, the motives for increased effort for employee welfare were different from those of the present. While today organizations are primarily concerned with maintaining reliable, educated and loyal employees as carriers of corporate development, in the past, it was essential to keep staff satisfied with the aim of eliminating the causes of social unrest and suppressing activities that were not in line with the socialist way of government management. In spite of this pejorative insight, the expectations of high-level employees remain high, and this is the cause of a large number of activities in the area of the internal social pillar as declared by the concept of CSR.

Similarly the imperfect implementation of the economic pillar can be explained. In today's business environment corruption is unfortunately still a common phenomenon, secondary insolvency destroys small entrepreneurs, and big companies are unable to get a big order without bribes. It is very difficult to appeal to the ethical principles of business when their (not) adherence is often dependent on existence and thus on jobs. Similarly, it is difficult to comply with the invoice maturity if the tax burden is so high and the maturity dates from customers are often shifted beyond the 3 months.

The ecological pillar seems to be the least interesting in this respect, as the vast majority of environmental measures commonly adopted by organizations are part of legislative codification. Organizations from ignorance of other options do not show an interest in further raising the level of the environmental pillar.

It is possible to claim that only one quarter (27.27%) of the surveyed organizations referred themselves as socially responsible fulfil all requirements of Czech national standard for corporate social responsibility ČSN 01 0391 in mutual balance.

In conclusion, it is a competitive advantage of the addressed organizations that many of the above-mentioned CSR requirements are already implemented, although not yet implemented in their complex form. This is a good starting point from which organizations can profit if they are willing to work equally to implement the requirements of all three pillars.

Acknowledgements

This paper was elaborated in the frame of the specific research project SP2018/97, which has been solved at the Faculty of Metallurgy and Materials Engineering, VŠB-TU Ostrava with the support of Ministry of Education, Youth and Sports, Czech Republic.

References

1. R. Petříková, Z. Hofbruckerová, P. Hercík, R. Lešingerová, *Společenská odpovědnost organizací* (DTO CZ, Ostrava, 2008)
2. A. McWilliams, D. Siegel, Strategic Management Journal, **21**, 5 (2000)
3. S. Maqbool, M. N. Zameer, Future Business Journal, **4**, 1 (2018)
4. A. G. Sahinidis, D. Daskalaki, E. Mantzari, I. Mantzaris, *Suistainability and Social Responsibility: Regulation and Reporting* (2017)
5. M. Costa, P. Torrecchia, Corporate Social Responsibility and Environmental Management, **25**, 2 (2017)
6. Z. Hofbruckerová, P. Halfarová, T. Divinová, Metal, **25** (2016)
7. Š. Markulík, L. Kamenický, J. Namešanská, A. Nagyová, *Advances in Physical Ergonomics and Human Factors* (2014)
8. Z. Hofbruckerová, T. Divinová, Współczesne problemy ekonomii – między teorią a praktyką gospodarczą, **1** (2016)
9. J.J. Rivera, E. Bigne, R. Curras-Perez, Spanish Journal of Marketing, **20**, 2 (2016)
10. J. Košťálová, L. Tetřevová, Metal, **25** (2016)

Problems in the Implementation of the Lean Concept at a Steel Works - Case Study

Adrian Maszke[1,*], *Renata* Dwornicka[2], and *Robert* Ulewicz[3]

[1]ArcelorMittal Warsaw, Poland
[2]Cracow University of Technology, Poland
[3]Czestochowa University of Technology, Poland

Abstract. The paper presents the experience of implementing selected instruments of the Lean concept at a steel works. The functioning of an enterprises in the currently prevailing conditions, i.e. in a turbulent environment, results in increased customer requirements. They expect, among other things, high quality of steel products delivered on time at a low price. Many organizations decide to implement the lean management concept without proper preparation. The purpose of this paper is to present the limitations occurring during the implementation of the lean management concept in a steel mill.

1 Introduction

Changes are a natural process occurring in each enterprise. They may be related to the transformation of the so-called "hard" components, e.g. technological changes, shop-floor workflow, purchasing of new technical equipment on the one hand, and the so called "soft" aspects, i.e. changes in the staff's attitudes or behaviour patterns on the other. In either case, one of the key factors behind successful implementation of the lean concept are- people [1,3]. Lean management is a Japanese concept comprising management methods and techniques including: waste elimination, process mapping, 5S, Just in Time, kanban, Single Minute Exchange of Die, Total Productive Maintenance, One Piece Flow, work standardization and Total Quality Management [4-8].

By implementing the lean management concept, an enterprise triggers a transformation of both the intangible and the tangible - technological aspects. Such transformation is expected to allow an enterprise, a steel works in this case, to attain a flexibility allowing it to quickly adapt to the turbulent environment and a springiness helping it to quickly react to changes [9-11]. The idea behind the lean concept is, among other things, to reduce the number of changes-corrections, eliminate the warehouse space, improve the work flow, lower the manufacturing costs (by reducing the stoppage time and increasing the general efficiency of using machines and devices) [12-19]. Unfortunately, the anticipated effects of the concept can be attained only on the condition that one avoids the mistakes that frequently result from oversight, neglect or simply lack of knowledge. Therefore, it is

* Corresponding author: a.maszke@ado-odblaski.pl

important to bring the mistakes occurring in the implementation of the lean concept to attention so as to avoid them [20-24].

2 Lean Concept at a steel works

The underlying activity of steel plant is the manufacture of liquid steel in different furnaces using different sources of heat necessary to carry out the production process and to further process steel in other departments. Currently, the Polish steel industry is controlled, among others, by the following companies: Arcelor Mittal, Cesla Huta Ostrowiec, CMC Zawiercie, ISD - Huta Częstochowa and Polish Steelworks.

These companies compete with each other, among other things, with respect to the technical manufacturing costs. The global competition significantly affects the headcount (social factor). One of the ways allowing to reduce the technical manufacturing costs is to implement the lean management concept to continually reduce losses. The starting point for a loss analysis should be the identification of the actual losses followed by determination of their causes, place and frequency of occurrence. The most important stage of the analysis is defining the present situation. Only then can improvements be proposed and verified. Elimination of waste is frequently impossible, so even its smallest reduction is a big achievement.

Each production process can be improved, and aiming at optimizing production by seeking any possible savings is the best illustration of that. No manufacturer tolerates any form of waste. Nowadays, however, this approach results in the production process being subjected to an in-depth analysis of its individual stages, especially with respect to the relevance and correctness of the current material and information flow of the customer service. This approach gives a company a unique opportunity to eliminate all redundant stages from the production process or to improve those which must not be eliminated due to their technological importance. In addition, each stage should be analysed in terms of its effectiveness to allow a company to produce only those products which can be used in a subsequent process (operation) as during these processes companies encounter barriers and limitations hampering proper implementation of the methods and techniques of lean management. According to research, the difficulty in implementing the lean concept results from the lack of identification with the company, its problems and needs, implementation of large, individual projects without following a long-term implementation plan, while ignoring the idea of continuous improvement, very high financial goals and expectations related to the lean department, the streams functioning as separate "companies within a company" and not as effect centres only, as well as from the lack of exchange of knowledge, ideas and information between the streams. An additional problem that frequently appears during the implementation is a lack of understanding of the Lean Concept by the entire management staff. Very often implementation of the concept is limited to the production area, which is perceived as an isolated element of the company. In the case of the analysed company, which is a part of a holding company, the key is to achieve financial objectives related to the lean department.

3 Case study

Implementation of the lean concept instruments is determined by attaining the anticipated financial goals by working toward optimizing the production by seeking potential savings. The problem to solve was to put individual costs into cost groups to facilitate the analysis and reduce waste, as no manufacturer tolerates any form of it.

Similar flow streams are applied at production departments of a given steel plant, such as the melt shop or the rolling mill. The melt shop uses a Danarc electric furnace to melt steel, which is subsequently cast in the continuous casting process known as COS and then put into ingot moulds at the ingots line. Production departments such as the melt shop or the rolling mill are primarily responsible for handling the production and control processes specified in process sheets and instructions in conformity with the customer's requirements, as well as the legal and environmental regulations to minimize their environmental impact. Another important element is regular recording, analysing and storing of data related to production, inspection of the products manufactured and monitoring of the environmental impact. Another important element of control of each department is ensuring identification and traceability of products, including non-compliant products, packing of semi-finished and finished products without causing damage to them, as well as analysing the causes behind significant, repeated, existing or potential discrepancies and determining corrective or pre-emptive actions necessary to remove such discrepancies to be taken in cooperation with the Quality Control Department. Fig. 1 presents elements of the value stream of the melt shop department. It is that part of the theory and practice of metallurgy which deals with the manufacture of steel in electrical furnaces and is referred to as electrometallurgy.

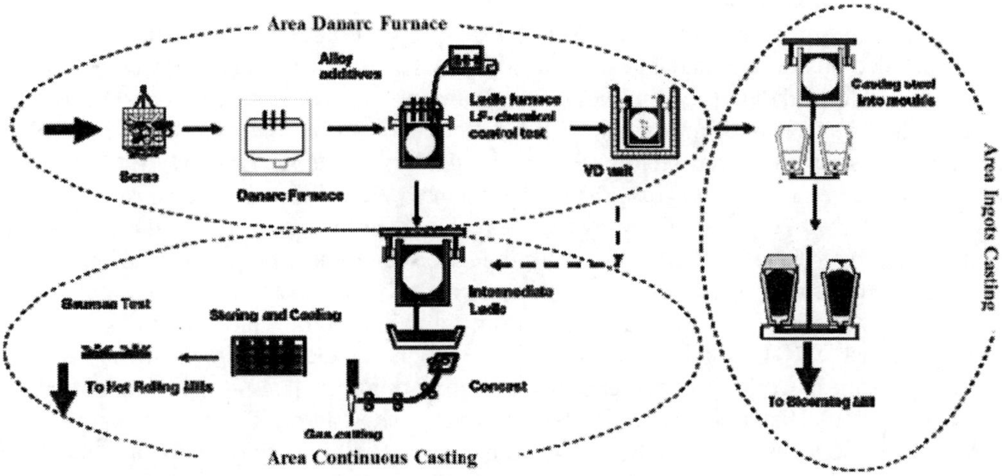

Fig. 1. Elements of the value stream at the melt shop department.

As the number of competitors who effectively use the direct labour force and machines increases, in addition to technical or process information, the management needs access to information about the costs of the production processes, products and customer service, such information being more precise than that obtained by means of the system used for the purposes of external financial statements. The simplest cost division is their division by resource type. All costs generated in the company in question are grouped by type of resource used and put into six groups of usage factors: C - charge factors, E - energy factors, M - material factors, S - service factors, L - Labour group factors, C2 - department common costs. Such cost division will allow to evaluate the effectiveness of the waste-eliminating activities or pro-lean activities introduced.

A preliminary research conducted to identify the obstacles in the implementation of the lean concept at the steel works indicated the following issues:

- lack of foundations for implementation of the lean concept. The organization does not have the so-called change and continuous improvement culture. Lack of teamwork ability. The employees seem not to identify with the company, its problems and needs. Another obstacle is the lack of conditions (lack of soft lean tools) to implement hard lean tools, such as standardization, TPM or 5S. Lack of employee's understanding of the lean concept.

- Focus on short-term projects without introducing long-term organizational changes. Implementation of individual, short-term projects without a long-term lean implementation plan while ignoring the idea behind continuous improvement.
- Financial requirements. The lean department is expected to keep cutting the operating expenses, as a result of which the lean processes are associated with reducing the headcount and total distrust of employees responsible for the implementation of the lean concept.
- Division of a company into streams. Streams function as separate enterprises within a company, rather than effect centres. Streams compete rather than cooperate with each other. This hampers exchange of information between particular streams and makes the company look more like a functional rather than a process-oriented enterprise.
- Procedures The procedures and instructions applied by the company are too complex and hamper work. Collecting data that is not analysed or difficult to interpret.
- Trainings. Lack of specific materials to be used during trainings and a failure to enforce implementation of particular stages of the lean concept. Lack of indicators for implementation of the lean concept.
- Unavailability of reliable data. Many lean projects have been conducted without a detailed observation and analyses, mostly based on a process-related documentation.

At the steelworks in question, the lean activities are evaluated based on cost groups by type of the resource used. In this study, the direct costs will be the following groups: C - charge factors, M - material factors, S - service factors, R-labour factors. Although they belong in the direct costs category, the charge factors (W) will be handled separately as an input cost as they constitute the biggest component. Both the costs and the production volume are time-related. Therefore, any production level must be determined relative to some general, average period of time. More importantly, the costs incurred by a company change in time. They may change both in a short and long term. Short-term costs may be fixed or variable. The total fixed cost (TFC) remains at the same level irrespective of whether a plant is working at full capacity or not working at all. As long as an enterprise has even a single fixed cost item, it operates over a short term. The total variable costs (TVC) increase together with the production costs. The total fixed cost and the total variable costs together yield the total cost. The total cost (TC) constitutes the sum of the fixed and variable costs at each production level. Following the classic cost division, each enterprise introduces small changes depending on its nature in order to best present its costs and provide information about them. The above-mentioned implemented system has resulted in an improved control over particular groups of cost factors. The system has contributed to a faster analysis of particular cost groups and details of a given group, which previously caused a lot of problems and required more time to identify a waste area.

4 Discussion and conclusions

Application of the cost-based approach in the analysis of effectiveness of lean activities has allowed the steelworks in question to:

- apply a quick and effective analysis in given cost groups,
- make faster decisions regarding areas needing improvement,
- reduce the consumption of utilities and auxiliary materials,
- shorten the calculation cycles for particular usage groups,
- improve the economic results.

At the stage of collecting the data relating to particular cost factors no problems were identified which could stop an already commenced grouping process, but only a discussion was noticed related to the location of particular cost groups. The applied model of division of costs into cost groups such as C - charge, E - energy factor, M - material factors, S-service factors, L-labour factors and K-2 common factors seems to be working and is verified on a monthly basis following balancing of data to verify its weaknesses and strengths. Each month a cost meeting is held with the team of a given department in order to carry out a costs analysis to be later reported to the management. Table 1 presents the percentage share of particular determinants of cost groups at the melt shop without taking into account the waste for particular sub-department such as: LSS - Liquid Steel Shop - the EAF furnace area, CCS - Continuous Casting Shop and CLS - Cast Line Shop - the ingots area.

Table 1. Percentage share of determinants of cost groups at the melt shop without taking into account waste.

		LSS	CCS	CLS
C	Charge factors	79.89	95.21	91.93
E	Energy factors	11.03	0.89	0.55
M	Material factors	6.10	2.14	3.93
S	Service factors	1.52	0.92	1.10
TVC	**variable costs**	**98.54**	**99.16**	**97.51**
L	labour factors	0.68	0.37	1.29
K2	common costs	0.78	0.47	1.20
TFC	**fixed cost**	**1.46**	**0.84**	**2.49**
TC	**total cost**	**100.0**	**100.0**	**100.0**

The results of a preliminary research of the value streams management as one of the elements of the lean concept indicates that the obstacle encountered at the steelworks in question is the culture of the organization in terms of implementing the lean instruments. A survey conducted among the shop floor staff and the managers shows that the most common problems include: a barrier in the contacts between the management and the employees, lack of standardization, short-term financial goals, referred to as short-termism, lack of information about the effects of activities and associating lean with headcount reduction. It should be pointed out, though, that application of the cost-based approach as an evaluation of the functioning of line activities in the company in question has yielded positive results.

Table 2. Waste in percentage at the melt shop for TC, TVC, TFC and particular determinant of cost groups for three production areas.

		LSS	CCS	CLS
Average monthly production (tonnes)		43,371	33,643	7,805
C	Charge factors	2.76	2.67	4.60
E	Energy factors	2.87	5.54	17.75
M	Material factors	2.01	6.42	14.17
U	Service factors	12.70	19.59	14.63
TVC	**variable costs**	**2.55**	**2.75**	**4.30**
R	labour factors	8.44	6.62	16.58
K2	common costs	9.80	9.79	16.92
TFC	**fixed cost**	**8.06**	**6.73**	**13.24**
TC	**total cost**	**2.59**	**2.74**	**4.41**

The impact of the waste costs factors on the company's economic result is complex in terms of the structure of the production and costs, while the presented results should be helpful in finding the optimal solution in the lean activities. The presented model functions in the company in question, however, an area has been identified that needs improvement in terms of a better tailored division of costs for an easier analysis.

References

1. W.L. Tate, L.M. Ellram, L. Bals, E. Hartmann, International Journal Production Economics, **120**(2), 512-524 (2009)
2. T.R. Holcomb, M.A. Hitt, Journal Operations Management, **25**(2), 464-481 (2007)
3. S. Wuyts, A. Rindfleisch, A. Citrin, Journal Operations Management, **35**, 40-55 (2015)
4. L.M. Ellram, W.L. Tate, C. Billington, Journal Operations Management, **26**(2), 148-163 (2008)
5. J.M., Hallikas, I. Karvonen, U. Pulkkinen, VM. Virolainen, M. Tuominen, International Journal Production Econmics, **90**(1), 47-58 (2004)
6. S.M. Handley, W.C. Jr. Benton, Journal Operations Management, **31**(3) 109-128 (2013)
7. H. Kenneth, J.B. Wathne, Journal of Marketing, **64**(4), 36-51 (2000)
8. S.M. Handley, J.V. Gray, Journal Operations Management, **22**(6) 1540-1556 (2013)
9. S. Borkowski, K. Knop, Production Engineering Archives, **1**, 25-28 (2013)
10. H. Golas, A. Mazur, P. Piasek, P. Czajkowski. Problemy Jakości, **2**, 10-14, (2017)
11. K. Mielczarek, K. Knop, Production Engineering Archives, **13**, 16-19 (2016)
12. K. Knop, K. Mielczarek, Technical Transactions Mechanics, **3**, 67-72 (2016)
13. J. Selejdak, T. Corejova, R. Ulewicz, *Total Quality Management* (Wz-PCz, 2016)
14. C.G. Drury, M.A. Sinclair, Human Factors, **25**, 391-399 (1983)
15. K. Stecuła, J. Brodny, SGEM 2016 Conference Proceedings, Book 1, **2**, 57-64 (2016)
16. J. Brodny, K. Stecuła, M. Tutak:. SGEM 2016 Conference Proceedings, Book1, **2**, 65-72 (2016)
17. J. Brodny, S. Alszer, J. Krystek, M. Tutak, Archives of Control Sciences, **27**(LXIII), 197-209 (2017)
18. J. Pietraszek, E. Skrzypczak-Pietraszek, Adv. Mat. Res.-Switz., **874**, 151 (2014)
19. J. Pietraszek, A. Gądek-Moszczak, T. Torunski, Adv. Mat. Res.-Switz., **874**, 139 (2014)
20. A. Czajkowska, R. Stasiak-Betlejewska, *Metal 2016: 25th Anniversary International Conference On Metallurgy And Materials,* 1763-1768 (2016)
21. M. Potkany, R. Stasiak-Betlejewska, R. Kovac, M. Gajdos. Polish Journal Of Management Studies, **13**(1), 145-156 (2016)
22. M. Ingaldi, S.T. Dziuba, *Metal 2016: 25th Anniversary International Conference On Metallurgy And Materials,* 1822-1828 (2016)
23. R. Ulewicz, M. Nowicka-Skowron. *26th Anniversary International Conference On Metallurgy And Materials,* 2338-2343 (2017)
24. M. Zasadzień, J. Zarnovsky, Management Systems In Production Engineering, **26**(1), 55-59 (2018)

Technological Resources Evaluation in the context of the micro-enterprise development

Renata Stasiak-Betlejewska[1,*], *Luminia* Parv[2], and *Justyna* Karolczyk[3]

[1]Ing. PhD., Czestochowa University of Technology, Faculty of Management, Department of Production Engineering and Safety, Al. Armii Krajowej 19B, Częstochowa, Poland
[2]Assoc. Prof., Transilvania University of Braşov, Faculty of Technological Engineering and Industrial Management Colina Universitatii, Colina Universitatii nr. 1, corp A, 500036 Brasov, Romania
[3]Student, Czestochowa University of Technology, Faculty of Management, Al. Armii Krajowej 19B, Częstochowa, Poland

Abstract. The object of the paper is technological resources analysis in the context of selection of the micro-enterprise development direction aimed at gaining the product quality improvement. Paper presents results of the study conducted in chosen micro-enterprise of the furniture industry with using ABC technology method to identify modernity technological level for chosen machine. Presented research results allows identifying technological development direction for the analysed furniture enterprise.

1 Introduction

The problem of the enterprise development is generally perceived in five characteristic dimensions: economic, organizational, personnel, information, technical and production [1]. These dimensions can be considered in the strict sense, but they will often be given an extended context. Dimensions are used as examples: economic and social, organizational and legal, production and logistics. All these dimensions, as well as their modifications, constitute a reference platform for management systems and processes shaping operational no tactivity and enterprise development. Generally widely understood development ability is a criterion for the enterprise assessment, which is a measure of projects implementing possibility at the high efficiency level. It expresses the ability to multiply assets and create progress in various fields of technological innovation activity in the context of enterprise development. Micro, small and medium-sized science and technology enterprises have become the important force to promote technological innovation and economical and social development [2].

One of an enterprise evaluation field is technological resources level. Products and processes require a number of technologies for its acquisition or realization. They may be of a key, while others are secondary. Technologies are characterized by the fact that they may be subject to minor changes over the decades, but they can also vary significantly in a short time [3, 16].

* Corresponding author: renata.stasiak-betlejewska@wz.pcz.pl

Technology portfolio include technologies needed to produce the product. Currently, effective technologies that require less time and costs for production, installation and use are gaining greater significance [4]. Innovative technologies included in the portfolio allow faster and more reliable manufacturing process that guarantees a high quality product obtaining. Technology portfolio contributes also to the production processes modernization. Technologies, depending on the level of their advancement, may in a different way affect the competitive advantage of the company. There are basic types of technological strategies used in micro, small and medium – sized enterprises [5, 6]:

1. Uniqueness - offering something that nobody else can offer.
2. Complexity - the difficulty of learning the technology.
3. Quick imitation of technology (the basic product, process).
4. Robust construction - the basic model of the product or process has a long life cycle.
5. Changing the rules of competition - changing the competitive rules of the game.
6. Continuous introducing of the technological innovation (product, process).
7. Niche-specialized know-how.
8. Binding of a major customer - the customer adoption of technological strategies.

The technology interdependence within the portfolio can decide on its choice. If there is not established interrelation between seemingly separate technologies, technological „bottleneck" can be a barrier in the product development [3]. The bottleneck is a vital element to the process realization, which despite the fact that it is used in 100% does not ensure maximum production [7, 9]. Therefore one should pay attention to the fact that the advancement in technology one can be inhibited by limitations occurring in the other. Proper selection of technology has a huge impact on the final product quality.

The object of the enterprise analysis in the context of its development is the technological resources evaluation with using ABC technology method aimed on selection of the micro-enterprise technological development direction [10-12].

2 Characteristics of the analysed micro-enterprise

The study object was a micro – enterprise called Drewdom operates in the Silesian province and it employs: designer, two editors and two production workers. It deals with the production of the kitchen furniture in accordance to own or commissioned projects. The product offer includes a wide range of non-upholstered furniture such as: kitchen furniture, chests of drawers, wardrobes, desks, tables, tables, chairs. The analysed enterprise was founded in 1985. Due to the market completion intensification in 2009, there was noted orders decrease in Drewdom, which prompted the enterprise owner to take actions to improve the activity aimed at kitchen furniture production and starting the design activity of manufactured products. Technological equipment park of the enterprise consists: multi-spindle drilling machine, upper spindle milling machine, edge banding machine, format saw and machine for cutting holes for hinges.

The subject of technological resources level evaluation is Felder FD 921 multi-spindle drilling machine for drilling holes for joints in solid wood and wood-based materials that has one drilling head. The machining can be carried out vertically and horizontally. Drilling from the bottom ensures that the working surface of the table is free of chips and the drilling depth is constant, which eliminates time-consuming adjustment when the thickness of the plate being drilled changes. Table 1 presents technical data for Felder FD 921 multi-spindle drilling machine.

Table 1. Technical data of the Felder FD 921 multi-spindle drilling machine

nominal power	1,5 kW
max. height of the work piece	65 mm
table dimensions	900 x 380 mm
number of spindles	21
spacing between spindles	32 mm
max. thickness element	55 mm
insertion depth	up to 300 mm
compensation for drilling depth	up to 8 mm
max. drilling depth	70 mm
spacing between extreme spindles	640 mm
weight	320 kg
number of pressure cylinders	2
min. air pressure	6 – 8 bar

2 Research results and discussion

Modernity level of machine parts can be classified with using the ABC technology analysis to assess their value and usefulness for the company in the period considered. As a criterion of evaluation, a competitive advantage may be used, expressed in the expensive leadership or in the product mix. The assortment's wealth can be achieved thanks to the flexibility of devices that allow the company to frequently change the type and quality of the product. The ABC analysis of the technology is based on the principle that each statistical group can be distinguished by several of its components (symbol A), which determine the main part of the results. At the other end of the distribution there are machine parts (symbol C) with a small contribution to the results of its activity [8, 10-15]. ABC analysis applies technology used to manufacture individual components in the machine and evaluate its modernity in a five-point scale Parker [10-13]:

1. Level 1 - simple parts that results from the craft techniques (e.g. a machine foundation).
2. Level 2 – parts based on unchanged and well - known technologies (e.g. standard engine cooling system).
3. Level 3 - parts produced by controlled technology, requiring appropriate technical knowledge (e.g. a standard electric motor).
4. Level 4 - parts that were produced using modern market technologies, e.g. displaying the diagnosis on the control panel computer.
5. Level 5 - parts resulting from the production of combined use of the state-of-the-art, patented technologies.

Results of the study affect decision on which machine parts should be upgraded or replaced. However, technologies are not isolated from each other, they often form constellations based on scientific principles. The basic subassembly consists of parts of category A, fundamental for the company's operations. They are also defined as elements implementing drive technologies, because they give special attributes (competitive advantage) to the product, such as a greater degree of surface smoothness. In this machine the primary component (Category A) is aimed at competitive advantage through its functions: control, input tools, programming, desktop support. The support sub-assembly (category B) is relevant to the drive system and the work piece handling system in the given case. Its role is to some extent dependent on the basic subassembly. The subassembly (category C) is associated with: the machine construction, foundation, energy supply and instrumentation. Parts of this subassembly are universal and, being used by any users, do not contribute to a competitive advantage. This does not mean that you can ignore them.

The ABC method has been carried out for multi-spindle drilling machines Felder FD921. Thanks to this method, the modernity degree of the machine and its individual components taking part in the production process can be determined in order to identify parts that require modernization. Table 2 presents technical characteristics of the analyzed machine.

Table 2. Technical characteristics of the machine

The machine element	Key technological aspects
Construction	steel table, on which the plate is placed which will be processed, after placing it is pressed and the spindles are drilled holes, stable side gears
Foundation	Rubber foot for machines
Energy supply system	Engines, starters, on/off switch, spindle inclination angle and drilling depth, compressor
Drive system	2 hp (1.5kW) engine with stable side gears, range and variability of drilling speed
Control system	Dialogue with operator, level of precision, safety limits
Instrumentation	The drill consists of 21 spindles that rotate one clockwise to the other, at an angle that has been set
The operating system of the work piece	Positioning, pneumatic clamping, spindles, side restrictions
System of support functions	The compressor that cools the motor and discharges chips

The most important machine elements are listed in Table 2. The machine elements include: construction, foundation, energy supply system, instrumentation, operating system of the work piece, support function. In addition, key technological aspects are described.

Table 3 presents the evaluation of the machine parts modernity level for a multi-spindle drilling machine.

Table 3. Evaluation of the level of modernity of a multi-spindle drilling machine

Subassembly	Symbol	Part / Element of the machine	The modernity level
A	A1	Control system	3
	A2	Measuring system	3
	A3	Drill head	3
	A4	Nominal power	3
	A5	Transmission feeds	2
B	B1	Drive system	2
	B2	Support	2
	B3	Handles	3
	B4	Pneumatic system	3
	B5	Clamp cylinders	3
	B6	Attaching the workpiece	2
	B7	Foundation	1
C	C1	Guideways	2
	C2	Lubrication system	3
	C3	Chip conveyor	3
	C4	Board	2
	C5	Regulating valve	2

Parts and components included in the drilling machine are presented in Table 3. Each of the analyzed machine components has been evaluated in terms of the modernity level. They are divided into three groups A, B and C. The first group are the parts responsible for the quality of the product that will be processed. The second support group contains the most important parts of the machine that determine its usefulness. The third group is responsible for safety within using the device. Research results of the evaluation of the machine parts modernity level for a multi-spindle drilling machine are presented in Figure 1.

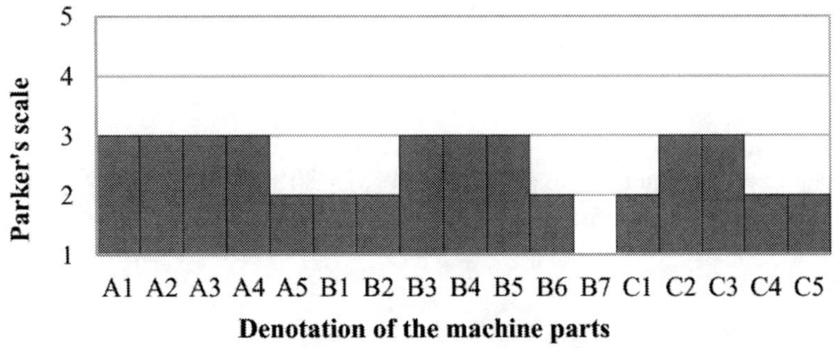

Fig. 1. Evaluation of the level of modernity of a multi-spindle drilling machine.

Data presented in Figure 1 show the research results of the machine parts modernity level analysis. The grades' average of the results for each subassembly group is following:

1. Category A: 2.8
2. Category B: 2.29
3. Category C: 2.4

The multi-spindle drill was evaluated according to the Parker scale (1 ÷ 5). The highest number of the highest grades (3) were granted in the case of the individual introduction of new technological solution based on own development of elements in the category A. Thanks to this, product obtained owing to this subassembly can be adapted to individual customer's needs. The highest grade (5) was not received. Grade 4 could be obtained for new applications. Also in this category the grade 4 was not granted. Grade 3 was is the most common evaluation level. The evaluation could be obtained for technology that already exists and is already known, but it is not obsolete. The grade 2 was obtained 7 times, which is the second most common grade that was awarded for obsolete technology.

Rating 1 has been given only for 1 element, which means that it is already an old technique (no new elements have been introduced). The rating was assigned only once. Thus, in subassembly group A received an average score of 2.8 the most common grade 3. In contrast, categories B and C show the greatest variation in assessments. The average rating for category B is at the level of 2.29 and category C at the average level of 2.4.

4 Conclusions

The object of the paper is technological resources analysis in the context of selection of the micro-enterprise development direction aimed at gaining the furniture product quality improvement. ABC technology method was applied to identify technological modernity level of the selected machine.

The analysed enterprise case confirms that applying of ABC technology method allows identifying modernity level of technological resources in the analyzed micro-enterprise what is one of the basic elements of decision process on the enterprise development. Presented research results allows identifying technological development direction for the analysed enterprise. It doesn't create additional costs and it can be effective tool for the eavluation of technological resources.

In the case of the analzyed micro – enterprise decision on the machine modernization with regard to the analzyed machine is connected with replacing individual elements that are responsible for the high quality of the product. Proposed solution should improve the machine's operation and increase its modernity level that affects the final product quality. The modern machine components will reduce the number of nonconformities, which will have a direct impact on the higher efficiency of the entire production process and lower costs associated with waste. Implementing advanced technology in the production equipment can affect the the competitive advantage of the company what result from implementing technological innovation strategy.

Acknowledgements. This work is related to the scientific program of the „Improving quality of processes, products and services" BW 615/201/07 supported by Polish Ministry of Science and Higher Education.

References

1. D.R. Gnyawali, D.S. Fogel, E.T. & P., 43-62 (1994)
2. X. Fan, YZ. Huang, *3rd International Conference on Education and Education Research* (Kuala Lumpur, Malaysia, 45-49, 2017)
3. P. Lowe, *Management of Technology: Perception and Opportunities* (Chapman & Hall, London, 1995)
4. S. Kildiene, E.K. Zavadskas, J. Tamosaitiene, J. C. E. & M., **20,** 280-290, (2014)
5. H. Mizgajska, *Aktywność innowacyjna polskich małych, średnich przedsiębiorstw w procesie integracji z Unią Europejską* (Wydawnictwo AE, Poznań, 2002).
6. M. Gancarzyk, J. Gancarzyk, E. M. J., **36**, 59-70 (2018)
7. R. Ulewicz, D. Jelonek, M. Mazur, Management and Production Engineering Review, **7**, 89-94 (2016)
8. R. Stasiak-Betlejewska, R. Ulewicz, *International Conference on the Path Forward for Wood Products: A Global Perspective* (Baton Rouge, LA, USA, 149-156, 2016)
9. A. Maszke, Production Engineering Archives, **17,** 45-48 (2017)
10. S. Borkowski S., J. Selejdak, Sz. Salamon, *Efektywność eksploatacji maszyn i urządzeń* (Wyd.Wydziału Zarządzania Politechniki Częstochowskiej, Częstochowa, 2006)
11. S. Borkowski, R. Ulewicz, *Instruments of production processes improvement*, (PTM, Warszawa, 2009)
12. M. Ingaldi, Independent Journal of Management & Production., **5**(4), 993-1003 (2014)
13. A. Pilkington, T. Teichert, Technovation, **26**, 288-299 (2006)
14. J. Pietraszek, E. Skrzypczak-Pietraszek, *Adv. Mat. Res.-Switz.,* **874**, 151 (2014)
15. K. Halicka, Technological and Economic Development of Economy, **22**, 4, 574-597, (2016)
16. R. Ulewicz, M. Nowicka-Skowron, *26th International Conference on Metallurgy And Materials*, 2338-2343 (2017)

Analysis of the 5S method in production enterprise - case study

Katarzyna Łyp-Wrońska[1], and *Bartłomiej* Tyczyński[1]

[1]AGH University of Science and Technology, Faculty of Non- Ferrous Metals, Poland

Abstract. One of the most popular system in production enterprise is Lean Management. 5S is a method which introduces workflows allowing to increase the efficiency. The aim of the research is to analyze the use of the 5S method on the example of a selected company. The scope of work covered the years 2013-2017. Research methods are: analysis of enterprise data, employee opinion survey, participation in audits. One of the ways to maintain a proper level of order and cleanliness according to the 5S method in the factory is to confirm compliance by conducting an audit. Inspections are carried out on two levels. One of them is an internal audit, which is carried out by a foreman who works at a given hall. The second level is an external audit, which is carried out by employees of other halls. Teams with the best results are rewarded with incentive bonuses. The first audit in 2013 at the enterprise showed that the state of order is unsatisfactory. Therefore, corrective actions have been developed and implemented. Analyzing the data in subsequent research periods - up to 2017, it can be stated that due to the introduced changes, the results of all audits, increased.

1 Introduction

The concept as ''lean'' have been used with growing frequency in management. It is a method to eliminate waste and create value at various types of enterprises. The aim of Lean Manufacturing is to reduce waste and make a quality of the product meet the customers expectation also defined a non-value activity consuming the resources. [1] Lean concepts can be put into place procedures and make better use of employee time and company resources. Lean manufacturing is an indispensable supplement to the class of experimental methods such as: TQM, TPM, 5S, SMED, Kanban, Kaizen, and Poka Yoke, 5 S etc. These philosohy, methods and tools are deployed by organizations for realizing consistent manufacturing performance improvements [2,3]. 5S has apotential to improve efficiency, quality, production, delivery adherence, safety, optimized cost, and acts as abaseline requirement for other programs [4,5,6,7]. The 5S method is a tool whose application will bring benefits to the enterprise, and at the same time will make work at a given position more pleasant. The application consists of 5 steps, whose names in Japanese begin with the letter "s", 1S- Selection (Seiri), 2S- Systematyka (Seiton) - 3S- Cleaning - 4S- Standardization, 5S- Self-discipline. 5S is valuable for enhancing the workplace, as well as that, 5S framework facilitates process and item quality guidelines, decrease lead time, and furthermore, obviously, lessen working expenses and improve overall performance [8] From literaturę perspective several studies have been reported by the practitioners about beneficial impacts of 5S practice in the different enterprises [9,10].

The aim of the research is to analyze the use of the 5S method on the example of a selected company. The scope of work covered the years 2013-2017. Research methods are: analysis of enterprise data, employee opinion survey, participation in audits

2 Implementation 5S

The researched object is a company - a factory in the cable industry, whose production is nearly 25,000. different cables and wires. The products of this company are used in many economic zones, such as construction, mining, renewable energy or telecommunications.

The company has quality certificates, but also uses modern management methods like lean management.

5S is one of the basic Lean Managmenet methods, which is implemented in the company. The meaning and location of this 5S method in the company's organizational structure is shown in Figure 1.

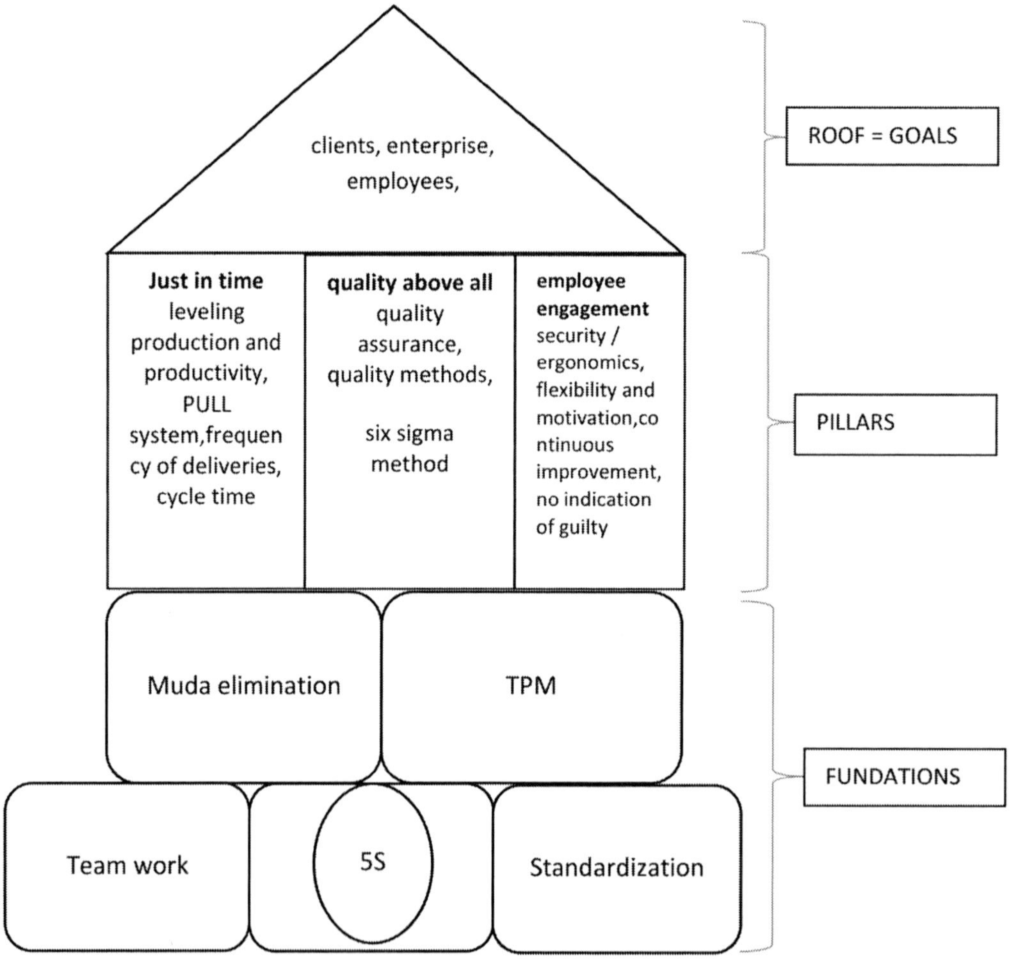

Fig. 1. The importance of the 5S method in the company's production system.

According to the illustrated graphics, 5S is included in the foundations of the entire production system. This system is aimed at meeting customer requirements while managing production, in which waste elimination and the TPM system play an important role. This means that the current operation of the plant could not be achieved without 5S, as well as team work and standardization.

One of the foundations of the 5S method is order and organization of work. In order for each employee to carry out his work in accordance with the assumptions, he must be trained. Then, the task of the management team is to supervise whether the employee / team fulfills his duties and this is done by conducting the audit. It is checking the condition of certain conditions (in this case it will check the cleanliness of machines and halls in the plant). Inspections are carried out on two levels. One of them is an internal audit consisting in checking the maintenance of order at work posts by individual brigades and employees. Then the person responsible for conducting such an audit is the master of change who works at the given hall. The second level is the external audit, whose purpose is the same as the internal audit, with the difference that it is carried out by employees of other halls. The idea of double control is to eliminate possible over-estimation, which would lead to a false state of affairs. In addition, brigades and halls compete with each other, which is an additional stimulus to take care of order. The grading scale is from 1-5. Teams with the best results are rewarded with incentive bonuses for conscientious performance of duties and are an incentive for further effort, resulting in continuous development.
In 2013 (year of implementation), the results of the first 5S audit were at a low level. The audit was carried out every day of the month. Employees were not informed which particular machine and team will be assessed. The company has 4 teams A, B, C, D and 3 halls X, Y, Z separated. The results of the audit are presented in Fig. 2.

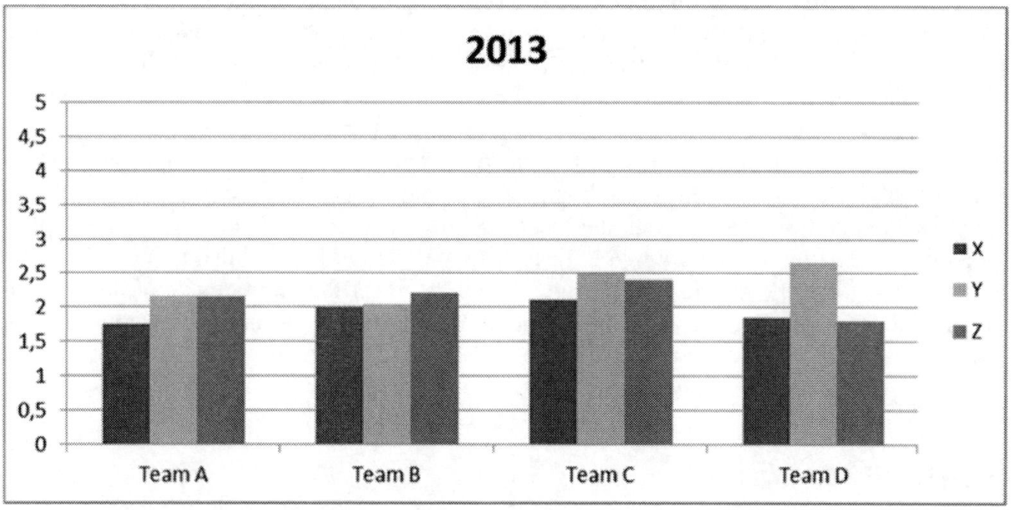

Fig. 2. Results of the audit of 4 teams of all halls.

Due to poor audit results, corrective actions were introduced and a system of division of responsibility for the divisions on individual machines and parts of the halls was developed, because it was believed that the employee / team who is responsible for a smaller area can more accurately take care of the order. The division of responsibility of 4 teams on a selected machine is shown in Fig. 3.

Team A	Team B
In charge of :	In charge of :
winder with extraction	wire guide elements
fence mesh	window sills on the reeds
window sills on the reeds	wire rewinder

Team C	Team D
In charge of :	In charge of :
delivery devices	caliber and its components
fence mesh	console
window sills on the reeds	window sills on the reeds
braider number 1	control cabinet
keeping the wire on the revier	braider number 2 and 3
	compensator

Fig. 3. Teams responsibilities.

Figure 3 presents a description of the responsibility of teams on individual departments. When an employee starts working on a new machine, he knows exactly where the elements he must take care of. Detailed determination of places and tasks to be carried out makes their implementation easier, which limits the possibility of a situation in which some part of the machine will not be cleaned for a longer period. This can lead to faster consumption of a given element, which generates costs related to repair or even replacement with a new one. When the part turns out to be so significant that the whole machine will have to be stopped, the plant will incur additional losses resulting from the reduction of the production potential. In smaller areas, not only is it easier to maintain order, but also requires less time to prepare it for work for the next change. You need to set the optimal size of the areas so that employees can spend the maximum amount of time on performing their duties, and the minimum amount to clean up and receive satisfactory results in one and the other time zone.

3 Effects 5S

After the implementation of corrective actions, an audit was carried out in 2015. The final evaluation of this audit, as a component of the grades obtained by each team - A, B, C, D is presented in Fig.4.

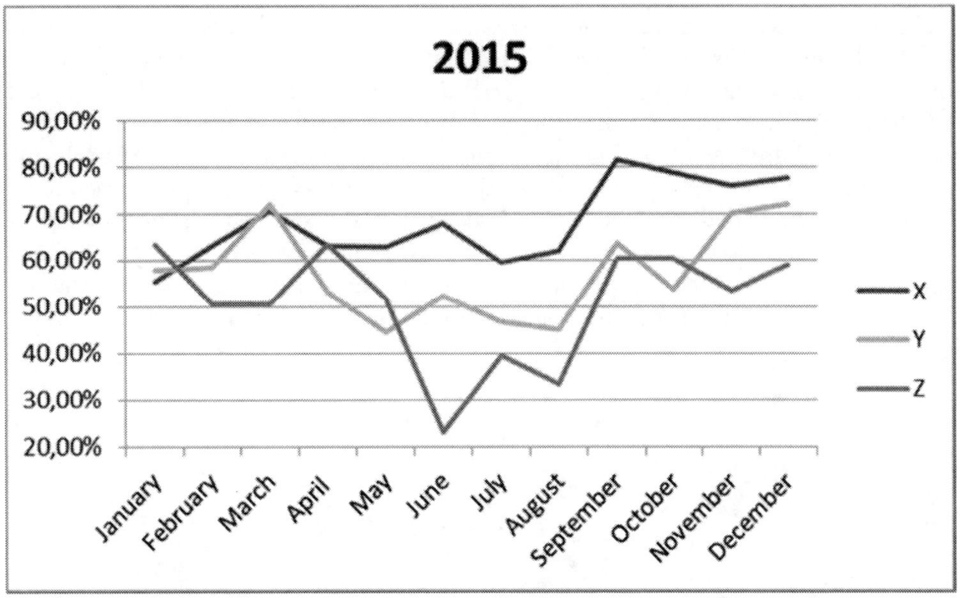

Fig. 4. Results of the 5S audit in 2015 of three halls each month.

The cleanliness of the machines and the area on which I find myself were checked. Almost, each month, the halls gained over 50% of possible points, which was a rarity in 2013 when they were introduced to the 5S plant (Figure 2). In fact, apart from the holiday period (from May to August), the ratings remained at a similar level or even increased. Weaker ratings resulted from the absence of employees responsible for 5S. However, despite this weaker period, the trend line is growing, which confirms that the state of the halls was getting better. Only the "Z" hall did not record an increase due to the highest absenteeism of employees. The next results of the 5S audit in 2016 are presented in Figure 5.

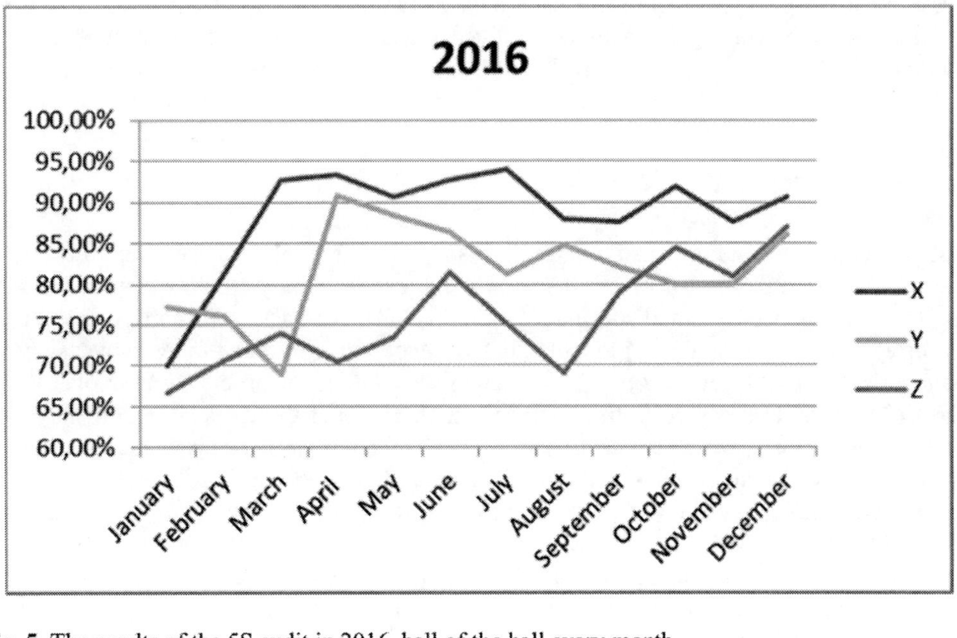

Fig. 5. The results of the 5S audit in 2016. hall of the hall every month.

The improvement of results during 5S audits is visible from year to year. In each month of 2016, better ratings were recorded than in previous years. Despite the clear and satisfactory increase in ratings and, consequently, also the improvement of cleanliness in the halls in 2015, the plant has not stopped improving in this area and 2016 has brought even better results in each hall. The trend line shown in the chart goes clearly sharply up. Even the "Z" hall, which struggled with staffing problems and separated from the other two halls, was better assessed. The company's management has analyzed the situation related to a clear decline and correct conclusions have been drawn. Even periods in which the plant has fewer employees at its disposal are able to maintain high quality. To illustrate all the activities, the results of the audit of all teams from 3 halls are presented, which is shown in Fig. 6.

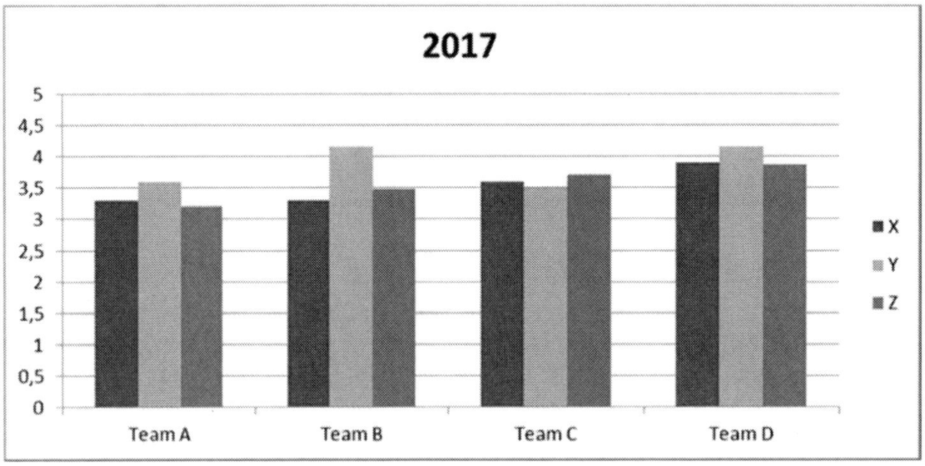

Fig. 6. Results of the audit of 4 teams of all halls.

In comparison to 2013, each hall improved its results. From year to year, audits showed an increasing improvement. An important element of the quest to improve the cleanliness was to plan the plan specifically for each halls separately. Due to the different size of the halls, the arrangement of the machines on it, the intensity of both foot and vehicle traffic, it was impossible to match one universal way that could provide improvement on each of the halls. Each machine and team in 2017 was rated higher than in previous years.

4 Conclusion

All tools at the workstations had a defined, easily accessible, place at any time. In case the situation required the use of more advanced tools, every employee knew exactly where to go. This management method introduced liquidity in the plant's operation process. These are effects from one side of good planning of work and training of employees by the managerial staff and, on the other hand, reliable execution of orders by employees. Few kinds of objects disappeared from the floor, which hindered the work of people staying in the hall. Situations in which the cable waste lying in the wrong place prevented the operator from re-arranging the machine no longer appeared, and protruding drums outside the designated storage area did not pave the passage of forklifts. Easy to implement changes have improved productivity and working conditions on many levels. Downtimes resulting from the lack of required equipment are no longer present, and machines due to improved technical condition spoil a lot less often. An employee who is not distracted from the tasks entrusted to him by removing unnecessary items and many other factors interfering with work is less frustrated and thus more efficient. According to the philosophy of lean management, in order for the company to generate more profits, it is necessary to reduce waste, i.e. all activities that do not bring benefits. While maintaining the order, the company is able to achieve these profits, but several aspects have to be taken into account. The first one is management in accordance with the possibilities of the plant and the people working there. Proper determination of the area in the hall and machine for the employee is an important issue for maintaining production continuity. Another important issue is the proper use of human resources. Each employee must know exactly what is required of him and how to perform the tasks ordered. Adequate training will increase the productivity of employees, their work comfort while maintaining high quality. Exact instructions will avoid many wastes and keep money in the company. In summary, the data collected should emphasize the importance of the human factor. Even the best management system will not yield satisfactory results if the employees or people who create this system will not perform their duties correctly. The inability to correctly perform tasks, for example due to the lack of appropriate tools or lack of commitment may be due to poor working conditions. Moreover, the 5S system will bring the desired results when all the above-mentioned factors are carried out reliably., what is similar results of another research as [9, 10].

References

1. R. Wolniak, B Skotnicka-Zasadzien, Metalurgija, **53** (2014)
2. K. Hemanand, D. Amuthuselvan, S.C. Raja, G. Sundararaja, International Journal of Lean Thinking **3**(2), (2012)
3. I. Madanhire, Ch. Mbohwa, *Total Productive Maintenance (TPM) Application for Cable Manufacturing Organisation* (6th Int'l Conference on Mechanical, Production & Automobile Engineering, 2014)
4. C. Jaca, E. Viles, L.P Galeano, J.Santos, R. Mateo, International Journal of Production Research, **52** (2014)
5. A. Pacana, L. Bednarova, I. Liberko, Przemysl Chemiczny, **7,** 93 (2014)
6. A. Pacana, L. Bednarova, J. Pacana, Przemysl Chemiczny, **12**, 93 (2014)
7. B. Main, M. Tabutiz, W. Wood, Professional Safety, **1** (2008)
8. J Liker, *The Toyota Way* (Simon & Schuster Inc., New York, 2004)
9. J. Randhawa, I. Ahuja, International Journal of Productivity and Performance Management, **67** (2018)
10. T. Bortolotti, S. Boscari, P. Danese, International Journal Production Economics, **160** (2015)

Selected problems of designing modern industrial databases

Dariusz Karpisz[1,*] and *Anna* Kiełbus[1]

[1]Cracow University of Technology, Faculty of Mechanical Engineering, al. Jana Pawła II 37, 31-864 Krakow, Poland

Abstract. The paper presents problems of designing databases for various branches of industry. The development of information technologies and in particular of object-oriented programming has caused a change from data modelling to the modelling of applications. The increase of unstructured Big Data in Industry 4.0 era and requirements of sharing data model between many applications needs a reversion to data analysis and design and it is presented in the article.

1 Introduction

Modern ICT systems (Information and Communication Technologies) to support the operation of enterprises are based on computing tools such as ERP, CRM as well as CAD/CAM modelling systems [1, 2]. The databases are a specific part of these applications. Both approaches to the design of any applications and the design of databases is associated with a well-recognized theory [3].

However, there are problems related to the development of the programming method itself from the method of data analysis earlier to OOP (Object-Oriented Programming) currently. The common belief of programmers in the possibility of creating any application without data modelling is unfortunately still a fact. Currently, most databases are generated from application class models. They are used for more complex ORM (Object-Relational Mapping) modules to generate and utilize data models. In addition, companies have entered the Industry 4.0 era, the essence of which is access to a huge amount of previously unavailable, unstructured data referred to as BigData. Some of this data may be stored in relational systems, others in databases known as NoSQL (referring to non relational data models).

The combination of classic databases and fast unstructured databases is often not implemented by ICT applications, which slow down the process of loading data and their subsequent analysis or is implemented in many separate applications. The authors' experience shows that the preparation of applications for the industry often begins from the preparation of data model. This also relates to ensuring adequate performance and safety level as shown in articles [4].

* Corresponding author: dkarpisz@pk.edu.pl

2 Methods

There are many methodologies available to design applications at different levels of specification. One of the first methodologies was Information Engineering Methodology (IEM) [5] shown as a simple diagram in Fig. 1. The IEM assumed that focusing on data can identify the information needed by business. Invention of the relational data model [3] together with the theory of normalization helped to simplify the process of searching for a minimal data set of data model. The methodology emphasized in particular the modeling of data and the functions of their processing, which then forced the design and implementation of the application. System analysts (SA) and Data Base Administrators (DBAs) have thus influenced functional requirements (steering by data) and non-functional requirements (e.g. performance and model adaptability to changes) to applications (today most common associated with Front-End).

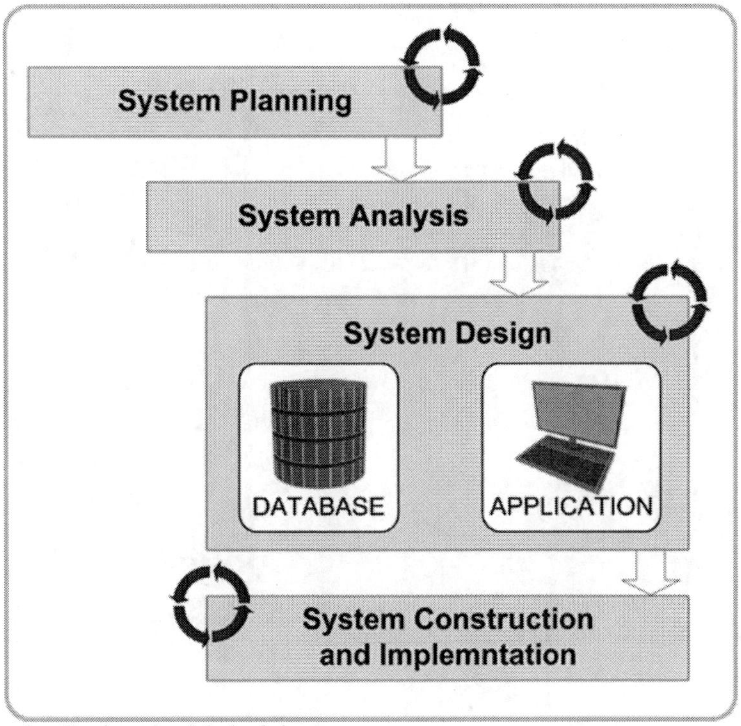

Fig. 1. Information Engineering Methodology.

It was important that the "System design" step assumed the parallel creation of both the data model and the application model by using multiple iterations. The database was often designed with overgrowth of tables, so that it contained tables and columns that can be used in subsequent versions of the application. The client application (as the presentation layer and also the logic layer) contained functionalities defined in functional and non-functional requirements defining appearance, available functions and their interrelationships (known as business logic). In order to support the design and implementation of software in IEM style, many tools and methods have been created under the common definition of CASE (Computer-aided Software Engineering) [6].

In response to the demand of the managerial staff, Business-Driven IEM variant was created. It gave the opportunity to better, proactively react to possible business changes in the organization. IEM has been completed with Strategic Planning to identify information required by management in strategic directions of organization changing.

As shown in Fig. 2, knowledge engineering was originally used to build ICT systems in many areas of the market according to the bottom-up design & development methodology. The evolution of programming languages and the acquisition by programmers competences of database designers, DBAs and data analysts, resulted in a general transition to the top-down application development model. It caused a drastic decrease in the quality of data. The data model created from the application class model in OOP style is compatible only with one application or its module. The resulting model is not optimized for data storage and search outside of the application. Correct data aggregation and analysis are also not possible due to the lack of proper data connections in the model - to correct data integrity, operation of the application from which the model was generated is always required.

Currently, especially in the *Industry 4.0* era, the importance of the data has grown again. When preparing modern production systems, the data model is analyzed again and this changes system design from top-down to bottom-up method.

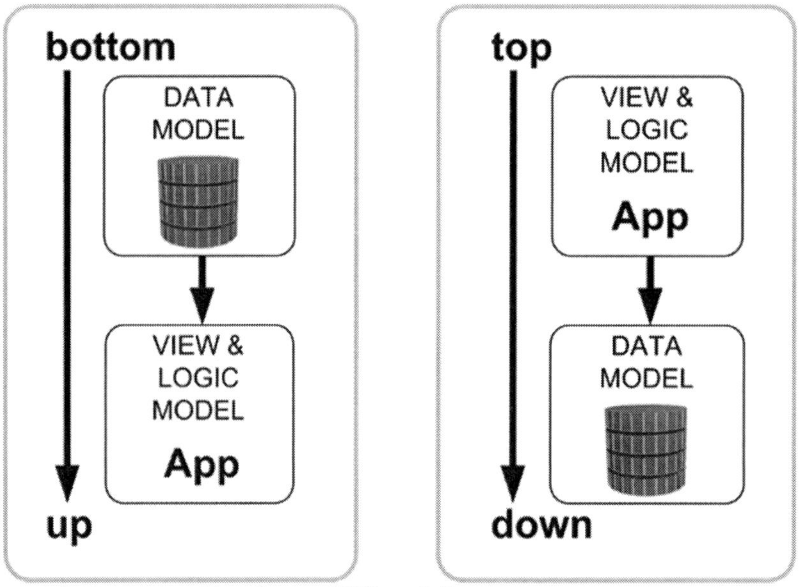

Fig. 2. Concepts bottom-up and top-down of ICT modeling.

3 Results

During the implementation of ICT systems projects for the machine industry, it turned out that precise determination of standards and factors are the most important ways of describing various technical and process objects. Both individual targets and their collections were analyzed.

The simplified ERD diagram from Fig. 3 shows the proposed way of storing information about any technical and non-technical objects. Each of the objects can be described by any number of associated standards with the joining entity "OBJECT Standards". The "Standards" entity can collect information related to the entire knowledge of the company e.g.:

- Official standards,
- Internal standards,
- Manufacturing experience,
- KnowHow.

In this way, management support was obtained using Knowledge Based Engineering (KBE) and also Knowledge Based (Expert) Systems. Entity "Standards" must be decomposed to the Physical Data Model (PDM) of database with these requirements:

- the possibility of cataloguing any standards, together with presenting both the vendor and the area of applying in manufacturing environment,
- the building of the hierarchy of standards/norms,
- the storing of the optional number of documents connected with standards.

A similar approach as presented for standards is necessary to determine the features of an object. These can be both descriptive and numerical features collected as a set of coefficients. The "Object" entity is described by a set of factors and it could be divided into many classes e.g. [4]:

- Nominal Factors - nominal monovalent,
- Range Factors - range from/to,
- Fuzzy Factors - described as fuzzy values (e.g. set: low, middle, high),
- Timed Changing Factors - time-varying values.

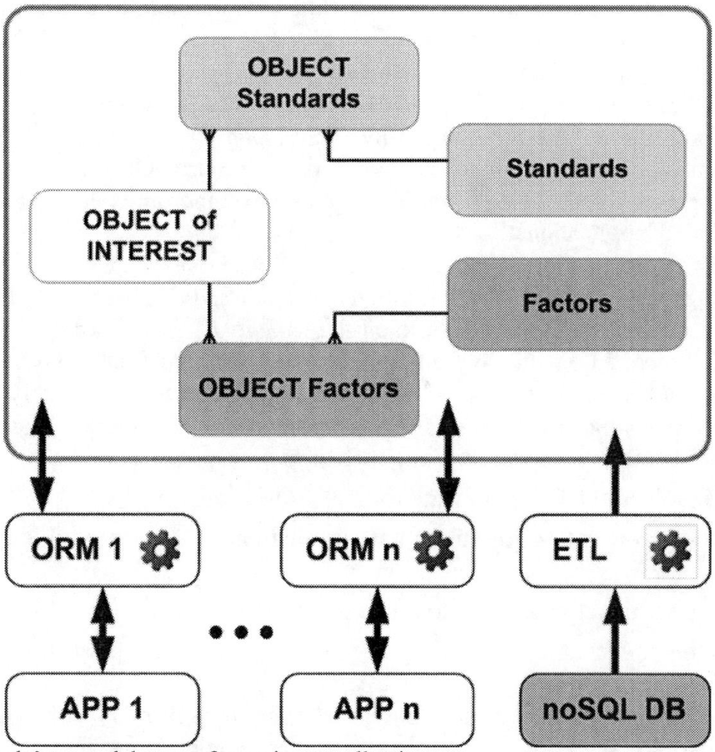

Fig. 3. Universal data model to use for various applications.

The "Factors" entity is a universal dictionary for gathering information about any factors. The "Factors" should implement the following requirements:

- the building of the hierarchy of factors,
- the possibility of cataloguing any factor with a set of units of measurement,
- the possibility of units conversion,
- the possibility of connecting factor with any norms/standards to it description in industrial environment.

The relational combination of a set of factors dedicated to the description of specific objects is accomplished by the entity "Object Factors" as shown in fig. 3. The diagram does not show additional entities for "Standards" and "Factors" and their connections.

The proposed data model has been used to build industrial databases for the application of experimental machining of hard-to-machine materials [7], databases of the distributor and manufacturer of tools for the furniture industry and for the construction of a data model for the application of a distributor of tools and clamping systems for machining metals. In all three databases, a similar physical model was used (PDM). The target databases have been connected to dedicated applications written in the OOP style and to production and inventory management modules of MRP and PLM systems of enterprises.

4 Discussion

The proposed common data model for various industries has given the opportunity to build ICT systems optimized for data, not for programmers or IT companies. In fact, the target physical model as a set of tables of relational databases are much more extensive. With the "Standards" entity, tables for storing any files are additionally connected. The "Factors" entity has been expanded with a dictionary of units and a table combining a coefficient with two units and a conversion rule. It is possible to use advanced database mechanisms such as virtual column used to automatic computing. This brings the possibility of using active database mechanisms [8].

Databases built on the basis of the model shown in Fig. 3 can cooperate simultaneously with many applications. The box market by ORM1 is a connection of the target database with the client application such as WWW portal, similarly ORM2 mapping data from database to macros for CAM System. This gives the opportunity to share data for very different technical applications.

An interesting application of the proposed model was to execute data insertion attempts from documents data model database (NoSQL) to relational database. Data from the sensor with fast data acquisition to the database on the MongoDB server, after the completion of the acquisition, were automatically transferred to the relational database (Oracle 12c) using the mapping marked ETL in Fig. 3. This is the application that directly addresses the problems of processing and storing BigData in *Industry 4.0* concept. The ETL refers to technology known from data warehousing meaning:

- extract – involves extracting the data form NoSQL database,
- transform – involves transformation data to relational structures with changing of data types,
- load – store the transformed data to relational database.

In this way, the collected data can be processed using any analytical tools (e.g. build-in analytical SQL or BI data tools).

The problem in using the bottom-up approach is the habit of IT companies to quickly create applications with OOP from well proven applications frameworks. A good application programmer will not be a good data analyst at the same time. It is difficult to reconcile contradictory goals of user interface and data model designing. The programmers' assignment is to take care of the quality of the application. This is not connected to the quality of the generated data model, because in the top-down style, Information Engineering does not apply. Therefore, it is important to support the IEM approach for industrial applications.

The well-designed and well-implemented database may be very useful tool for the users located far from the software industry: image analyzers with their huge datasets [9, 10], especially in materials science [11-13] and medicine [14], the phytochemistry and the biotechnology [15, 16] or large-scale companies activity [17-20].

References

1. U.S. Access Board, *Information and Communication Technology (ICT) Standards and Guidelines, Notice of Proposed Rulemaking*, 36 CFR Parts 1193&1194, United States Access Board (2015)
2. R. Tongia, E. Subrahmanian, V. S. Arunachalam, *Information and Communications Technology for Sustainable Development, Defining a Global Research Agenda* (Allied Publishers, Bangalore, 2005)
3. E. F. Codd, *Commun. ACM* **13**, 6 (1970)
4. D. Karpisz, *Technical Trans.* 113, 73 (2016)
5. C. Finkelstein, *An Introduction to Information Engineering: From Strategic Planning to Information Systems* (Addison-Wesley, 1989)
6. A. F. Case, *ACM SIGMIS Database*, **17**, 35 (1985)
7. J. Gawlik, A. Kiełbus, D. Karpisz, *Solid State Phenom.* **223**, 35 (2015)
8. D. Karpisz, *Technical Trans.* **108**, 179 (2011)
9. A. Gadek-Moszczak, *Image Anal. Stereol.* **36**, 151 (2017)
10. A. Gadek-Moszczak, P. Matusiewicz, *Image Anal. Stereol.* **36**, 207 (2017)
11. J. Pietraszek, A. Gadek-Moszczak, A. Szczotok, *Solid State Phenom.* 197, 162 (2013)
12. J. Korzekwa, M. Bara, J. Pietraszek, P. Pawlus, *Int. J. Surf. Sci. Eng.* **10**, 570 (2016)
13. R. Dwornicka, N. Radek, M. Krawczyk, P. Osocha, J. Pobedza, *Proc. 26rd Int. Conf. on Metallurgy and Materials*, Brno, Czech Republic, May 24-26, 1252 (2017)
14. A. Gadek-Moszczak, J. Pietraszek, B. Jasiewicz, S. Sikorska, L. Wojnar, *Studies Comp. Intell.* **572**, 127 (2015)
15. E. Skrzypczak-Pietraszek, J. Pietraszek, *Acta Biol. Crac. Ser. Bot.* **51**, 62 (2009)
16. E. Skrzypczak-Pietraszek, I. Kwiecień, A. Goldyn, J. Pietraszek, *Phytochem. Lett.* **20**, 443 (2017)
17. P. Osocha, *Adv. Mat. Res.-Switz.* **874**, 89 (2014)
18. R. Ulewicz, J. Selejdak, S. Borkowski, M. Jagusiak-Kocik, *Proc. 22nd Int. Conf. on Metallurgy and Materials*, Brno, Czech Republic, May 15-17, 1926 (2013)
19. T. Lipinski, *Solid State Phenom.* ***163***, 183 (2010)
20. T. Lipinski, A. Wach, *Proc. 23rd Int. Conf. on Metallurgy and Materials*, Brno, Czech Republic, May 21-23, 738 (2014)

Tribological characteristics of Al/SiC/Gr hybrid composites

Sandra Veličković[1*], *Slavica* Miladinović[1], *Blaža* Stojanović[1], *Ružica* Nikolić[2] *Branislav* Hadzima[2], *Dušan* Arsić[1], and *Jozef* Meško[3]

[1]University of Kragujevac Faculty of Engineering, Kragujevac, Serbia
[2]University of Žilina, Research Center, Žilina, Slovakia
[3]University of Žilina, Faculty of Mechanical Engineering, Žilina, Slovakia

Abstract. Metal matrix composites (MMCs) are considered as important engineering materials due to their excellent mechanical, as well as tribological properties. When the metal (or alloy) matrix is reinforced with two or more reinforcements, those composites are the so-called hybrid composites. The aluminum metal matrix composites, reinforced with silicon carbide (SiC) and graphite (Gr), are extensively used due to their high strength and wear resistance. The tribological characteristics of such materials are superior to characteristics of the matrix. This research is presenting influence of the load and the graphite and silicon carbide contents the composites' wear rate and the friction coefficient.

1 Introduction

Composite materials consist of the matrix, content of which is much greater with respect to other materials, and the reinforcer(s), i.e. material(s) whose properties result in desired properties of the composite. Materials that can be used for the substrate, i.e. the matrix of composites, can be metals, polymers and ceramic materials. If the composite matrix is a metal, then one speaks about the metal matrix composites (MMC); if the matrix is made of the polymer material, then such composites are the so-called polymer matrix composites (PMC); the composites whose matrix is a ceramic material are the ceramic matrix composites (CMC). Metal matrix composites (MMCs) are the materials with the superior properties with respect to the conventional materials [1].

Because of the raising necessity for producing materials with better characteristics, than the existing ones, the investigation of the MMCs is increased in last decades. Usually, reinforcements of the MMCs are ceramics, which can be in form of particles, whiskers or fibers. When two or more different reinforcements are incorporated in the metal matrix the new type of the MMC is developed, called the hybrid metal matrix composite (HMMCs). [2, 3].

Characteristics of hybrid composites depend on properties of their reinforcements (e.g. their hardness, size of particles, volume fractions), distribution of the reinforcing phases, as well as the manufacturing technique. Development of the hybrid MMCs is performed so

* Corresponding author: sandrav@kg.ac.rs

that their performances would be better than those of the conventional MMCs [2, 4]. The high resistance-to-weight and strength-to-weight ratios, hardness and wear resistance, good creep behavior, small weight, design flexibility and other good characteristics are the reason why the HMMCs' application is rapidly growing [5-7].

The most used MMCs are the aluminum matrix composites (AMCs). The fact that aluminum and its alloys have low specific (volume) weight, high corrosion resistance and easy processing, is the reason for aluminum to be widely used in industry, in general [4, 8].

Compared to conventional alloys, the AMCs provide many advantages, such as: high strength, superior high temperature properties, low density, better stiffness, good damping characteristics, abrasion and wear resistance [8, 9].

2 Types of reinforcements

Characteristics of composites are determined by: type of reinforcement, particle sizes of reinforcements and volume fractions of reinforcements, microstructure, homogeneity and isotropy of the system.

As reinforcements in aluminum matrix composites (AMCs) usually are used graphite (Gr), silicon carbide (SiC), titanium carbide (TiC), titanium diboride (TiB_2), alumina (Al_2O_3), tungsten carbide (WC), boron carbide (B_4C), fly ash, zircon (Zr), silicon nitride (Si_3N_4) and some organic wastes such as the rice husk ash, coconut shell ash, palm oil fuel ash and the sugar cane bagasse. With addition of hard reinforcements (silicon carbide, alumina, titanium carbide), with respect to the matrix alloy, the hardness, strength and wear resistance of the composites are improved [10, 11].

Compared to conventional alloys, the SiC particle reinforced AMCs have higher mechanical strength and wear resistance, due to the fact that SiC particles have high modulus of elasticity and yield strength, and the base alloy has the low yield strength and good plasticity [4, 10]. Because of those characteristics, in machining of the composites, matrix deforms plastically and the SiC particles only deform elastically or break. For the better tribological characteristics of aluminum matrix composites, as a reinforcements, commonly are used both SiC and Gr [12, 13].

By combining these two reinforcements with aluminum matrix the hybrid composites are develop. Those hybrid composites are known as the Al/SiC/Gr hybrid composites.

3 Manufacturing methods

Manufacturing process has a big effect on tribological properties of the Al/SiC/Gr composite materials. Microstructure, distribution of reinforcement particles and bonds between the matrix and the reinforcement are influenced by the fabrication process. It can cause porosity formation, poor wettability and improper distribution of reinforcement.

At the industrial level, there are two groups of basic processes for fabrication of the aluminum matrix composites: processing in the solid state and processing in the liquid state.

The solid state fabrication of Metal Matrix Composites is a process, in which the MMCs are formed due to bonding of the matrix metal and the dispersed reinforcement phase, which is a result of the mutual diffusion the in solid state, at elevated temperature and under pressure. The fabrication processes in the solid state include the powder metallurgy technology, diffusion metallization and the vapor deposition technology. Sintering fabrication of the Metal Matrix Composites is a process, in which a powder of a matrix metal is mixed with a powder of the dispersed phase (in the form of particles or short fibers), followed by compacting and sintering in the solid state (sometimes with presence of a liquid).

<u>The liquid state fabrication</u> of Metal Matrix Composites is a process where the dispersed reinforcement phase is incorporated into the molten matrix metal, what is followed by solidification. Fabrication processes in the liquid state include stir-casting, compo-casting, squeeze casting, spray casting and reactive – in situ process.

Selection of the fabrication process depends on numerous factors, like the type and level of reinforcement load, desired degree of micro-structural integrity etc. [14-16].

Based on review of literature on HMMCs, it was noticed that the beginning of the Al/SiC/Gr hybrid composites research points to application of the higher weight/mass % share of SiC particles, as compared to research results of the same composites of the more recent times. The higher %/mass share of the SiC particles leads to more difficult manufacturing and processing of the material, since increased content of the ceramic particles causes increase of the composite's hardness. In the last few years, research of these composites shows that the content of the SiC particles did not exceed 10 wt %, while the minimum content was 3% of SiC particles, what improves the manufacturing and machining of the Al base hybrid composites, without compromising their mechanical properties.

Researchers have also studied tribological behavior of the Al/SiC/Gr composites, by application of the statistical methods with smaller number of experiments like the Taguchi method, the Analysis of Variance (ANOVA), the Full Factorial Method, Response Surface Methodology, Center Composite Design, Taguchi Grey Analysis and Artificial Neural Network.

4 Influence of the load and reinforcement content on composites' tribological properties

Infiltration of the reinforcement particles into the metal matrix affects the tribological and mechanical properties of the composite materials. Depending on various requirements of the industry, the volume shares percentages of SiC and Gr are being combined, as well as sizes of the reinforcing particles.

Generally, the friction and wear properties of composites also depend on the shape and distribution of hard or soft particles filled in the matrix. Hard particles increase the strength and wear resistance of composites, but decrease their ductility, whereas the soft particles acting as lubricant decrease the coefficient of friction [17, 18].

Tribological behavior of specimens made of the Al/SiC/Gr composites were studied by conducting the dry sliding wear test, mainly on the pin-on- disc tribometer. The wear test parameters were within the following intervals: load between 10 N to 60 N, sliding speed from 1 m/s to 3 m/s, the sliding distance up to 1000 m. The size of the reinforcement particles varied from nm for SiC to mm for Gr.

The effect of load on the wear rate of the Al/SiC/Gr composites, with different weight percentage of SiC particles, in the presence of graphite, is shown in Figure 1.

Based on the diagram in Figure 1, one can conclude that the wear rate increases with the load increase. Increase of the SiC reinforcing particles volume share in the hybrid composite, on the other hand, causes reducing of the wear rate, since the hardness of the composite material is increased. The sliding speed increase, for the same experimental conditions (the same load and the sliding distance), causes decrease of the wear rate.

The effect of the load on coefficient of friction of the Al/SiC/Gr composites, with different weight percentage of SiC particles, in the presence of graphite, is shown in Figure 2.

In addition, the effect of load on coefficient of friction of the Al/SiC/Gr composites with different weight percentage of Gr particles and fixed 5 wt % SiC is shown Figure 3.

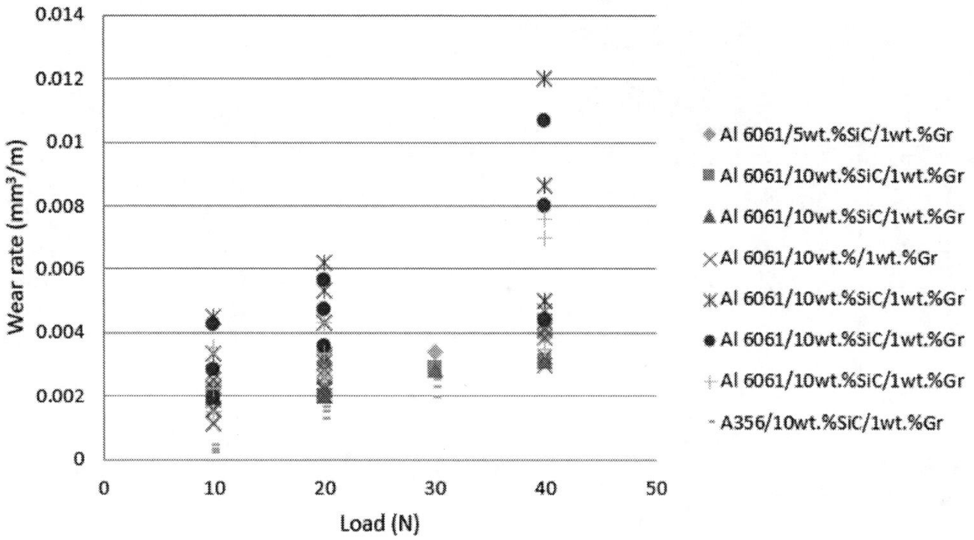

Fig. 1. The effect of load on the wear rate of the Al/SiC/Gr composites with different wt % of the SiC particles and fixed 1 wt % Gr [13, 19-20].

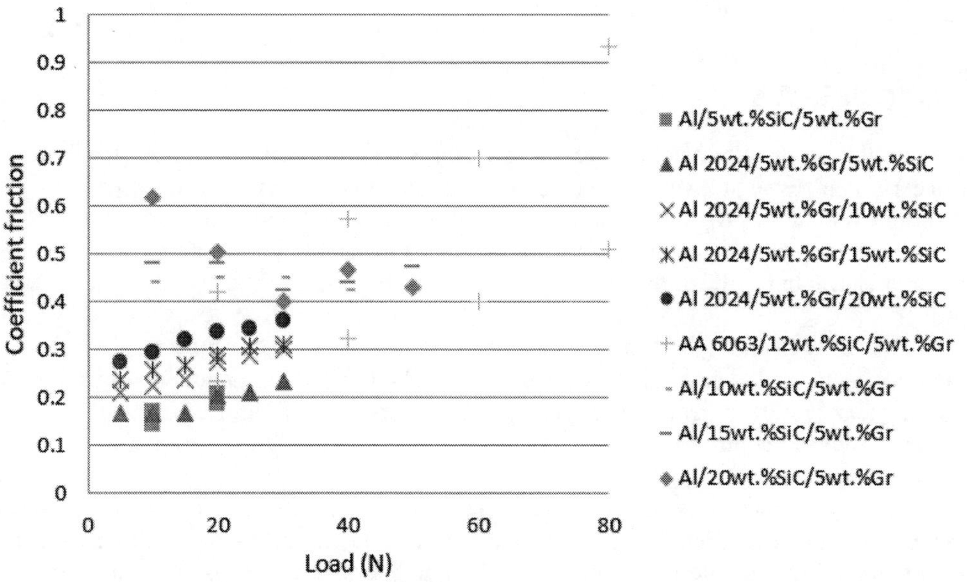

Fig. 2. The effect of load on coefficient of friction of the Al/SiC/Gr composites with different weight percentage of SiC particles and fixed 5 wt % Gr [21-24].

From diagram in Figure 2, one can notice that the friction coefficient increases with the load increase, as well as with the SiC particles volume share increase. The sliding distance has the same effect, as well, while the sliding speed increase causes decrease of the friction coefficient of the composite material.

From diagram in Figure 3 one can notice that the coefficient of friction increases with increase of the load, as well as that the sliding distance increase cause increase of the coefficient of friction. It can also be concluded that the friction coefficient increases with increase of the Gr reinforcing particles.

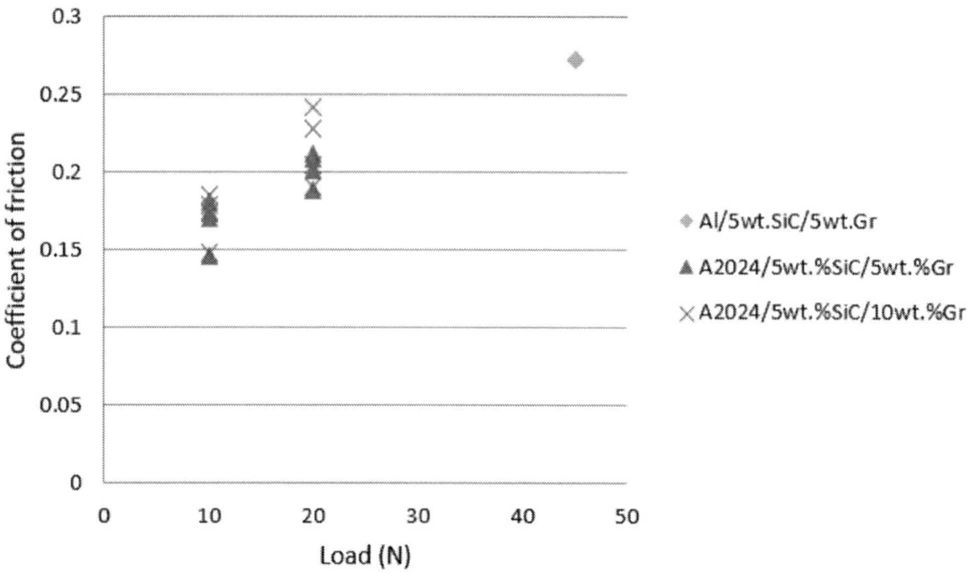

Fig. 3. The effect of load on coefficient of friction of the Al/SiC/Gr composites with different wt % of the Gr particles and fixed 5 wt % SiC [15, 25].

5 Conclusions

Though the composite materials were known for a long time and were used for thousands of years, their research is intensified and well invested in only in recent times.

Improvement of properties of the Al/SiC/Gr hybrid metal matrix composites (HMMCs) is strongly dependent on the nature of the reinforcement, its hardness, particles sizes, volume share and uniformity of dispersion within the matrix and the method of hybrid production.

The quality of the hybrid metal matrix composites (HMMCs) can also be improved by the manufacturing process itself, with the proper choice of the production parameters (the pouring temperature, stirring speed, preheating temperature of reinforcement etc.).

The hybrid composites generally have better tribological properties than the conventional composites.

Adding of the hard reinforcer – ceramic materials (SiC) – causes increase of composite's mechanical and tribological properties with respect to the same properties of the base aluminum alloy (hardness, tensile strength, toughness, wear resistance and corrosion resistance).

Adding the second, soft reinforcer – graphite (Gr) – causes decrease of the friction coefficient and improves the machinability of the hybrid composites.

Thus, one can conclude that the adequate combination of the manufacturing process, reinforcement material, its content and size of its particles, can result in optimal tribological properties of the aluminum based hybrid composites that are much better than properties of the conventional composites.

Acknowledgement

This research was financially supported by European regional development fund and Slovak state budget by the project "Research Centre of the University of Žilina" and by the Ministry of Education, Science and Technological Development of Republic of Serbia through grant TR 35021

References

1. B. Stojanovic, S. Milojevic, *Characterization, Manufacturing and Application of Metal Matrix* Composites/*Advances in Materials Science Research* (Nova Science Publishers, Ed. Maryann C. Wythers, New York, 2017)
2. A. Zeren, Ind. Lubr. Trib., **67**(3), 262 (2015)
3. S.O. Adeosun, L.O. Osoba, O.O. Taiwo, Int. J. Chem. Mol. Nucl. Mats. Metall. Eng., **8**(7), 731 (2014)
4. S. Basavarajappa, J.P. Davim, Mater. Res., **16**(5), 990 (2013)
5. G. Dixit, M.M. Khan, Jordan J. Mech. Ind. Eng., **8**(6), 7 (2014)
6. B. Stojanovic, L. Ivanović, Tech. Gazz., **22**(1), 247 (2015)
7. B. Stojanovic, J. Glisovic, J. 2016. *Automotive Engine Materials. In: Saleem Hashmi (ed.) Reference Module in Materials Science and Materials Engineering* (Elsevier, Oxford, 2016)
8. S. Bansal, J. S. Saini, Def. Sci. J., **65**(4), 330 (2015)
9. P. Suresh, K. Marimuthu, S. Ranganathan, Austral. J. Bas. Appl. Sci., **7**(9), 341 (2013)
10. M. Naveed, A. R. Anwar Khan, Int. J. Sci. Res., **4**(9), 146 (2015)
11. L. Lancaster, M.H. Lung, D. Sujan, World Acad. Sci. Eng. Techn., **73,** 1136 (2013)
12. P. Suresh, K. Marimuthu, S. Ranganathan, G. Arunkumar, Int. J. Adv. Eng. Appl., **6**(3), 38 (2013)
13. M. Zakaulla, A.R. Anwar Khan, World Acad. Sci., Eng. Techn. Int. J. Envir. Chem. Ecol. Geol. Geoph. Eng., **9**(6), 697 (2015)
14. B. Stojanovic, *Tribological behavior of hybrid composites with the A356 matrix, PhD Dissertation, Faculty of Engineering* (University of Kragujevac, Serbia, 2013)
15. S. Suresha, B.K. Sridhara, Mat. Des., **34,** 576 (2012)
16. P. Ravindran, K. Manisekar, P. Rathika, P. Narayanasamy, Mat. Des., **45,** 561 (2013)
17. A. Vencl, I. Bobic, S. Arostegui, B. Bobic, A. Marinkovic, M. Babic, J. All. Comp., **506,** 631 (2010)
18. N. Radhika, R. Subramanian, S. Venkat Prasat, B. Anandavel, Ind. Lub. Trib., **64**(6), 359 (2012)
19. S. Şahın, N. Yüksel, H. Durmuş, S.G. İrızalp, Mat. Techn., **48**(5), 639 (2014)
20. B. Stojanovic, J. Blagojevic, M. Babic, S. Velickovic, S. Miladinovic, Ind. Lub. Trib., **69**(6), 1005 (2017)
21. B. Pavithran, J. Swathanandan, Int. J. Techn. Enhan. Emerg. Eng. Res., **3**(04), 60 (2015)
22. A.N. Varghese, P.V. Pradeep, Int. J. Theor. Appl. Res. Mech. Engin., **3**(3), 2319 (2014)
23. B. Stojanovic, M. Babic, S. Veličković, J. Blagojević, Tribol. Trans., **59**(3), 522 (2016)
24. S. Veličković, S. Miladinović, B. Stojanović, R. R. Nikolić, B. Hadzima, D. Arsić, Production Engineering Archives, **14**, (2018).
25. P. Ravindran, K. Manisekar, P. Narayanasamy, N. Selvakumar, R. Narayanasamy, Mat. Des., **39,** 42 (2012)

The influence of simultaneous action of the aggressive environment and loading on strength of RC beams

Jacek Selejdak[1], *Roman* Khmil[2,*], and *Zinoviy* Blikharskyy[1]

[1]Czestochowa University of Technology, The Faculty of Civil Engineering, Department of Concrete Structures and Geotechnics, Akademicka 3, 42 200 Czestochowa, Poland
[2]Lviv Polytechnic National University, Department of Building Constructions and Bridges, Karpinskogo Str. 6, Lviv, 79013, Ukraine

Abstract. The article is devoted to an experimental research of the strength of reinforced concrete beams, and its dependence on a simultaneous influence of a corrosion environment and a loading factor. The tests have been carried out upon reinforced concrete specimens of 2100×200×100 mm size, with a regular reinforcement. The beams are of a span equaling to 1,9m with different reinforcing ratio of beams. The acid environment, namely 10 % H_2SO_4, was taken as a model of an aggressive environment. Reinforced concrete beams have been tested with and without the co-action of the aggressive environment and loading factor. Beams, which underwent a simultaneous action of the corrosive environment and loading, were loaded to a level 0.7 of its load-carrying capacity. The load-carrying capacity in aggressive environment in all the beams of all the series was achieved in 46-60 days. The influence of the simultaneous action of the aggressive environment and loading on the strength of reinforced-concrete beams has been described in the following work. It is necessary to note that the design code of Ukraine does not allow determining load carrying capacity of the beams affected by corrosion with simultaneous influence of loading with adequate accuracy. The analysis of experimental data has been done and the main directions of the design code's correction have been formulated.

1 Introduction

It is known that reinforced concrete is one of the most widespread building materials. This material has found its application in production and in a constructional engineering [1, 2, 3]. Its carrying functions are successfully applied in structural designs under a wide range of power load and other influencing factors in different environments. It is known that all existing constructions are influenced in a variety of ways by corrosion processes. Around 20 to 70% of constructions in nonferrous metallurgy, chemical, wood-pulp and paper industry, and energy-generating factories is affected by various corrosive environments. A corrosion process can last longer when in an aggressive environment of small

* Corresponding author: roman_hl@ukr.net

concentration whilst, when there is a high concentration corrosive environment, a construction can collapse very quickly.

Durability and load-carrying capacity generally depends not only on loading but also on the influence of the environment [4-7]. Every year losses resulting from corrosion are estimated by different specialists in boundary values of 4-6% of gross domestic product (GDP) [4]. Constructions of buildings of galvanic workshops, chemical enterprises and other elements of buildings which have a direct contact with reactive material undergo a corrosive processes to the greatest extent [4, 5]. The outcomes of this process can be the decrease of a cross section of elements, decrease of crack resistance and strength, build-up of deformations.

When constructions which are affected by corrosive environment are exploited, the effect of this interaction should be taken into consideration on the stage of design as well as controlled by specialists. An estimation of the load-carrying capacity of reinforced concrete constructions which have corrosion - mechanical injuries of concrete can be given on the basis of results of theoretical and experimental research.

2 Experimental research and discussion

For the experimental research there were made three series of reinforced concrete beams by the sizes of 2100×200×100 mm with reinforcement in the stretched working area 2∅12, 14 and 16 bars A400C accordingly, and in upper compressed area 2∅5 Bp-1. The transverse reinforcement was from ∅5 bars Bp-1 with step of 75-100 mm. The concrete mix of series is accepted with stone aggregate 10-20 mm. The concrete compressed strength made 44 MPa. In general 12 beams (four specimens in each series) were tested.

Firstly, in order to define strength characteristics of experimental beams without the effect of the corrosive environment, they were tested in terms of bending by means of a short term load applied till they fail. The beams were marked BO. Applying of a short term load was achieved with two concentrated forces in one-third of their span. The loads were applied in steps. Thus, deflection, presence and evolution of cracks, deformations of concrete and the reinforcement were fixed. Ultimate limit state (ULS) of beams passed due to the reinforcement's yield in the stretched area of the greatest bending moment with the following failure by crushing of the compressed zone of concrete.

For the research of effect of simultaneous acting of the corrosive environment and load on strength, two beams of each series was investigated, while one of the beams of each series were subjected only by corrosive environment. The beams were marked BD. Schemes of special equipment on which the experimental research was carried out presented on Fig. 1. The testing time of the corresponding beams - twins in the corrosive environment under loading and without loading was accepted as equal. After the beams stayed in the corrosive environment, they were tested in terms of a short term loading till their failure.

Beams undergoing a simultaneous action of the corrosive environment and loading were loaded to a level 70% from load-carrying capacity of ULS of beam marked BO. The loading's value was controlled by special ring-balance dynamometers. The loading was put with special spring devices which provided a static loading (Fig. 1b). As corrosive environment of 10% solution of sulphuric acid H_2SO_4 was used. Such environments occur in separate chemical productions, electroplating shops, smoke pipes etc. High concentration of solution also resolved for rather small period simulation effect of simultaneous acting of the corrosive environment and loading.

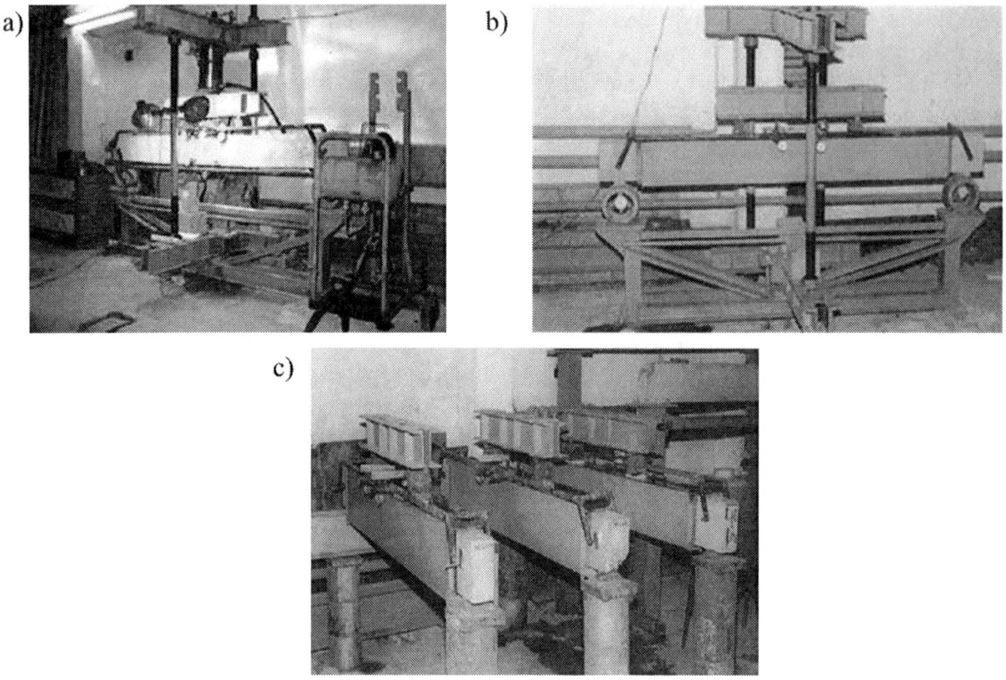

Fig. 1. The experimental specimens during testing: a) on short-term acting of loading; b) on long-term acting of loading and corrosive environments; c) on separate acting of a corrosive environment and loading after one.

The solution of acid was put into special baths, processed with corrosion-resistant coatings, timbered on the beams. The constant supervisory control of concentration of the corrosive environment was carried out. At decrease of concentration of acid more than 1%, according to standards [8], the solution was renewed. During the research, the reinforcement ratio of all beams was changed. It occurred because of corrosion cross section of the concentre. The reinforcement at given loadings level did not corrode. The corrosion of the reinforcing bars was avoided in order to exclude its impact on the stress-strain state of the beams. This effect is more complex and requires a separate study [5, 9]. At achievement by experimental samples of ULS the reinforcing ratio made for a series 3 (2∅12) - 1.93-1.99%, for a series 1 (2∅14) - 2.6-2.8 %, for a series 4 (2∅16) - 3.43-3.54%.

Achievement of the reinforcement yield of beams of all series was comprised almost equally - for 46-60 days (Table 1).

The final cross section of the concrete of beams at failure after simultaneous acting of the corrosive environment and loading and a beam without acting of loading were approximately equal (Table 1). It does not enable to compare strength of beams at achievement of ULS by them; nevertheless, the comparison of strength at failure of beams was possible (Table 2).

Experimental values of the bending moments which related to load-carrying capacity by ULS and physical failure and also their correlation are presented in Table 2. The theoretical bending moments according to design codes [10, 11, 12] are also presented in Table 2. It is necessary to note that the technique of standards [11] allows determining load-carrying capacity of the beams tested without of simultaneous effect of the corrosive environment and loading with adequate accuracy. The deviation between theoretical and experimental values is 1.1 % for a beam of the BD-4.4c of a series of 4 and -0.7 % for a beam of the BD-1.4c of a series 1 towards excess of experimental value above theoretical.

Table 1. Overall parameters of experimental research.

Series of testing	Specimens	Concrete cross section, b×h, mm			Testing time, days		Type of destruction
		before influence	during influence		when reinforcement yield	when failure	
			at reinforcement yield	at failure			
1	BO 1.1	101×202	without corrosive influence		short-time act of loading		bending
	BD 1.2-0.7c	99×200	74.4×174.3	65.0×166.0	46	56	bending
	BD 1.3-0.7c	102×200	70.2×169.6	61.0×166.0	49	60	shear
	BD -1.4c	100×206	60.0×167.0		short-time act of loading in 60 days after corrosion		bending
3	BO 3.1	99×208	without corrosive influence		short-time act of loading		bending
	BD 3.2-0.7c	101×208	70.4×177,2	60.0×176.0	46	54	shear
	BD 3.3-0.7c	100×207	68.2×176,9	58.0×174.0	47	56	bending
	BD -3.4c	98×204	60.0×169.0		short-time act of loading in 56 days after corrosion		shear
4	BO 4.1	100×202	without corrosive influence		short-time act of loading		bending
	BD 4.2-0.7c	102×200	71.3×171,0	66.5×172	50	54	bending
	BD 4.3-0.7c	99×201	72.9×175.0	65.0×173.0	48	57	bending
	BD -4.4c	99×202	60.0×171.0		short-time act of loading in 57 days after corrosion		bending

Table 2. Strength of experimental specimens.

Series of testing	Specimens	Experimental bending moments, kN×cm		Theoretical bending moments, kN×cm	Correlation, $\frac{M_u^{norm}}{M_u^{exp}}$
		at reinforcement yield M_u^{exp}	at failure M_f^{exp}	at reinforcement yield M_u^{norm}	
1	BO 1.1	2539.7	2924.7	2253.2	0.887
	BD 1.2-0.7c	1759.4		1956.9	1.112
	BD 1.3-0.7c	1759.4		1921.2	1,092
	BD -1.4c	1773.3	1959.5	1785.6	1.007
3	BO 3.1	1931.7	2084.9	1768.6	0.916
	BD 3.2-0.7c	1369.9		1558.1	1.137
	BD 3.3-0.7c	1369.9		1550.6	1.132
	BD -3.4c	-	1318.0	1467.8	-
4	BO 4.1	3166.7	36600	2809.1	0.887
	BD 4.2-0.7c	2141.9		2176.9	1.016
	BD 4.3-0.7c	2141.9		2246.3	1.049
	BD -4.4c	1933.3	2122.0	1910.5	0.989

The analysis of results shows, that simultaneous acting of the corrosive environment and loading for research of load-carrying capacity of beams is a worse possible combination. For beam which was only in the corrosive environment (BD-1.4c) the bending moment at failure was greater (1959.5 kN×cm) than for beams (BD 1.2-0.7c; BD 1.3-0.7c) tested under simultaneous acting of corrosive environment and loading (1759.4 kN×cm). For beam BD -4.4c the bending moment at failure was approximately equal (2122.0 kN×cm) for beams (BD 4.2-0.7c; BD 4.3-0.7c) tested under simultaneous acting of corrosive environment and loading (2141.9 kN×cm). Physical failure of a beam BD-3,4c passed from shear and earlier from load-carrying capacity by ULS. It happened at 1318 kN×cm when the theoretical ULS are 1467.8 kN×cm. The reason of failure was full corrosion of transvers bars for this beam that was not anticipated.

In general, the results of the beams tested under simultaneous acting of the corrosive environment and loading was worse. It is explained by the fact that the cracks revealed by the microscope in concrete at long acting of loading, being passed round come together at a macro crack. It increases deformability concrete, influences its strength and results in its premature failure. At short-term acting of loading this effect has no time to reveal itself.

The scheme of failure types of beams samples is presented in Fig. 2.

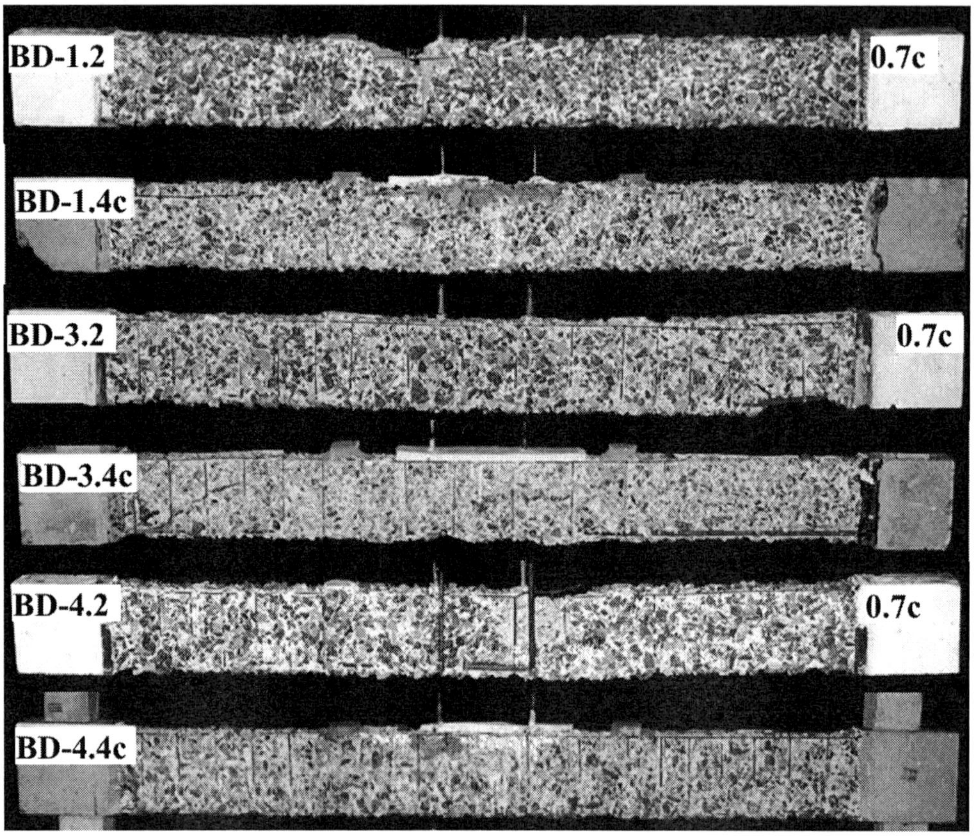

Fig 2. The scheme of typical failure of experimental beams.

This research shows that failure of reinforced concrete beams in environment of sulphuric acid has passed owing to corrosion of concrete, sometimes together with corrosion of the reinforcement. Corrosion of concrete was caused by chemical reactions of acid with constituent parts of cement stone which contains calcium. Thus, corrosion proceeded gradually from external sides of concrete in depth of experimental samples with formation of a contact layer on a surface of concrete. It is necessary to notice, that corrosion of concrete of beams occurred continually from the moment of putting beams into aggressive environment. The quartz sand and granite crushed stone practically did not give in to corrosion, and fell out from the surface of a beam during the corrosion of a cement stone.

The corrosion processes resulted in the decrease of the size of cross section beams over time on linear dependences. The reduction of the area of the compressed zone of concrete, as well as effective height of section, resulted in the increase of stresses in concrete and in the reinforcement. Failure of beams happened with the achievement of stress in the reinforcement of yield line with a further destruction of the compressed zone of concrete. Due to progressive corrosion of practically all of transverse bars, the beam failures caused by shear were appeared.

The analysis has also shown, that the technique of standards [11] in all cases of the design of bending reinforced-concrete elements with the corrosion damage taking at long simultaneous acting of a corrosive environment and loading does not allow to determine a load-carrying capacity of such elements with the real accuracy. The deviation between theoretical and experimental values increased to 13.7% towards the excess of theoretical values above experimental. It is explained by the effect of the corrosion micro cracks, as stress concentrators in the compressed concrete and as consequence the decrease of strength of concrete at presence of the corrosion processes and long acting of loading. It is necessary to take into account the presence of a contact concrete layer with the lowest physical-mechanical characteristics and strength. In the control process, testing beams and beams affected by the corrosive environment without simultaneous acting of loading, the technique of standards [11] give sufficient convergence of results. Therefore, the research shows that presence of simultaneous act of corrosive environment and loading significantly influences load carrying capacity of beams.

3 Conclusions

1. The analysis of results shows, that simultaneous acting of the corrosive environment and loading for research of load-carrying capacity of beams is a worse combination.
2. The deviation between theoretical and experimental values made up to 13.7% towards excess of theoretical above experimental ones for beams tested with simultaneous acting of the corrosive environment and loading. At the same time for beams without simultaneous acting the deviation made – 0.7-1.1%.
3. These obtained features should be taken into account when beams applied to real operating conditions are designed.

References

1. K. Brózda, J. Selejdak, P. Koteš, Procedia Engineering, **192**, 64 (2016)
2. M. Mazur, Prod. Eng. Arch., **13**(4), 41 (2016)
3. M.A. Fernandez-Ruiz, L.M. Gil-Martin, E. Hernandez-Montes, Int. J. of Concr. Struct. and Mat., **12**, 6 (2017)
4. S.N. Alekseev, F.M. Ivanov, S. Modry, P. Shysel, *Durability of reinforced concrete in corrosive environment* (316, 1990)
5. D.B. McDonald. D.W. Pfeifer. M. R. Sherman. *Corrosion evaluation of epoxy-coated, metallic-clad and solid metallic reinforcing bars in concrete* (124, 1998).
6. J. Selejdak, R. Ulewicz, M. Ingaldi, *23rd Int. Conf. on Metal.&Mat.* (1882-1888, 2014)
7. G.M. Kotsovou, D.M. Cotsovos, Eng. Struct., **160**, 44 (2018)
8. *National Standard of USSR, The methods of corrosion tests. General demands* (GOST 9.905-82, 5, 1983)
9. C.G Berrocal, I. Lofgren, K. Lundgren, Eng. Struct., **163**, 409 (2017)
10. *Eurocode 2: Design of concrete structures – Part 1-1: General rules for buildings* (EN 1992-1-1:2004 (E), 225, 2004)
11. *National Norms of Ukraine, Concrete and reinforced concrete structures* (DBN B.2.6 - 98: 2009, 84, 2011)
12. *National Standards of Ukraine, Concrete and reinforced concrete structures with heavy concrete* (DSTU B.2.6 - 156:2010, 118, 2011).

The computational analysis of the crack width of beams reinforced with CFRP and steel bars

Kinga Brózda[1,*], and *Jacek* Selejdak[1]

[1]Department of Concrete Structures and Geotechnics, Faculty of Civil Engineering, Czestochowa University of Technology, Poland

Abstract. The aim of the paper is to explain the particular problems related to the use of FRP reinforcement in bending beams. The paper compares the aspect of Serviceability Limit State (SLS) of simply supported concrete beams reinforced with Carbon Fiber Reinforced Polymer (CFRP) bars and three various types of steel bars (the steel grades are 35G2Y, 20G2VY-b and B500SP). The computational analysis of change in crack width under the influence of various level of service live load was investigated. The calculations of beam reinforced with FRP bars was based on Italian design guideline (CNR-DT 203/2006) and the conventional reinforced concrete beams (RC) were analysed in accordance with the European standard Eurocode 2 (EN 1992-1-1:2004).

1 Introduction

The low corrosion resistance of the steel reinforcement is a typical limitation of RC (reinforced concrete) structures, especially if the structures are exposed to the aggressive environmental factors [1-7]. To avoid such problems, the alternative FRP (Fiber Reinforced Polymer) rebar could be used as an internal longitudinal flexural reinforcement. This relatively new solution provides high protection against corrosion damage of RC structures. In additional, very good physical and mechanical properties of the FRP as low density and what is connected with it the low dead load, low thermal expansion coefficient, high tensile strength-to-weight ratio and very good fatigue properties cause the FRP an attractive alternative for steel reinforcement [7-11]. Unfortunately, the disadvantage of this solution is low value of modulus of elasticity of FRP materials, which has a significant impact on Serviceability Limit State (SLS) as deflection or crack width [11-14].

2 Computational analysis

The main purpose of the paper is an assessment of the impact of load intensity to the crack width increasing. The computational analysis concerned the beams reinforced with steel and polymer rebar. The crack width increasing of beams reinforced with selected steel rebar and CFRP bar were compared.

* Corresponding author: kbrozda@bud.pcz.pl

2.1 Assumed calculation model

The calculations of rectangular-shaped, simply supported beams were done. The static scheme and localization of reinforcement A_f are presented in the Fig. 1. The length of the span was assumed as $L_{eff} = 4.0$ m and the dimensions of cross-section are: $b = 0.18$ m, $h = 0.35$ m. The investigated beams were under various static stress of service live load $w_L = \{0.5; 2.0; 3.5; 5.0; 6.5; 8.0; 9.5; 11.0\}$ kN/m, dead load $w_D = 1.58$ kN/m and service dead load of $w_{SDL} = 3.0$ kN/m.

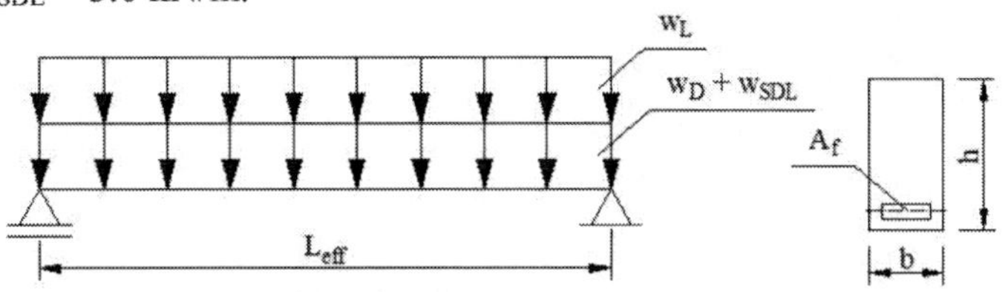

Fig. 1. The assumed static scheme and cross-section of beams.

The beams were reinforced with 4Φ12 of polymer rebar CFRF (Carbon Fiber Reinforced Polymer) with a relatively high modulus of elasticity and "C" steel grade reinforcement: 35G2Y, 20G2VY-b and B500SP, which have the highest values of tensile strength. The concrete is specified by C20/25 compressive strength class: the characteristic compressive strength $f_{ck} = 20$ MPa, the mean value of axial tensile strength of concrete $f_{ctm} = 2.2$ MPa, the ultimate compressive strain in the concrete $\varepsilon_{cu3} = 0.0035$, the modulus of elasticity $E_{cm} = 30 \cdot 10^3$ MPa. The concrete cover thickness is assumed as $c_{nom} = 35$ mm. The material characteristics of rebar are presented in the Table 1. The strength properties of the polymer rebar are reported by the manufacturer's material data [15].

Table 1. The material characteristics of reinforcement [15,16].

Type of rebar	The characteristic yield strength [MPa]	The characteristic tensile strength [MPa]	The modulus of elasticity [MPa]	The design strain [–]
CFRP	-	2 300	$130 \cdot 10^3$	0.018
35G2Y	410	550	$200 \cdot 10^3$	> 0.0675
20G2VY-b	490	590		
B500SP	500	575		

2.2 Results

The results of calculated beams are presented in the Table 2.

Table 2. The crack widths w_k depend on type of used bars and live load level w_L.

The live load level [kN/m]	0.5	2.0	3.5	5.0	6.5	8.0	9.5	11.0
Type of rebar	w_k *[mm]*							
CFRP	0.202	0.217	0.225	0.229	0.232	0.234	0.235	0.236
35G2Y	0.133	0.144	0.156	0.193	0.231	0.262	0.290	0.316
20G2VY-b	0.151	0.163	0.174	0.197	0.240	0.276	0.307	0.334
B500SP	0.155	0.167	0.178	0.197	0.242	0.278	0.310	0.338

In the computational analysis, the calculations of beams reinforced with steel rebar were done in accordance with the EC2 standard [16]. There is no available standard to design the polymer reinforcement. However, the guidelines for design the structural elements reinforced with FRP bars were elaborated as a result of work experience of researchers and designers. The calculations the crack width of beams reinforced with CFRP was done in accordance with the Italian design recommendations [17], due to the similarity to EC2 standard [16].

The crack width w_k of the beams reinforced with steel rebar (Fig. 2) depending on the value of the mean strain allowing under the relevant combination of loads for the effects of tension stiffening, shrinkage, etc., seems to increase linearly. Assuming according to the EC2 standard that limitation of the crack width $w_{k.lim}$ is equal 0.4 mm, none of investigated beams exceeded the limit value.

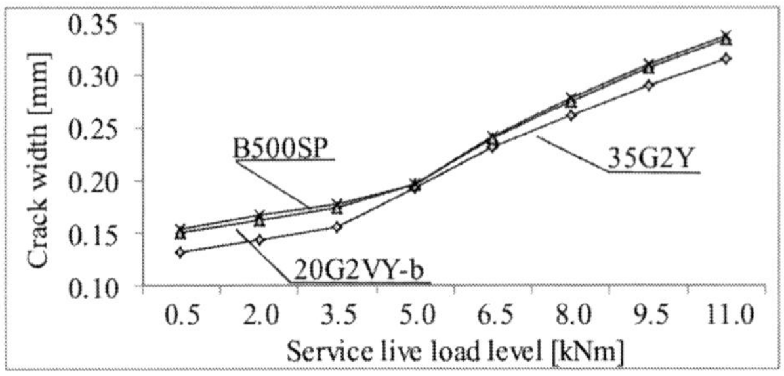

Fig. 2. The crack widths of steel rebar depending on service live load level.

In contrast to steel reinforcement, the w_k of beam reinforced with CFRP increases logarithmically. It means that the crack width increase is more dynamically at the initial load. The graph in the Fig. 3 presents the relation between crack width and service live load level.

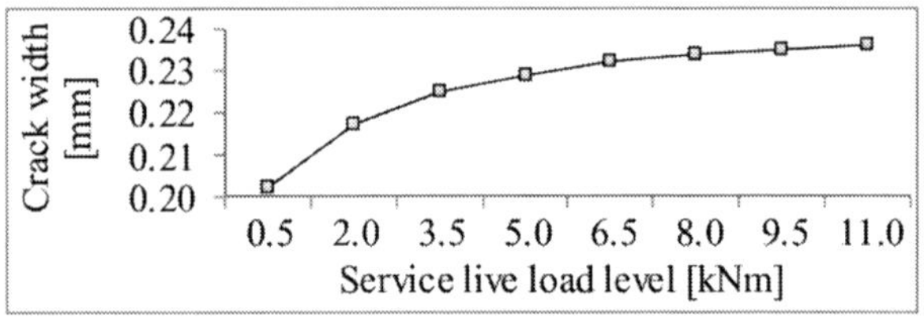

Fig. 3. The crack widths of CFRP rebar depending on service live load level.

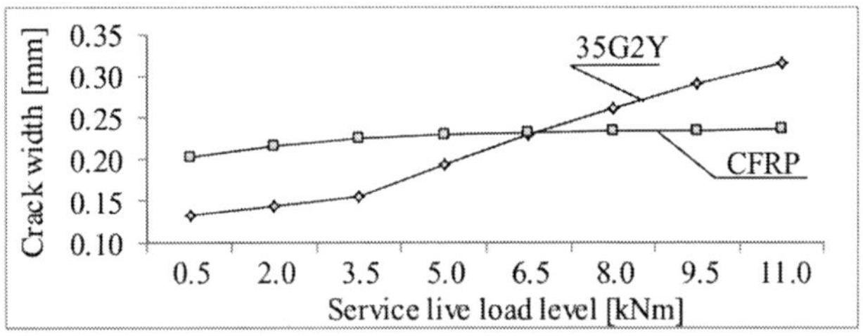

Fig. 4. The comparison of crack widths of CFRP and 35G2Y rebar.

In the Fig. 4 the increase of the crack widths of beams reinforced with CFRP and steel 35G2Y rebar were combined and compared. The steel 35G2Y were selected because the achieved values of crack width were the lowest (Fig. 2). As shown in the graph (Fig. 4) the increase of the crack width due to level of the service live load is more dynamic in case of the steel reinforcement. The component lines intersect almost at the point of 6.5 kN·m value. It means that above that level of load the beam reinforced with CFRP rebar has higher flexural capacity (due to higher tensile strength then steel's) and lower crack width. This phenomena makes that beam more effective in high levels of loading.

3 Summary

The crack widths depend on various level of load of the beams reinforced with steel and CFRP rebar were analysed and compared. During the computational analysis based on available standards and recommendations [16, 17], the high influence of relatively low modulus of elasticity on the SLS were concluded. Moreover, the differences between graphs of the crack width increase were noticed. The relationship between service live load level and crack widths of beams reinforced with steel rebar is linear, while the graph of CFRP reinforced beam increases logarithmically. It leads to conclusion that the CFRP reinforcement is more effective when the service live load achieves level higher than 6.5 kN·m. That phenomena in combination with very good flexural capacity due to the high value of the tensile strength, makes the CFRP rebar an attractive solution for RC structures which are exposed to aggressive environmental factors.

However, the only selected issue of the Serviceability Limit State was examined. It is recommended to limit the SLS deflections. The high impact of the low value of modulus of elasticity on the increased deflection is expected.

References

1. T. Lipinski, *16th Inter. Sci. Conf. Eng. for Rural Devel.*, (1069-1074, 2017)
2. I. Pliszka, N. Radek, Proc. Eng., **192**, 707-12 (2017)
3. C. Barris, L. Torres, C. Miàs, I. Vilanova, J.of Civ. Eng. Man., **18**(6), 843-857 (2012)
4. A. Szczotok, J. Nawrocki, J. Pietraszek, Arch. of Metal. Mat., **62**(2), 587-593 (2017)
5. J. Selejdak, R. Ulewicz, M. Ingaldi, *23rd International Conference on Metallurgy and Materials, Conference Proceedings* (Brno, Czech Rep., 1882-1888, 2014)
6. J. Pietraszek, A. Szczotok, N. Radek, Arch. Metall. Mater., **62**, 235 (2017)
7. H.A. Abdalla, Composite Structures, **56,** 63-71 (2002)
8. K. Brózda, J. Selejdak, Production Engineering Archives, **12**(3), 2-4 (2016)
9. K. Brózda, J. Selejdak, P. Koteš, Proc. Eng., **192**, 64-68 (2016)
10. C. Bywalski, M. Drzazga, M. Kamiński, Build. Mat., **6**, 72-73 (2014)
11. K. Brózda, M. Major, J. Selejdak, *23th Int. Conf. EM2017* (218-221, 2017)
12. R.J. Gravina, S.T. Smith, Eng. Struct., **30**, 2370-2380 (2008)
13. H. Toutanji, Y. Deng, Constr. Build. Mat., **17**, 69-74 (2003)
14. J. Pietraszek, E. Skrzypczak-Pietraszek, Adv. Mat. Res.-Switz., **874**, 151-155 (2014)
15. Sireg Geotech S.r.l. prod. data sheet, available at: www.sireggeotech.it/en (22.12.2016)
16. *Design of concrete structures. General rules and rules for buildings* (EN 1992-1-1:2004)
17. *Guide for the Design Cons. of RC Struct. Reinf. with FRP Bars* (CNR-DT 203/2006)

Quasi-optimization of the slender geometrically nonlinear supporting system with non-prismatic element under the specific load

Janusz Szmidla[1], and *Anna* Jurczyńska[2,*]

[1]Institute of Mechanics and Machine Design Fundamentals, Częstochowa University of Technology, Częstochowa, Poland
[2]Institute of Mechanics and Machine Design Fundamentals, Częstochowa University of Technology, Częstochowa, Poland

Abstract. This paper is devoted to the issue of free vibrations of a geometrically nonlinear column with a nonprismatic element modelling the supporting structure subjected to the specific load. The boundary problem was formulated on the basis of the Hamilton's principle and its solution was obtained using the Lindstedt – Poincare's small parameter method (a perturbation method). Within the range of the kinetic criterion of the static equilibrium, an influence of the shape of a rod of variable cross-section on the values of free vibrations and the bifurcation load of the system was determined. Presented quasi-optimization issue has been reduced to defining the ranges of physical and geometrical parameters where an increase of the bifurcation load is the biggest in comparison with the reference system.

1 Introduction

Slender systems are commonly used as a components of many structures, among others in machinery construction, civil engineering or mining industry. Widespread application forces researchers to predict their behaviour and strength (stability). Studies on controlling their dynamic properties and the issue of optimization are of particular importance.

Work [1] concerns the influence of the size of the amplitude parameter on the natural frequency of the partially compressed slender column under the Euler's load. The problem of optimization of a similar system with regard to the occurrence of cracks is described in [2], where genetic algorithms were used. The topic of work [3] is the solution of the problem of a dynamic stability in stepped cantilever beam with attached discrete elements obtained by applying the mode summation method. The original optimization algorithm was proposed in the works [4, 5].

This paper proposes a quasi-optimization algorithm in relation to the control of natural vibration frequencies of the geometrically nonlinear slender system under the follower force directed towards the positive pole (a specific load).

2 Boundary problem formulation

A geometrically nonlinear column **N** built of three prismatic rods connected at the free ends presented in figure 1b is considered in the paper. The relation between the flexural stiffness of the particular rods is given by the coefficient of the flexural stiffness asymmetry μ while the total flexural stiffness is assumed to be constant:

$$\mu = \frac{(EJ)_{III}}{(EJ)_{I} + (EJ)_{II}}, \quad \sum_{k=I}^{III}(EJ)_k = const. \tag{1, 2}$$

In the target system **NN** (fig. 1a) the internal element was replaced by the nonprismatic rod of the same volume, resulting from the value of the coefficient μ. The variable cross-section was modelled by the division of the rod into the finite number of prismatic segments of the same length and thick, which widths were described by the approximating functions (fig. 1c). The calculations were limited to the shape approximation by means of the polynomial of degree I and II.

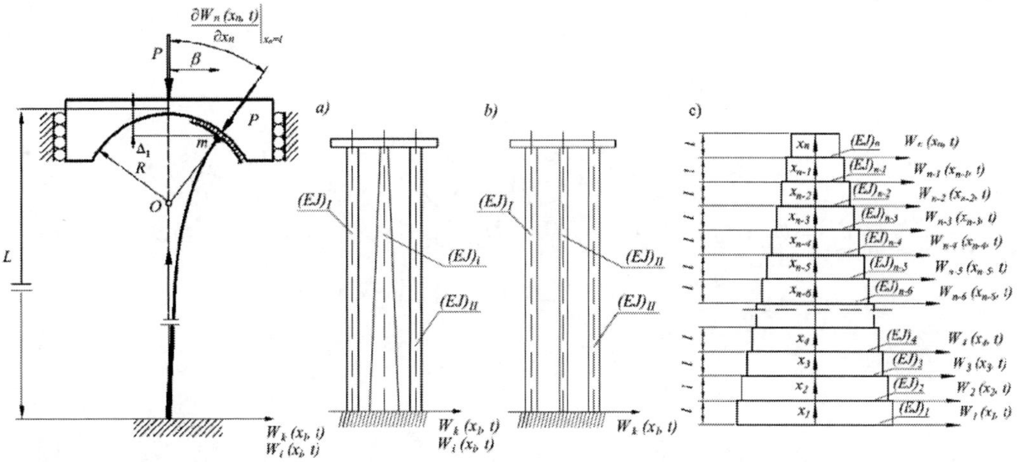

Fig. 1. The physical model of the geometrically nonlinear column a) column with the nonprismatic element, b) the comparative column, c) the model of the rod of variable cross-section.

The columns were subjected to the follower force directed towards the positive pole, realized through the loading structure built of elements of circular outlines (parts of the rolling bearing). The applied load is a case of the specific load defined by L. Tomski [6] and it combines features of the follower force (nonconservative) and the force directed towards the pole. The load is transmitted from the loading head to the receiving head and then via a rigid element (enabling correct mounting of the column) on the system. The advantage of using systems under the specific load as a supporting structure is obtaining higher values of the bifurcation and critical loads in comparison with the classical Euler's load.

Taking into account the Bernoulli-Euler's theory and described physical models, the total potential energy V as well as the kinetic energy T (comp. [7, 8]) of the analyzed system were defined. The boundary problem was formulated on the basis of the Hamilton's principle (principle of stationary action) using the calculus of variations:

$$\int_{t_1}^{t_2} (\delta T - \delta V)\,dt = 0. \tag{3}$$

The differentia equations of motion in the transverse (eq. 4) and longitudinal (eq. 5-6) directions obtained based on the relation (3) may be written as follow:

$$(EJ)_i \frac{\partial^4 W_i(x_i,t)}{\partial x_i^{\,4}} + S_{III}(t)\frac{\partial^2 W_i(x_i,t)}{\partial x_i^{\,2}} + (\rho A)_i \frac{\partial^2 W_i(x_i,t)}{\partial t^2} = 0, \quad i = 1,\ 2,\ ...,\ n. \tag{4}$$

$$(EJ)_k \frac{\partial^4 W_k(x_k,t)}{\partial x_k^{\,4}} + S_k(t)\frac{\partial^2 W_k(x_k,t)}{\partial x_k^{\,2}} + (\rho A)_k \frac{\partial^2 W_k(x_k,t)}{\partial t^2} = 0, \quad k = I,\ II, \tag{5}$$

$$U_i(x_i,t) - U_i(0,t) = -\frac{S_i(t)}{(EA)_i} x_i - \frac{1}{2}\int_0^{x_i}\left(\frac{\partial W_i(x_i,t)}{\partial x_i}\right)^2 dx_i, \quad i = 1,\ 2,\ ...,\ n, \tag{6}$$

$$U_k(x_k,t) = -\frac{S_k(t)}{(EA)_k} x_k - \frac{1}{2}\int_0^{x_k}\left(\frac{\partial W_k(x_k,t)}{\partial x_k}\right)^2 dx_k, \quad k = I,\ II. \tag{7}$$

The geometrical and natural boundary conditions describing the column, including the continuity conditions relate to the particular segments of the nonprismatic rod, are given by the following formulas:

$$W_I(0,t) = W_{II}(0,t) = W_1(0,t) = 0, \tag{8}$$

$$W_I(L,t) = W_{II}(L,t) = W_n(l,t), \tag{9}$$

$$\left.\frac{\partial W_I(x_I,t)}{\partial x_I}\right|_{x_I=0} = \left.\frac{\partial W_{II}(x_{II},t)}{\partial x_{II}}\right|_{x_{II}=0} = \left.\frac{\partial W_1(x_1,t)}{\partial x_1}\right|_{x_1=0} = 0, \tag{10}$$

$$U_I(0,t) = U_{II}(0,t) = U_1(0,t) = 0, \quad U_I(L,t) = U_{II}(L,t) = U_n(l,t), \tag{11}$$

$$\left.\frac{\partial W_I(x_I,t)}{\partial x_I}\right|_{x_I=L} = \left.\frac{\partial W_{II}(x_{II},t)}{\partial x_{II}}\right|_{x_{II}=L} = \left.\frac{\partial W_n(x_n,t)}{\partial x_n}\right|_{x_n=l}, \tag{12}$$

$$W_n(l,t) - (R - l_0)\left.\frac{\partial W_n(x_n,t)}{\partial x_n}\right|_{x_n=l} = 0, \tag{13}$$

$$\begin{gathered}\sum_{k=I}^{II}(EJ)_k \left.\frac{\partial^3 W_k(x_k,t)}{\partial x_k^{\,3}}\right|_{x_k=L} + (EJ)_n \left.\frac{\partial^3 W_n(x_n,t)}{\partial x_n^{\,3}}\right|_{x_n=l} + \\ +(R - l_0)\left[\sum_{k=I}^{II}(EJ)_k \left.\frac{\partial^2 W_k(x_k,t)}{\partial x_k^{\,2}}\right|_{x_k=L} + (EJ)_n \left.\frac{\partial^2 W_n(x_n,t)}{\partial x_n^{\,2}}\right|_{x_n=l}\right] - m\frac{\partial^2 W_n(l,t)}{\partial t^2} = 0,\end{gathered} \tag{14}$$

$$W_j(l,t) = W_{j+1}(0,t), \quad U_j(l,t) = U_{j+1}(0,t), \tag{15}$$

$$\left.\frac{\partial W_j(x_j,t)}{\partial x_j}\right|_{x_j=l} = \left.\frac{\partial W_{j+1}(x_{j+1},t)}{\partial x_{j+1}}\right|_{x_{j+1}=0}, \tag{16}$$

$$(EJ)_j \left.\frac{\partial^2 W_j(x_j,t)}{\partial x_j^2}\right|_{x_j=l} = (EJ)_{j+1} \left.\frac{\partial^2 W_{j+1}(x_{j+1},t)}{\partial x_{j+1}^2}\right|_{x_{j+1}=0}, \tag{17}$$

$$(EJ)_j \left.\frac{\partial^3 W_j(x_j,t)}{\partial x_j^3}\right|_{x_j=l} = (EJ)_{j+1} \left.\frac{\partial^3 W_{j+1}(x_{j+1},t)}{\partial x_{j+1}^3}\right|_{x_{j+1}=0}, \quad j = 1, 2,\ ...,\ n-1. \tag{18}$$

Due to the geometric nonlinearity of the analyzed system, the perturbation method (small parameter of the amplitude method) was used to solve the boundary problem. By substituting the solution of the differential equations of motion into the boundary conditions

(8-18), the system of homogeneous equations is obtained which solution is a transcendental equation describing free vibrations of the column.

3 Results of numerical computations

Taking into consideration the mathematical model of the analyzed system, an original computational algorithms were developed. Results are presented in the dimensionless form after prior definition of the parameters of external λ and bifurcation load λ_b, frequency of vibrations Ω^* and the parameters describing the shape of the nonprismatic rod:

$$\lambda = \frac{PL^2}{(EJ)_{por}},\ \lambda_b = \frac{P_b L^2}{(EJ)_{por}},\ \Omega_i^* = \frac{(\rho A)_{por}\,\omega_i^2 L^4}{(EJ)_{por}},$$
$$Z^* = \frac{b_1 - b_n}{L},\ p^* = \frac{p}{L},\ q^* = \frac{q}{b_{pr}},\ R^* = \frac{R - l_0}{L}. \tag{19-25}$$

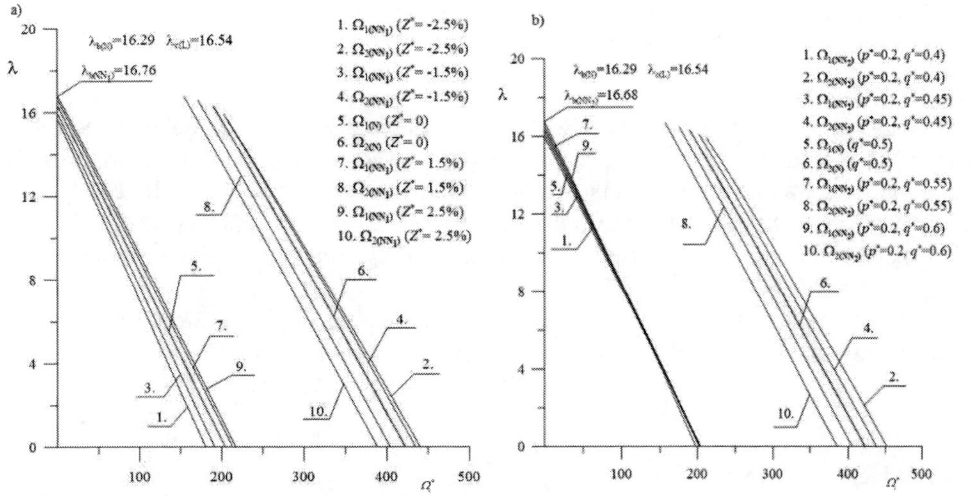

Fig. 2. The first two frequencies of vibrations as a function of external load parameter for selected values of the approximating parameters ($R^* = 0.35$, $\mu = 0.25$).

A change of two first natural frequency parameters as a function of the external load for selected geometry of the loading structure was presented in figure 2. The values of load for which the frequency reaches the zero value, refer to the bifurcation load of the system (the structure lost its rectilinear form of the static equilibrium). Taking into account the element of the variable cross-section in the physical model has got an influence on the value of the natural frequency parameter and it may be used for control of the dynamic properties of the column. Properly selected parameters of the approximating functions coefficients may also cause an increase in the bifurcation load compared with the reference system made of prismatic rods (compare fig. 3).

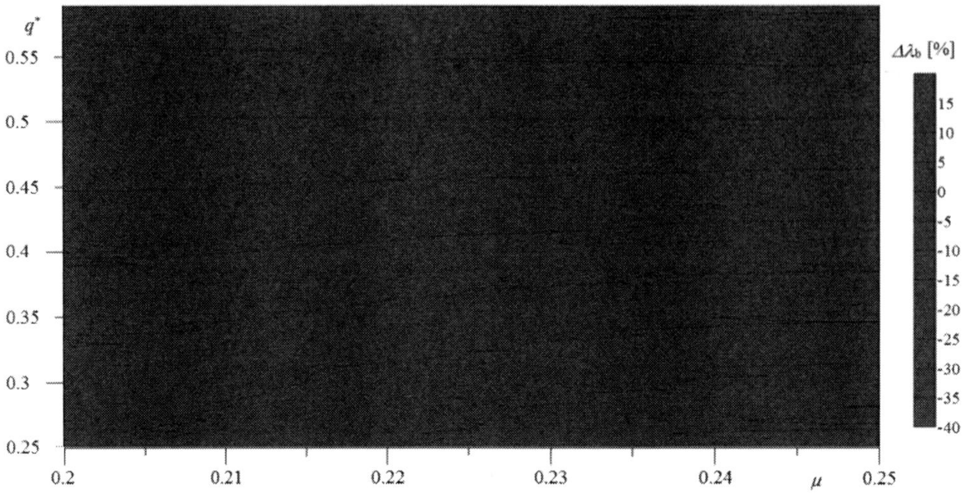

Fig. 3. An increase of the bifurcation load of the geometrically nonlinear column with nonprismatic rod in relation to the asymmetry of the flexural stiffness coefficient (p^*=0.4, R^*=0.087).

4 Conclusions

After taking into account in the model of the geometrically nonlinear column the rod of variable cross-section, a change in the value of the natural frequency of vibrations and bifurcation load was noted in comparison with the systems built of prismatic elements. It was proved that the appropriately chosen values of the parameters describing the nonprismatic rod outline can be used for to control the dynamic properties and increase the strength of the system.

The calculation model proposed in this work and the algorithm developed on its basis is universal and enables a quick analyze od complex beam systems with nonprismatic elements described by any approximating functions, realizing different types of external loading.

Acknowledgements
The study has been carried out within the framework of project BS/MN 1-101-302/17/P realized at Częstochowa University of Technology.

References

1. S. Uzny, K. Sokół, M. Osadnik, *Eng. Mech Conf. Proceedia* (562-565, 2016)
2. K. Sokół, A. Kulawik, J. Appl. Math. Comput. Mech., **13**(1), 115-124 (2014)
3. W. Sochacki, J. of Vibroeng. **15**(1), 280-290 (2013)
4. J.T. Katsikadelis, G.C. Tsiatas, Intern. J. of Non-lin. Mech., **42**, 164-171 (2007)
5. J.T. Katsikadelis, G.C. Tsiatas, Intern. J. of Mech. Sc., **49,** 1204-1212 (2007)
6. R. Bogacz, S. Imiełowski, L. Tomski, *Dyn. Of Continua Intern* (Symp., 1996)
7. L. Tomski, J. Szmidla, S. Uzny, Thin-Wall. Struct., **45,** 945-949 (2007)
8. L. Tomski, S. Uzny, Intern. J. Sol. Struct., **45**, 87-112 (2008)

Possibilities of using 3D printing for surface layers

Agnieszka Dulska[1,*]

[1]Silesian University of Technology, Department of Foundry Engineering, Towarowa St. 7, 44-100 Gliwice, Poland

Abstract. This article describes how to improve the tribological properties of cast iron by local strengthening by titanium layer, directly in the casting process. The research comprised test molds with a titanium (composite) layer. The insert with titanium was made basing on proprietary development of geometric assumptions. The innovation relies on application the 3D printing insert obtaining in SLM (selective laser melting) method. The tests included measurement of resistance to abrasive wear and carried out metallographic, microhardness measurements as the presents detailed geometry of the insert as well. On the basis of obtaining results was stated that there is a possibility of reinforcing surface layer of the grey cast iron casting by using 3D printing scaffold insert in the method of mould cavity preparation. Moreover there was a local increase in hardness and abrasive wear resistance in spite of the precipitation of titanium carbides in surface layer of grey cast iron.

1 Introduction

The market requirements affecting the development of industry in the world are constantly growing, what increases the demand for new, better production technologies. The quality and durability of machine components or structures must also be improved. Traditional materials are increasingly unable to meet the technological requirements. Increasing interest around surface layers and elements reinforced with the core caused their very dynamic developement. Improving the usable properties, extending the life of tools and subassemblies, and also reducing the production time and production costs are just a few of the advantages of using surface layers [1-4].

One of the methods of obtaining the surface layers is pouring a liquid metal into the mold with the perform (insert, cover) inside. This is a casting method for obtaining surface layers. After pouring the liquid metal into the mould, the covering create a layer on the surface of the cast. The processes which take a place inside the mold cause the creation of a cast with a permanent combination of two chemically different materials with different properties. The big advantage of using this method is that the obtained layer is created in the one process - casting, and the technology does not differ from the production of castings - it requires only pre-preparation of the insert (preform) [1-7].

* Corresponding author: agnieszka.dulska@polsl.pl

The methods of preparing the mould:
- layers formed as a result of pouring by liquid metal the inserts into the mold (for example: bimetals),
- layers obtained by melted the alloy components on the mold surface by the heat of the poured liquid metal,
- grained layers created by infiltration of a liquid metal into a layer formed on a mold.
- The preform should have appropriate properties such as:
- adequate wettability,
- melting temperature lower than the melting temperature of the base material,
- a certain thickness,
- non-gaseous binder material.

The inserts used to create the surface layers most often consist of alloy components such as: Cr, Ni, Co, Fe-Cr, Fe-Mn, Fe-W, Fe-Mo, as well as binders formed from surface-active compounds [1,3,4].

A new way to prepare inserts for the production of surface layers is the 3D printing method. It allows to produce inserts with a complicated shape, fully recreating the shape of the place in the mold cavity, where higher properties, eg higher abrasion resistance are needed to obtain [3, 8-10].

2 Methods and Materials

The purpose of the research was to create an innovative material based on the cast iron with titanium reinforcement.The research included observation of the structure on the optical and scanning microscope as well as the measurement the microhardness and abrasion resistance.

The gray eutectic cast iron (base cast) and the insert made of titanium alloy (working part) were used to make the composite cast. The modular insert (50x15x15 mm) was made using the method of 3D printing in the process of Selective Laser Melting (SLM) from titanium alloy (TiAl6V4 - table 1).

Table 1. Chemical content of the alloy TiAl16V4.

Element	Fe	C	N	O	H	Al	V
Content [%]	≤0,30	≤0,08	≤0,05	≤0,20	≤0,015	5,5-6,75	3,5-4,5

A modular insert with a geometrical form of elementary cells (Fig. 1a) was designed and printed. The cell is formed by a circular section connectors that combined in the nodes. In this way, elementary cells in the shape of a regular body (cube) were obtained. The single cell model is shown in the Figure 1b and 1c. The distance between individual nodes is 2 mm, the diameter of the connector is 1.5 mm. In addition, the inside diameter of the connector was set at 1 and 0 mm, depends on the connector :"empty" or "full" (Fig. 1b and 1c). The shape of the cast with the place of location of the insert is shown in Figure 2.

a) b) c)

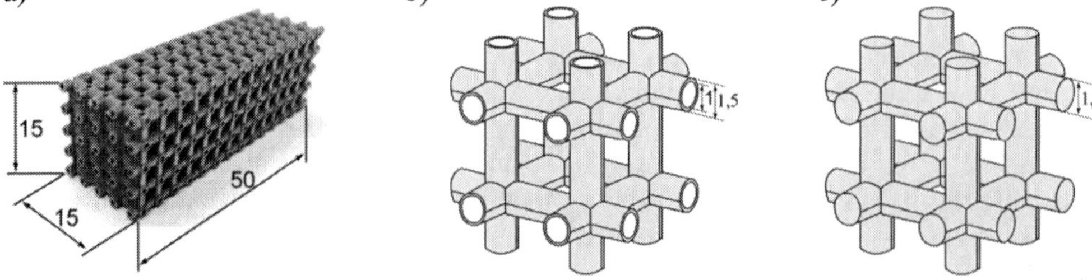

Fig. 1. 3D model of a titanium insert; a) a real insert; b) an insert with an empty connector; c) an insert with a full connector.

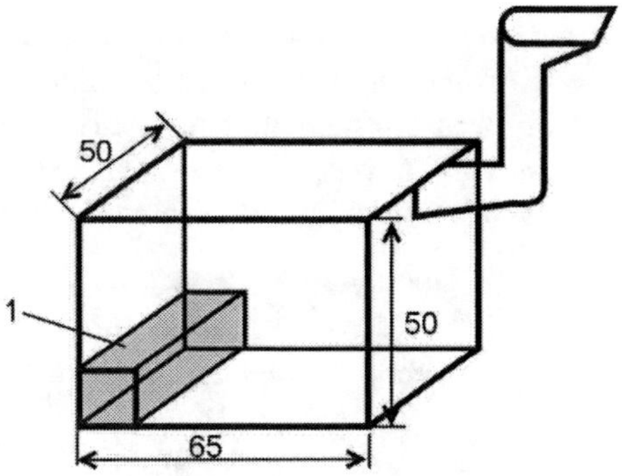

Fig. 2. The shape of the cast with dimensions; 1 – modular insert.

3 Results of the research

The research included observation of the structure of the materials connection on the optical and scanning microscope as well as the measurement of the microhardness and abrasion resistance.

Metallographic examinations of the resulting layer were performed on a microscope Nikon Eclipse LV 150N (fig. 3) to determine the type of connection in the structure. The result of combining the two materials was a layer of gray cast iron and titanium reinforcement (with many carbide precipitates). The formed carbides have an irregular and different shape, different size, and their concentration is highest at the place of dissolution of the insert.

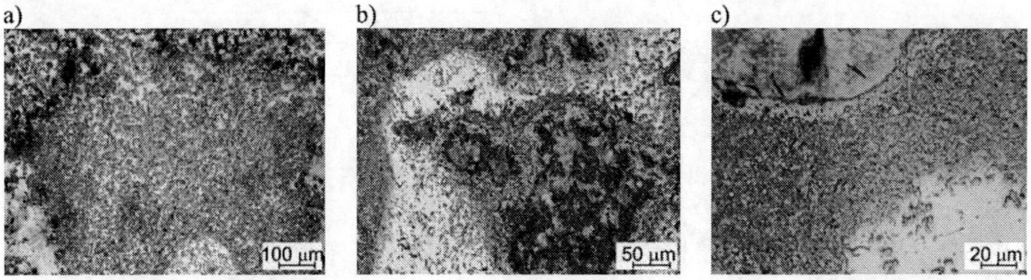

Fig. 3. Areas of Titanium carbides precipitates: a) zoom 100x; b) zoom 200x; c) zoom 500x.

The pictures from the Phenom scanning microscope are presented in Figure 4. Determining the approximate size of the precipitated titanium carbides, which ranged from 1.5 to 6 µm was based on the pictures from the microscope. Figures 4a and 4b present different concentration zones of the carbides. The concentration zones depend mainly on the distance from the skeletal insert - closer to the insert the concentration is bigger.

Additionally, Figure 4c shows photos of the breakthrough of the sample, on which the shape of carbide precipitates is visible.

The research included also the point analysis of chemical composition. The purpose of this examination was to determine the elements occurring in the structure of the resulting material (Figure 5).

The microhardness test were executed on the hardness tester FUTURE TECH FM-700 with a load 1000 g. The testing area included zones with significant concentrations of titanium carbide precipitates. Figure 6 shows an example of a measurement place and obtained results of microhardness.

Diversified values of microhardness measurements are caused of inhomogeneous concentration of carbides, therefore it can be concluded that a higher measurement value means a higher concentration of carbides. At very low values, it can be concluded that the measurement covered a larger part of the warp of the material than the carbides. It was noted that the results of the warp measurement were slightly higher than for typical gray cast iron. Therefore, the warp was strengthened not only in the place of the material connection, but in the region of the precipitations. The results were compared with materials exposed to abrasive wear (table 2) and presented in table 3.

a) b) c)

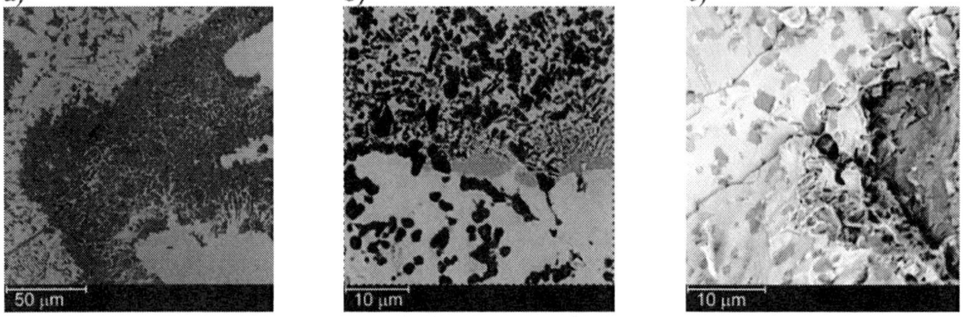

Fig. 4. Microstructure of the surface layer; a), b) upper view; c) breakthrough.

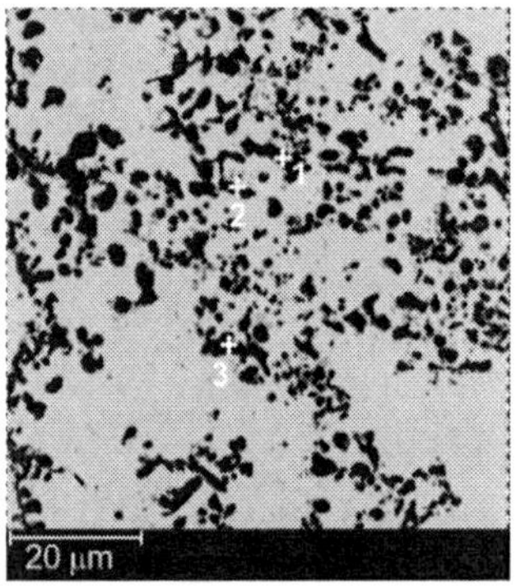

Test	Ti%	C%	Fe%
1	72,1	21,4	6,5
2	71	21,9	7,1
3	71	22,8	6,3

Fig. 5. The point analysis of chemical composition.

a) b)

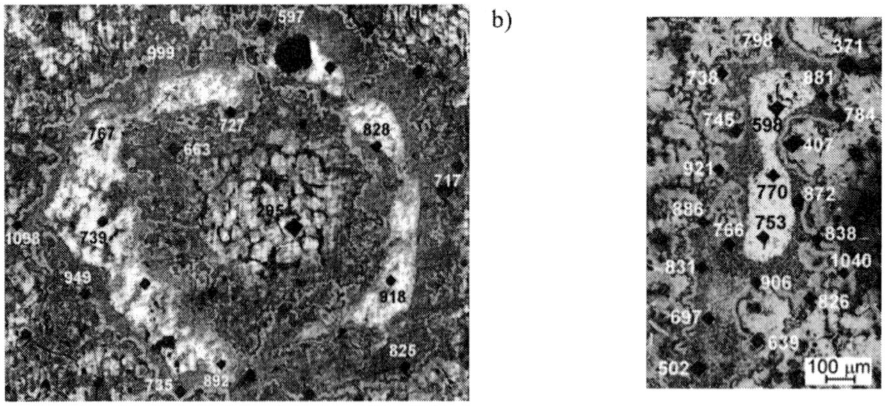

Fig. 6. The microhardness measurement area for the surface layer with an insert with a connector: a) empty; b) full.

Tabel 2. Chemical content of comparative materials.

Material	C [%]	Mn [%]	Si [%]	Cr [%]	Ni [%]	Mo [%]
High-chromium cast iron	2,85	0,385	0,657	20,4	1,48	0,565
Low-alloy steel	0,22	1,09	0,75	0,65	0,296	0,25
GJL-250	3,00-3,25	0,40-0,70	1,85-2,10	-	-	-

The obtained surface layers were tested to abrasion resistance. The test device consists of a table on which the counter-sample is mounted, the motor generating the rotary motion and the arm which generates a reciprocating movement of the sample on the surface of the counter-sample. The device allows to use different values of a load by using weights of a given mass. The silicon carbide abrasive paper (SiC) with a granularity of P50 was used in each of the tests. The grinding cycles were carried out with a load of 10 kg. One measurement consisted of 2,500 repetitions (ie 500 m). Four measurements were made for each of the samples. The exact loss of material was determined by weighing the sample before and after each measurement. Average losses in mass are showed in Table 4.

Tabel 3. Average value of the microhardness test for materials (Sm – average value of the microhardness).

Test number	**1**	**2**	**3**	**4**	**5**	**6**	**7**	**8**	**S_m**
High-chromium cast iron	463	459	540	450	449	506	410	558	**479**
Low-alloy steel	594	559	753	524	623	590	655	713	**626**
Gray cast iron, base cast	308	295	329	324	318	283	321	333	**314**

Tabel 4. Losses of the materials under the load 10kg/10cm².

	Test number	**Weight [g]**	**Loss [g]**
Titanium layer - empty connector number 1	0	192,437	-----
	1	191,880	0,557
	2	191,459	0,421
	3	190,916	0,543
	4	190,470	0,446
Average			0,492
Titanium layer - empty connector number 2	0	191,088	-----
	1	190,960	0,128
	2	190,844	0,116
	3	190,736	0,108
	4	190,621	0,115
Average			0,117
Titanium layer – full connector	0	187,871	-----
	1	187,746	0,125
	2	187,663	0,083
	3	187,580	0,083
	4	187,497	0,083
Average			0,094

The obtained results were compared with measurements for higher abrasion resistanced materials (high-chromium cast iron) and a base material (gray cast iron GJL-250). The tested alloys were characterized by the chemical composition shown in Table 2 and abrasion resistance presented in Table 5. Tests made for the reference materials were carried out under the same conditions as cast iron castings with a titanium insert.

Tabel 5. The results of the abrasion resistance test under the load 1kg/1cm².

Sample	**Average loss [g]**
Titanium layer - empty connector number 1	0,049
Titanium layer - empty connector number 2	0,012
Titanium layer – full connector	0,009
High-chromium cast iron	0,018
GJL-250	0,221

Comparing the results, it was found that the use of 3D printed titanium alloy inserts greatly increases abrasion resistance in comparing to the "base" material – gray cast iron GJL-250. Compared to high-chromium cast iron, the weight loss value is comparable. However, the main advantage of the tested castings with a titanium layer is the possibility of reinforcing only in a place exposed to wear (local reinforcement) and not as in the case of chromium cast iron - in the entire volume. The largest loss in mass was observed in the tested sample No. 1 (casting with a titanium insert, empty connector). It probably means that local melting of the insert appeared (too high pouring temperature).

4 Conclusions

The inserts used during the research were made using the SLM method. The advantage of these inserts is that they can be produce in the shape of a flat skeleton or any shape corresponding to the work surface of the cast part (machine).
The conclusions were made on the basis of the conducted research:

1. It is possible to strengthen the surface layer of a gray cast iron casting using a modular insert printed in the SLM method.
2. As a result of the reaction between the insert (TiAl6V4) and the metal (gray cast iron) locally numerous carbides (mainly TiC) were created. Carbides guarantee the increase of hardness and resistance to abrasive wear.
3. The functional properties depend mainly on the size of the insert (the internal dimensions of the connector) and on the temperature of the pouring the liquid metal into the mold.

References

[1] T. Wróbel, Archives of Materials Science and Engineering, **48**(2) (2011)

[2] M. Dojka, R. Dojka, A. Studnicki, Archives of Foundry Engineering, **17**(1) 37-40 (2017)

[3] N. Przyszlak, A. Dulska, T. Wróbel, J. Szajnar, Archives of Foundry Engineering **18**(1), 99-102 (2018)

[4] A. Dulska, A. Studnicki, J. Szajnar, Archives of Metallurgy and Materials, **62**(1), 365-367 (2017)

[5] E. Çalikoğlu, M. Çetin, *Production of bi-metal material, Materials of International Iron and Steel Symposium* (Karabük, Turkey, 156-161, 2012)

[6] S. Ertürk, O. Çakir, L. Kumruoglu, A. Ozel, Acta Physica Polonica A, **125**(2), 452-453 (2013).

[7] P. Jonšta, P. Váňová, S. Brožová, P. Pustějovská, J. Sojka, Z. Jonšta, M. Ingaldi, Archives of Metallurgy and Materials, **61**(2), 709-711 (2016)

[8] M. Cholewa, M. Dziuba-Kałuża, Archives of Foundry Engineering, **9**(3), 29-34 (2009)

[9] M. Król, T. Tański, Archives of Metallurgy and Materials, **61**(3), 1291-1296 (2016)

[10] Patent Application No. P.417657

Surface properties in S500MC steel with zinc coating

Agata Dudek[1,*], *Katarzyna* Strzelczak[1], and *Barbara* Lisiecka[1]

[1]Institute of Material Engineering, Faculty of Processing Engineering and Materials Technology, Czestochowa University of Technology, Armii Krajowej 19, 42–200 Czestochowa, Poland

Abstract. Zinc metal has a number of characteristics that make it a well-suited corrosion protective coating for iron and steel products. Due to the zinc's corrosion resistance on most environments, it provides an effective protective coating on many products and in many exposure conditions. Zinc, like all metals, corrodes when exposed to the atmosphere. However, because of its ability to form dense, adherent corrosion byproducts, the rate of corrosion is considerably lower than ferrous materials (10 to 100 times slower depending on the environment). In this study, a comparison of two series of specimens of alkaline electrochemical galvanizing steel S500MC (HSLA) was made. One series of samples was not stripped of the oxide layer from the substrate. The analysis was made using optical microscopy, electron scanning microscopy and X-ray diffraction.

1 Introduction

Zinc- and zinc-alloy-coated steel products are finding numerous applications in various branches of industry [1,2] as shown in Table 1.

Table 1. Applications of zinc-coated steel products [3].

Coating	Application
Coatings by continuous process	
Zn-Fe	Autobody panels and structural components
Zn and Zn-Al	Roofing, siding, doors, culvert, housing, autobody panels and structural components
Zn-Ni	Autobody panels, housings, appliances, fasteners
Zn-Co	Automotive small parts and fasteners
Batch galvanized coatings	
Zn	Structural steel for power-generating plants, petrochemical facilities, heat exchangers, cooling coils, bridge structural members, culverts, marine pilings, rails, reinforcing steel for concrete structures

* Corresponding author: dudek.agata@wip.pcz.pl

The prevalence of zinc coatings resulting from the availability, low price, good protective and decorative properties of the metal. Zinc, as a more electronegative metal than iron, forms anodic coatings on steel (Fig. 1). In moist air, corrosion products formed on the surface create a tight layer that insulates the substrate from the environment [4-5]. As additional corrosion protection of zinc coatings, the production of chromate or phosphate layers on them, and the lastly recommended chromite ones are used [6-8].

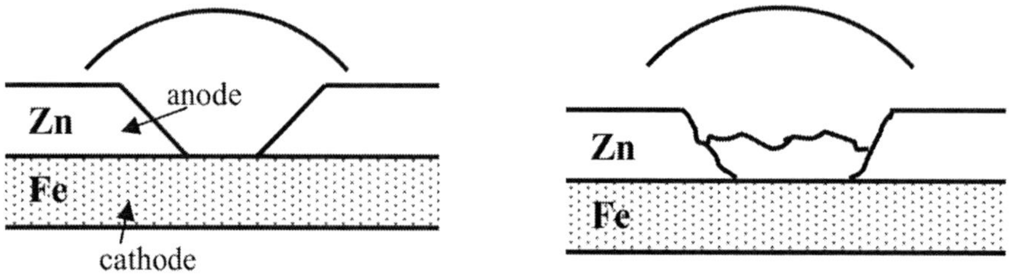

Fig. 1. Diagram of corrosion of anodic coatings.

Several types of galvanic baths are used for galvanizing, depending on the required properties and application of the coating. These are acidic, weak acidic and alkaline baths – cyanide and non-cyanide. New generation galvanizing is based on alkaline non-cyanide electrolytes [9].

Galvanic coatings should be characterized by good adhesion of the coating to the substrate, external appearance, fine-crystalline structure, tightness and appropriate minimum coating thickness for given conditions of use. The main factors affecting the quality of the galvanic coating are: concentration and intensity of electrolyte mixing, temperature, presence of surfactants, cathodic current density, type of electrolyte, the property of the metal on which the coating is deposited [10-11].

In this study S500MC (HSLA) was subjected to alkaline electrochemical galvanizing. The S500MC is a hot rolled, high-strength low-alloy steel (HSLA) characterized by reproducible quality and excellent engineering properties. The well-balanced chemical composition together with the fine-grained microstructure created in the thermomechanical rolling process creates steel with the highest class of mechanical properties and cold formability. Typical applications of S500MC include a wide range of steel components and structures, e.g. truck chassis and crane booms [12-17].

2 Material and Methods

The research was carried out for specimens of alkaline electrochemical galvanizing steel S500MC (HSLA). As additional corrosion protection of zinc coatings, the production of chromate layer on steel was used. Compositional specification of S500MC steel is included in Table 2.

Table 2. Chemical composition of S500MC steel (HSLA) (in wt. %).

Mn (max %)	C (max %)	Si (max %)	P (max %)	S (max %)	Al (max %)	Nb (max %)	V (max %)	Ti (max %)
1.60	0.10	0.21	0.025	0.010	0.015	0.09	0.20	0.15

Two series of samples were prepared, which differed in the way of preparing the surface before the zinc coating process. Both series were degreased in acetone, left to dry at room temperature and washed with de–ionized water. The thickness of the zinc layer of each sample series was 10 μm.

Analysis of microstructure of specimens after galvanizing process was carried out using microscope OLYMPUS GX41 with stereo digital image recording and electron scanning microscope Jeol JSM–6610LV. Identification of phase composition of zinc coatings was made using X–ray diffractometer Seifert 3003 TT with cobalt lamp with characteristic radiation wavelength of $\lambda_{cok\alpha} = 0.17902$ nm.

3 Results and Discussion

The microstructure were observed using the metallographic sections etched with Nital. Figure 2 illustrates the results obtained for the zinc coating by the optical microscope Olympus GX41. In Figure 2a the sample has a partially recrystallized ferrite structure and shows effects of cold work after plastic working resulting in elongated ferrite grains with irregular borders. While the sample in the Figure 2b was characterized by the presence of numerous polygonal ferrite grains, but also by the carbide phase disclosed in the pearlite. For this series of samples, a full recrystallization of the structure is noticeable. Probably after the annealing process, the surface of the sample with poor quality of steel substrate (Fig. 2b) was characterized by the presence of an oxide layer that had not been thoroughly removed before zinc coating.

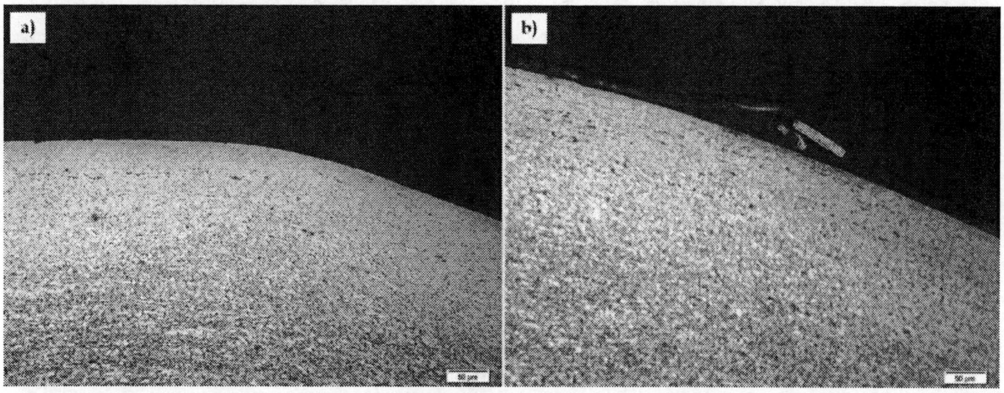

Fig. 2. Microstructure of zinc coating with: a) good, b) poor quality of the steel substrate.

The zinc coating deposited using the electrochemical method on the S500MC steel had thickness of around 10 μm. Figure 3 presents the cross–section of zinc coating microstructure obtained from the scanning microscope Jeol JSM–6610LV.

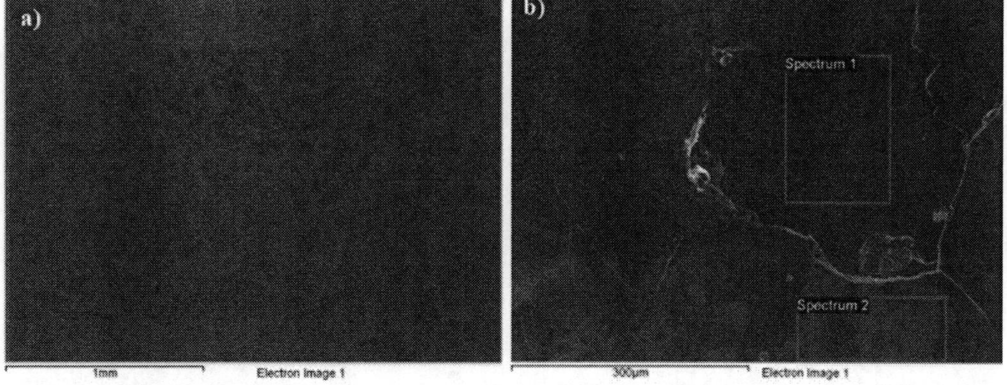

Fig. 3. Microstructure of the cross–section of zinc coating with: a) good, b) poor quality of the steel substrate.

Figure 4 presents the microstructure of the zinc coating surface obtained by the scanning microscope Jeol JSM–6610LV. Furthermore, in the pictures are marked location of the measurements EDS spectra in zinc coating identified by EDX analysis areas. The analysis of chemical composition of the zinc coating surface are presented in Table 3.

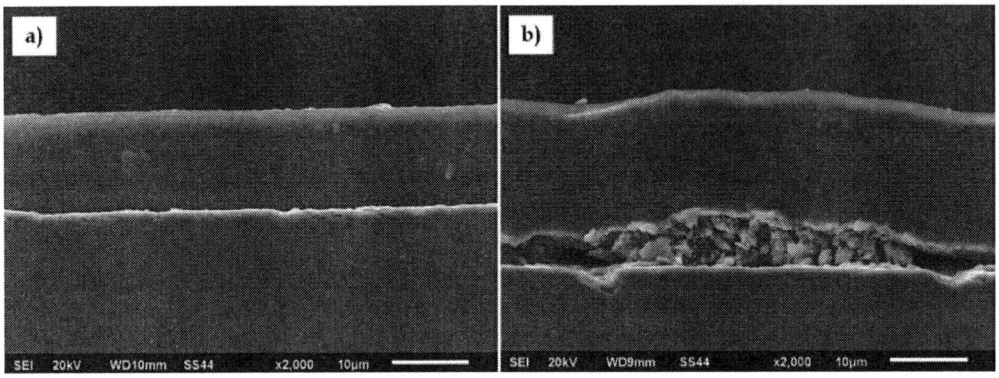

Fig. 4. Microstructure of zinc coating with: a) good, b) poor quality of the steel substrate.

Table 3. EDX–analysis of chemical composition of the surface in zinc coating (location of the measurements is marked in Fig. 5).

	Element	Weight [%]		
		Good quality of the steel substrate	Poor quality of the steel substrate	
		Spectrum 1	Spectrum 1	Spectrum 2
Zinc coating	C	17.05	6.33	13.56
	O	9.56	4.77	5.86
	Al	–	0.31	0.12
	Si	0.26	0.73	0.53
	S	1.98	0.27	0.96
	Cr	1.00	–	0.47
	Mn	–	1.66	–
	Cl	–	–	0.23
	Fe	0.45	83.12	0.48
	Zn	69.70	2.81	77.79

The X–ray phase analysis was conducted to identify the phase composition of the zinc coating. The results of the analysis are presented in Figure 5.

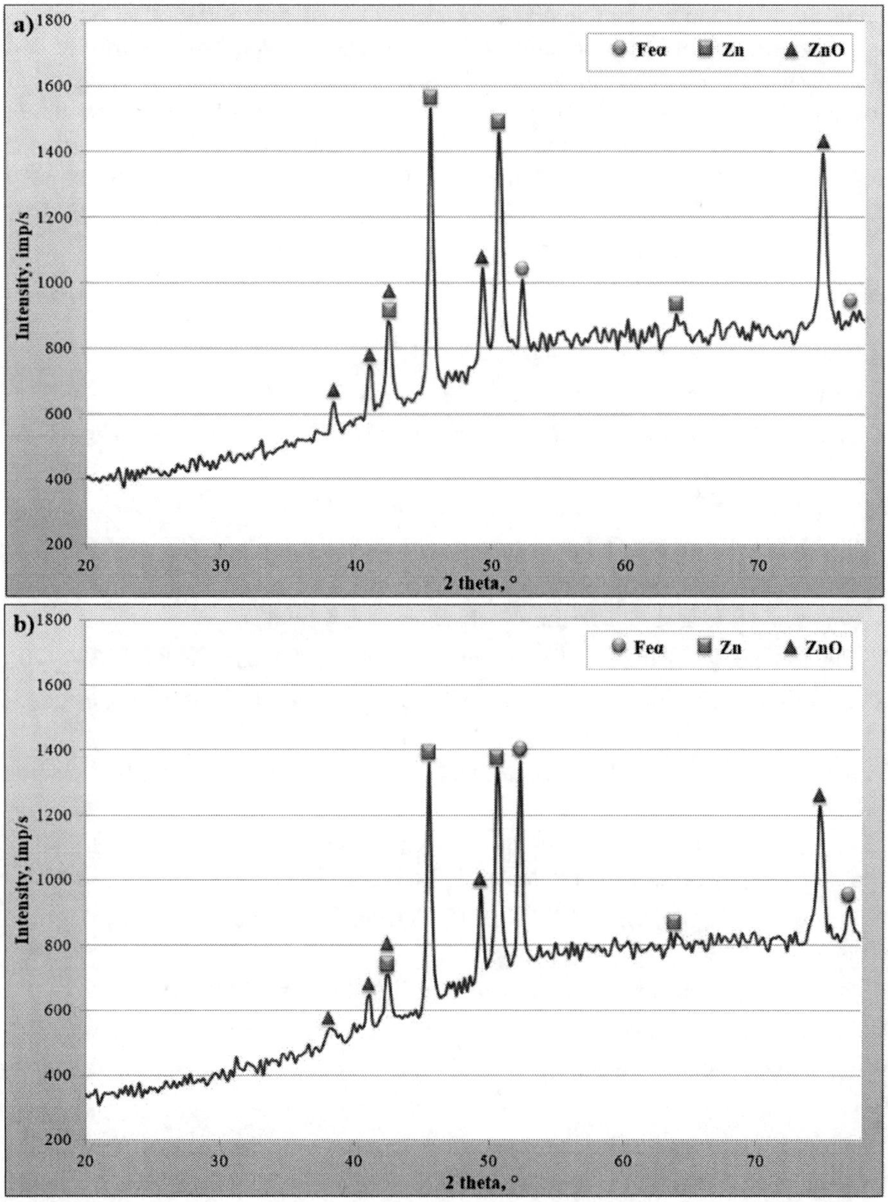

Fig. 5. Diffractogram of the zinc coating with: a) good, b) poor quality of the steel substrate.

The phase composition analysis for individual specimens revealed presence of the ferritic phase (cell parameters: a = b = c = 0.286 nm, $\alpha = \beta = \gamma = 90°$) which it is crystallizes in cubic cell. Additionally, analysis revealed presence of zinc crystallizes in hexagonal cell (cell parameters: a = b = 0.266 nm, c = 0.495 nm, $\alpha = \beta = 90°$, $\gamma = 120°$) and zinc oxide crystallizes in hexagonal cell (cell parameters: a = b = 0.325 nm, c = 0.521 nm, $\alpha = \beta = 90°$, $\gamma = 120°$).

5 Conclusion

Steel galvanizing process is widely used in the automotive industry by means of a component protection method against corrosion. The galvanizing S500MC steel presented in this study is used for the production of truck chassises. As results from the conducted research, preparing the substrate before the galvanizing process is of great importance. Microstructural examinations have proven that the series of samples were subjected to a recrystallizing annealing process under different conditions. For this reason, the series of samples subjected to the test were characterized by a different thickness of oxides formed as a result of the heat treatment.

References

1. S. Rashmi, L. Elias, A.C. Hegde, JESTECH, **20**, 3 (2017)
2. S. Ghaziof, W. Gao, Appl. Surf. Sci., **311** (2014)
3. X.G. Zhang, *Corrosion and Electrochemistry of Zinc* (Springer-Verlag, New York Inc., 6, 2013)
4. A. Amirudin, D. Thierry, Prog. Org. Coat., **28**, 1 (1996)
5. V. Kuklik, J. Kudláček, *Phenomena on galvanized coatings, Hot-Dip Galvanizing of Steel Structures* (67-92, 2016)
6. F. Simescu, H. Idrissi, Sci Technol Adv Mater., **9**, 4 (2008)
7. N. Jantaping, C.A. Schuh, Y. Boonyongmaneerat, Surf. Coat. Technol., **329** (2017)
8. R. S. Lillard, C. Valot, M. A. Hill, P. D. Dickerson, R. J. Hanrahan, Corros. Sci., **60**, 11 (2004)
9. M. Kobya, E. Demirbas, F. Ozyonar, G. Sirtbas, E. Gengec, Process Saf Environ Prot., **105** (2017)
10. K.A. Yasakau, I. Giner, C. Vree, O. Ozcan, R. Grothe, A. Oliveira, G. Grundmeier, M. G.S. Ferreira, M.L. Zheludkevich, Appl. Surf. Sci., **389** (2016)
11. Z.F. Lodhi, W.J. Hamer, J.M.C. Mol, H. Terryn, J.H.W. de Wit, Mater. Corros., **59**, 4 (2008)
12. R. Oyyaravelu, P. Kuppan, N. Arivazhagan, J. Adv. Res., **7** (2016)
13. N. Wint, J. Leung, J.H. Sullivan, D.J. Penney, Y. Gao, Corros. Sci., **136** (2018)
14. R. Ulewicz, F. Novy, Journal of The Balkan Tribological Association, **22**, 1147-1155 (2016)
15. N. Radek, J. Pietraszek, B. Antoszewski, Adv. Mat. Res.-Switz., **874**, 29 (2014)
16. P. Szataniak, F. Novy, R. Ulewicz, *23rd International Conference on Metallurgy And Materials* (778-783, 2014).
17. T. Lipinski, A. Wach, *23rd International Conference on Metallurgy And Materials* (778-783, 2014).

Arrangement of the foundations under the new hotel in Lviv

Andrii Kramarchuk[1], *Borys* Ilnytskyy[1], and *Oksana* Lytvyniak[2*]

[1]Department of Building Constructions and Bridges, Institute of Building and Environmental Engineering, Lviv Polytechnic National University, Ukraine
[2]Department of Civil Safety, Viacheslav Chornovil Institute of Sustainable Development, Lviv Polytechnic National University

Abstract. The paper discusses a renovation technique of a historic housing system used in Lviv. The technique presented involves an arrangement of foundations including conservation, as well as displaying exposed fragments of ancient buildings. The foundations of a hotel complex are accomplished owing to drilled injection piers. There are also full-scale tests of drilled injection piers on axial static pressed load which are carried out on the construction site.

1 Introduction

The building of a hotel complex on the corner of Krakivska and Virmenska streets in Lviv was supposed to begin in 2004, when the executive committee of Lviv borough council adopted the solution concerning the reproduction of the lost historical housing system of north commercial city block within the area of these streets. Later, this area was allotted for a housing system of a hotel complex. The 868 m^2 space under the building and the plot of land under the hotel is located in the area of protection of cultural monuments by UNESCO.

In 2010 Lviv borough council and company «Budinvest» organized international architectural contest for a project of a hotel complex in terms of regeneration of historical city block. The winner of the contest was an Austrian architect German Gubert. His project, however, was approached critically by the citizens of Lviv because of the uncharacteristic architectural solution for the urban core and the form of object. After some public protests and public hearings, the investor refused the realization of Gubert's architectural project. In 2010, a new project of hotel was launched by architects of a research-and-development laboratory, which operates under the Department of Restoration of Architectural and Artistic Heritage of Lviv Polytechnic National University. The resources for the project were provided by the investor. At the same time, the realization of project became complicated because of fragments of building foundations dating back to the XVI – XVII century which were discovered during archeological digs on the corner Krakivska and Virmenska street (Fig. 1).

* Corresponding author: lytvyniak.oksana@gmail.com

Fig. 1. The discovered pieces of foundation of a building.

The new project, which is developed by main architect of Lviv - Yurij Dubyk - in co-authorship with colleagues from Lviv Polytechnic National University, is a reproduction of a lost historical building belonging to a hotel complex in terms of a renovation project of this city block. It complies to the rules of the construction concernig a regeneration of a historical city block with elements form the mid-XIX century used in the elevation (Fig. 2, 3). When Yurij Dubyk presented this project he said, «During the development of this project we used existing documents in historical archives about architectural special features of buildings in this part of city and the parceling, in other words planning out site of model of XVI-XVII centuries. Primary at the corner streets Krakivska-Virmenska were four buildings with superficiality in three or four floors. Then the planning out of site under the object will focus on the parceling of XVI-XVII centuries. We took the solution to use in elevation of future hotel architectural models by middle of XIX century to save the interior of historical epoch».

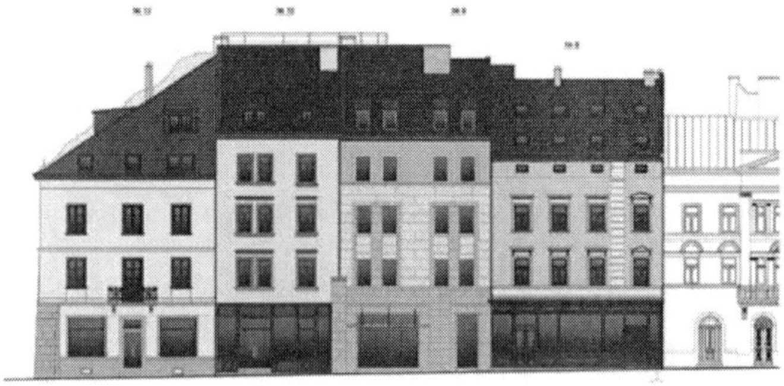

Fig. 2. The elevation of hotel (the look of Krakivska Str.).

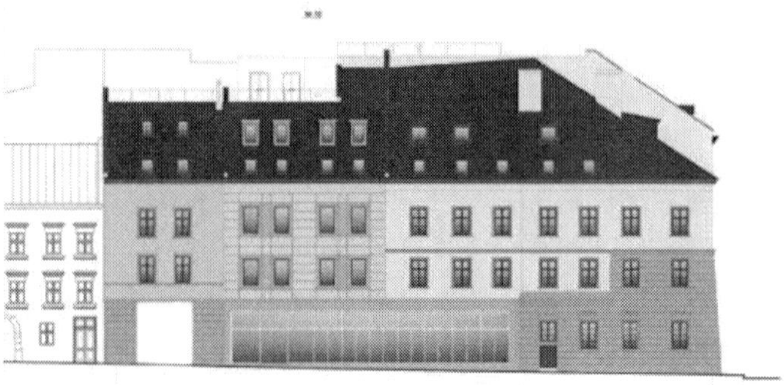

Fig. 3. The elevation of hotel (the look of Virmenska Str.).

2 Experimental investigations

2.1 Basic information

The methodology of renovation of a lost housing system in a historical city center is actively discussed among architect-restorers, guards of monuments, historians and the community. They cannot give the right solution whether to restore or to keep the places of lost housing system as free orderly urban territories. All in all, global city planning and architectural practice find different approaches to decide about similar problems in terms of requirements and the strategy of city development. Particularly, the methods of renovation of historical edifices which were destroyed or damaged as a result of war operations were interrogated by researchers of states. Especially, after Second World War in Warsaw, Wroclaw, Elblong not just individual edifices were built, but the whole city blocks in places where lost city housing system was located. Accordingly, these architectural transformations found their arguments in theoretical researches. Famous Polish explorers Edmund Małachowicz described several methods, which are used for similar models of transformation, that is reintegration, reconstruction, reproduction, and restitution. Which one to chose depends on the status of a territory, the technical condition of protected building fragments, the building standards [1].

Besides, the dictates of [2] foresee the use of the regeneration principle in architectural projecting for protected territories.

The main aim of this investigation was to design the arrangement of the foundations with conservation and displaying exposed fragments of ancient buildings. The complication which arose was that the foundation fragments occurred on different benchmarks and now they have different physical-mechanical characteristics because of the use of different materials for their considerable age.

2.2 Additional information

It was determined in the project that the arrangement of the foundations ought to be done with compliance to the law, and, at the same time, maximum conservation should be done and historic fragments of the building should be exposed (Fig. 1). However, geological conditions of the building site complicated the solution of this problem. The existent foundations had different physical-mechanical characteristics because different materials was used for them over time, besides, they had different depth, and they were built on man-made grounds. At the same time, natural grounds, which could become a base (semi-solid clay and semi-rocky ground - marl under clay), dropped by 4.5 m in terms of building site limits. In the other words, the grounds came under the base of separate foundations in one part of the building site; however, they did not come under the base of foundations by 5.4 m in another part of the building site.

Thereby, a solution was created to drill injection piers with diameter 600 mm under on marl. Afterwards, high grillage was made on the drilled injection piers in the form of monolithic reinforced concrete slab. There also was floor slab over basement, which was constructed from restored exposed fragments of ancient buildings. The choice of the piers by diameter 600 mm was executed of following conditions: physical-mechanical characteristics of base's grounds and possibility to use building technique under conditions of historical housing system. The scheme of pile foundations was shown in Fig. 4.

Calculated load on piers was up to 2500 kN. The reinforcement of piers was a spatial framework. It was constructed with twelve longitudinal steel reinforcements of 25 mm diameter and transverse (circular) steel reinforcement of 6 mm diameter with a pitch of 200 mm.

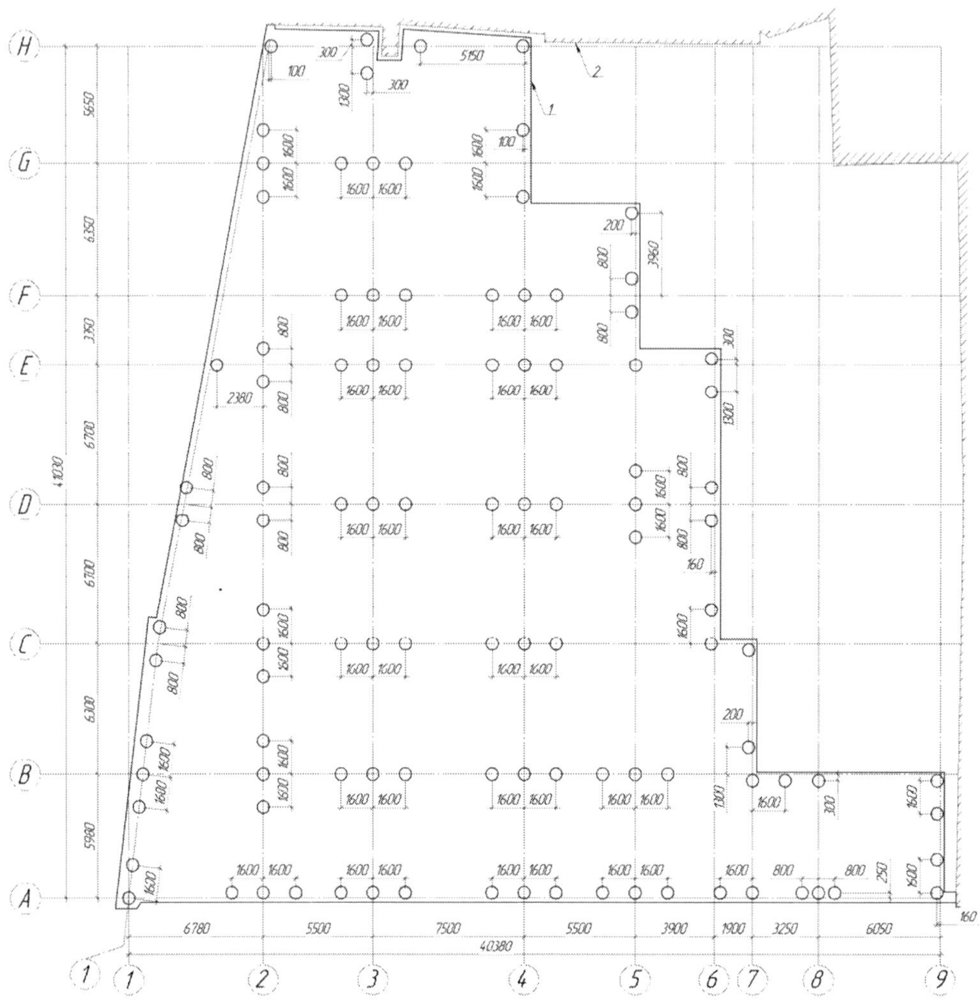

Fig. 4. The scheme of pier foundations: 1 – outer contour of hotel complex; 2 – boundary of existing house system.

3 Results and discussion

Full-scale test of piers was conducted with a maximum loading of 3000 kN on a pier on the construction site. The experiments were carried out according to the standards of [4] test of piers on axial static pressed load. The pier PK-1 during the test and test bench are shown in Fig. 5.

The test of the pier was carried out with the scheme for pier trial [4] by means of twelve anchor piers, two control traverses and one power traverse, which were supported by two hydraulic jacks (Fig. 4). The two hydraulic jacks were used for loading on piers which allowed developing the force 4000 kN. The tests were carried out stepped load of pier trial to proof of pressed load 3000 kN.

Fig. 5. The test bench for trial of a pier.

There were provided experimental piers, which were subjected to static load, eleven step loads (a range of 240…350 kN) according to [4] (Table 1).

Table 1. Static test of piers.

Step	1	2	3	4	5	6	7	8	9	10	11
Display of manometer, kilogram force/cm^2	40	70	100	130	160	190	220	250	280	310	340
Loading, kN	290	560	800	1080	1340	1600	1860	2120	2380	2650	3000
Total loading, kN	290	850	1650	2730	4070	5670	7530	9650	12030	14680	17680

The movement of pier was measured by deflection indicator. Real movement of a pier was calculated as a value of index of two deflection indicators.

The reading was carried out by all measuring instruments in the following sequence:

- zero reading - before loading on pier;
- first reading – immediately after application of loading, then consistently with an interval of 30 minutes to stabilize the deformation, which did not exceed 0.1 mm;
- following readings - a similar procedure, as described earlier.

The results of a field testing of pier PK-1, PK-2 with a static pressed load are shown in Table 2.

Table 2. Static pressed load of piers by loading.

Step	Loading, P, kN	Average pier settlement PK-1, S, mm	Average pier settlement PK-2, S, mm
1	290	0.00	0.00
2	560	0.12	0.12
3	800	0.22	0.21
4	1080	0.41	0.40
5	1340	0.66	0.65
6	1600	0.90	0.88
7	1860	1.28	1.26
8	2120	1.58	1.56
9	2380	2.02	2.00
10	2650	2.46	2.44
11	3000	3.37	3.36

The experimental piers were loaded after 28 days their arrangements (the demand of [4]. For example, the experimental pier PK-1 was loaded on 22nd January 2016 (the pier PK-1 on 03 November 2015), the experimental pier PK-2 was loaded on 29th January 2016 (the pier PK-2 on 17 November 2015).

The stabilization of experimental piers' deformation was taken by value increase of deformation not more 0.1 mm for last two observations [4]. The values of average pier settlement by maximum pressed load are shown in Table 3.

Table 3. Average piers settlement by maximum pressed loading.

Number of test	Number of pier by project	Length of pier, m	Diameter of pier, mm	Final test of pier by loading, kN	Average settlement, mm
1	PK-1	11.0	600	3000	3.37
2	PK-2	11.9	600	3000	3.36

4 Conclusions

The results of the 600 mm diamerer pier test, which were performed with the use of the drilled injection method on the building site of a hotel complex at the corner Krakivska and Virmenska Street in Lviv showed the value of average settlement of test piers with a maximum loading which exceeds 16.85% of limit value of an average pier settlement by 20 mm (according to [3]). Thus, the undertaken constructive solution allow to keep the fragments of ancient buildings while retaining the neccessary bearing capacity of their structure.

References

1. Małachowicz E. *Konserwacja i rewaloryzacja architektury w środowisku kulturowym* (Oficyna Wydawnicza Politechniki Wrocławskiej, Wrocław, 2007)
2. DBN 360-92, *Town planning. Planning and development of urban and rural settlements* (Ukraina, 2002)
3. DBN V.2.1-10:2009, *Basements and foundations of buildings* (Ukraina, 2009)
4. DSTU B.V.2.1-95, *Soils. The methods of field trials by piers* (Ukraina, 1995)

MATEC Web of Conferences **183**, 02008 (2018) https://doi.org/10.1051/matecconf/201818302008
QPI 2018

Experimental study of the strengthening effect of reinforced concrete columns jacketed under service load level

Pavlo Krainskyi[1] , *Yaroslav* Blikharskyy[2,*], *Roman* Khmil[1], and *Zinoviy* Blikharskyy[1]

[1]Lviv Polytechnic National University, Department of Building Constructions and Bridges, 79013, Lviv, Ukraine
[2]Lviv Polytechnic National University, Department of Highways and Bridges, 79013, Lviv, Ukraine

Abstract. The need of structural retrofitting and strengthening of different buildings, engineering structures or their elements is always present. Among the main reasons are demages and material deterioration due to aging, improper maintenance or physical damages; planed repairs; reconstruction or extension of the building; technical modification or complete change of operations inside the building or the structure, etc. In some cases operation of the building during retrofitting or strengthening of its structures has to be partially or fully stopped. In other cases the strengthening process takes place while the building is still operational which means that structures are strengthened under service loads. The main goal of this research is to determine the strengthening effect of reinforced concrete jacketing applied to columns under service load level. For that the experimental study of six reinforced concrete columns were carried out: four reference columns, both strengthened by jacketing and unstrengthened and two strengthened under service load. The main results of the research are presented.

1 Main goal of the research

Reinforced concrete jacketing is one of the traditional and widespread methods of RC structures strengthening. RC jacketing can be applied to both flexural and compressed members and has some significant advantages. Jacketed member usually do not need additional corrosion or fire protection.

A lot of research have been carried out to determine the stress-strain state of the jacketed RC members [1-3], bonding and surface preparation influence on the performance of jacketed structure [4] and seismic retrofitting using RC jacketing [5, 6]. Nevertheless the further study of this type of strengthening is relevant, since there is little known about the performance of the structures strengthened under loads of different magnitude.

In this study we try to experimentally determine the effect of initial loading, which is present during strengthening process, on further work of jacketed RC column. Reinforced concrete columns strengthened under service load are the main object of this research.

* Corresponding author: yaroslav.z.blikharskyy@lpnu.ua

2 Research methodology

2.1 Test specimen description and material properties

For this study six test specimen of reinforced concrete columns were prepared. The overall dimensions of the test specimens presented on Fig. 1. The length of the columns was 2.2m and a cross section dimensions before jacketing was 0.14m x 0.18m. Columns were tested as compressed-bent elements and were manufactured with corbels on both ends for that purpose. Cross section dimensions of the jacketed columns were increased to 0.2m x 0.26m.

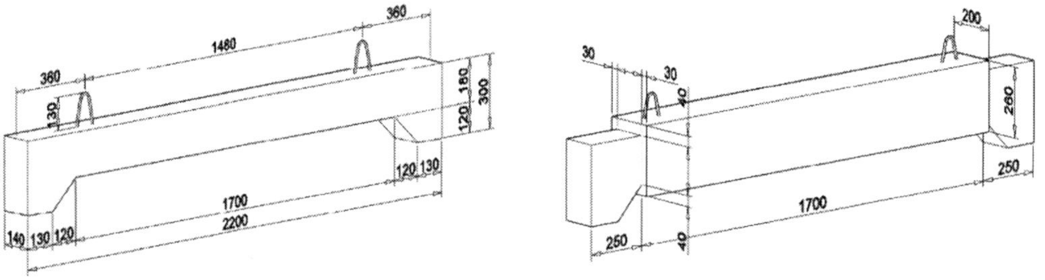

Fig. 1. Dimensions of the reinforced concrete columns test specimens.

All six columns were manufactured from the same materials. Material specimens were selected and tested to determine the mechanical properties. For cross section dimensions and reinforcement details see Fig. 2 a, b. Reinforcement ratio of unstrengthened column equaled 1.8%. Four 12mm rebar were used as longitudinal reinforcement with 6mm rebar used for ties. The rupture stress of main rebar was 720MPa; yield stress – 630MPa; Young's modulus – 210GPa. Average concrete cube strength equaled 37MPa which corresponds to C25/30 concrete class.

Reinforced concrete jacketing length equaled 1.7m. Reinforcement ratio equaled 1.17%. Four 10mm rebar were used as jacketing longitudinal reinforcement (Fig. 2, 3) with 6mm rebar used for ties. Mechanical properties of concrete and rebar were the same as previous. The preparation of the concrete surface before jacketing consisted only from dust removal and ample hydratation.

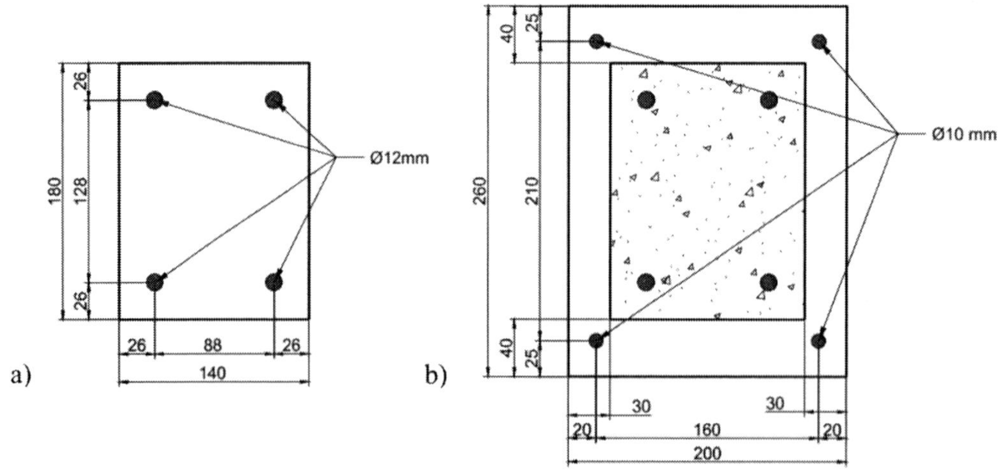

Fig. 2. Cross section dimensions and reinforcement details of unstrengthened (a) and strengthened (b) column specimen.

Fig. 3. Reinforcement details of strengthened column specimen.

2.2 Experimental program

The first two columns C-1 and C-2 were reference columns. They were tested without strengthening to determine the bearing capacity of the specimen and to adopt the service load value for the next tests.

Columns C-3-0 and C-4-0 were also reference columns. They were tested after strengthening but without any initial load. Bearing capacity of this specimens was considered as maximum possible and was used for comparison of the strengthening effect of the following jacketed columns.

Columns C-5-S and C-6-S were initially loaded to a service load which was adopted as 65-70% of C-1 and C-2 reference columns bearing capacity. After that they were strengthened and tested to failure.

Columns were tested with pinned supports. Load was applied incrementally as a force on one of the supports. The following data was acquired during the tests of the columns: strain in the rebar; strain on the compressed face of the concrete; deflection of the midheight of the column.

3 Results of the experimental study

3.1 C-1, C-2 reference columns

C-1 and C-2 columns were tested to failure without strengthening. Specimens failed in flexural manner. Average bearing capacity of two specimens equaled 175kN. From Fig. 4

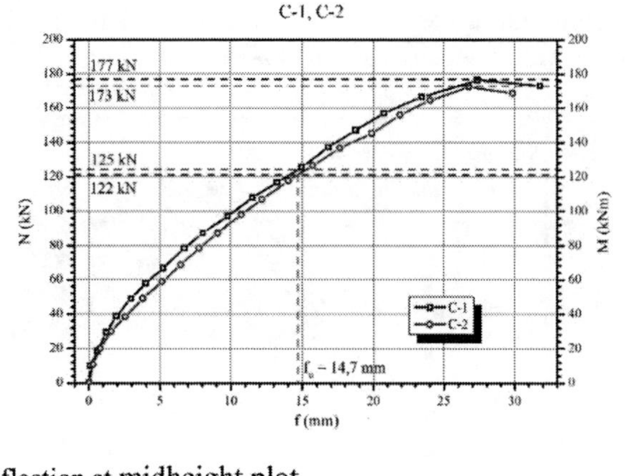

Fig. 4. C-1 and C-2 deflection at midheight plot.

we can see that the loads that corresponded to a maximum allowed deflection of 14.7mm (calculated according to Ukrainian codes for serviceability limit state) were 122kN and 125kN. Considering these values we adopted a service loads for following tests of 120kN.

3.2 C-3-0, C-4-0 reference columns

C-3-0 and C-4-0 columns were strengthened without initial loading and tested to failure. Specimens failed in flexural manner. Average bearing capacity of two specimens equaled 475kN. It is 2.7 times bigger than the bearing capacity of C-1 and C-2 (see Fig. 5). We consider it to be a maximum possible effect from strengthening since there were no initial stresses or deformations in the columns before strengthening.

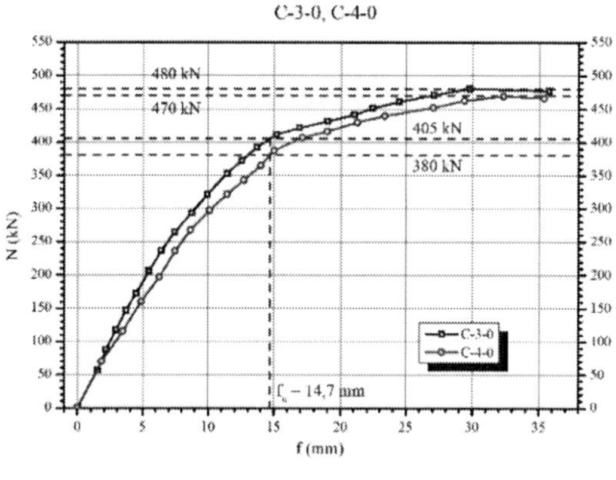

Fig. 5. C-3-0 and C-4-0 deflection at midheight plot.

3.3 C-5-S, C-6-S columns

C-5-S and C-6-S columns were initially loaded to a service load of 120kN, strengthened an then tested to failure. Specimens failed in flexural manner. Average bearing capacity of two specimens equaled 430kN. It is 90% of a maximum possible effect and 2.6 times bigger than the bearing capacity of C-1 and C-2 (see Fig. 6). Although the application of the reinforced concrete jacketing was almost at the verge of exceeding the serviceability limit state requirements, we still achieved some positive effect in deflection manner of the columns. on Fig. 5 we can see the increase in stiffness of the columns immediately after strengthening. This resulted in a very slow increase in deflections on first steps of loading after jacketing and 1.5 times increase in serviceability capacity.

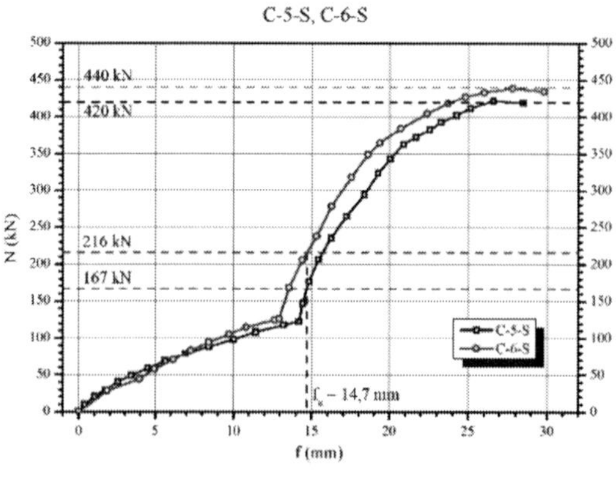

Fig. 6. C-5-S and C-6-S deflection at midheight plot.

For comparison of all specimens the experimental data was summarized in the table 1.

Table 1. Results of research.

Specimens	Initial load, kN		Bearing capacity, kN		Strengthening effect, %	
	specimen	average	specimen	average	specimen	average
C-1	-	-	177	175	-	-
C-2	-		173		-	
C-3-0	0	0	480	475	175	172
C-4-0	0		470		169	
C-5-S	118	120	420	430	142	146
C-6-S	122		440		151	

According to table 1 the strengthening effect for specimens C-3-0 and C-4-0 was 172%, for specimens C-5-S and C-6-S – 146%. The strengthening effect decreases when the structure is under load during strengthening process what can be seen from the test results.

4 Conclusions

The effect of strengthening by reinforced concrete jacketing proved to be still significant even when the columns were strengthened under service loads (70% of bearing capacity). Bearing capacity increase was 2.6 times while serviceability capacity increase was 1.5 times. The effect was also less then maximum possible determined from the tests of the columns strengthened without initial loading.

References

1. E. Julio, F. Branco, V. Silva, Prog. in Struct. Eng. and Mat., **9** (2003)
2. A. Meda, G.A. Plizzari, Z. Rinaldi, G. Martinola, Con. Rep., Rehab. and Ret., **II**, 6 (2009)
3. Y.M.K. Vijaya, G. Siddappa, S. Chandra, IJRET, **4** (2015)
4. C. Talbot, M. Pigeon, D. Beaupre, D. Morgan, ACI Mat. J., **91**, 7 (1994)
5. D. Stoppenhagen, J. Jirsa, L. Wyllie, ACI Str. J., **92**, 11 (1995)
6. S. Sugano, New Zeal. Nat. Soc. for Earthq. Eng., **14**, 14 (1981)

Fracture toughness of RC beams on the shear, strengthening by FRCM system

Zinoviy Blikharskyy[1], *Pavlo* Vegera[2 *], *Rostyslav* Vashkevych[2] and *Taras* Shnal[2]

[1]Lviv Polytechnic National University, Department of Bridges and Structural Mechanics, Karpinskogo Str. 6, Lviv, 79013, Ukraine
[2]Lviv Polytechnic National University, Department of Building Constructions and Bridges, Karpinskogo Str. 6, Lviv, 79013, Ukraine

Abstract. This research paper present the experimental study of reinforce concrete beams strengthened by FRCM system in the shear area. All samples are without transverse reinforcement at the shear distance. First beams tested as control sample, with shear distance a/d=2. Another three beams tested with strengthening by FRCM system. Variable parameter is the level of initial load, before strengthening. The initial load values were 0, 0.3, 0.5 from carrying capacity of control samples. All RC beams were designed to fail in shear, even strengthened samples. In the result of this research we determined that fracture toughness of RC beams increases with a decrease of span to the effective depth ratio. Results are comparable with the increasing of bearing capacity. The first inclined crack opend in mid-height cross section at the load 50 kN at an angle equal to 45^0 and it does not depend on the shear span. The maximum width of inclined cracks decrease with decreasing of shear span. The maximum width of inclined cracks decrease with decreasing of the shear span. The maximum values of width of inclined cracks are fixed at the value of 0.55 - 0.85 mm. Using the limit crack opening width as criteria of exhaustion of serviceability (SLS) we saw that about 16 - 29% of bearing capacity still remains before the danger of shear failure of the beam.

1 Introduction

In influence of time, environment and different factors most structures need repairing and strengthening. There are a lot of different type of strengthening. Some of them are jacketing with self-consolidating concrete [1], application of steel plates for and others.

Most popular materials nowadays are different composites called FRP (Fiber Reinforce polymer) or FRCM (Fiber Reinforce Cement Polymers) [2].

There are interest for using different polymer plates for shear strengthening of RC beams by many researchers [3,4]. There are some works with different computer modeling of shear crack [5].

* Corresponding author: Pavlo.I.Vehera@lpnu.ua

The goal of this article is to experimentally determine shear strength and serviceability of reinforced concrete beams strengthened with FRCM. Main variable in the experiment was the initial load applied to the beam before strengthening.

2 The testing program

Four reinforced concrete beams were tested. All samples are without transverse reinforcement at the shear distance. First beams tested as control sample, with shear distance a/d=2. Another three beams tested with strengthening by FRCM system. Variable parameter is level of initial load, before strengthening. Initial load was 0, 0.3, 0.5 from carrying capacity of control samples. All RC beams were designed to fail in shear, even strengthened samples. For the research, a technique has been developed for testing each beam twice, which was described in our article [8].

Tested beams had 2100 mm length, 100 mm width, and 200 mm height of cross section. As the beam's tension reinforcement, A400C Ø18 mm rebar was chosen, A400C Ø10 mm – as compressed reinforcement. Transverse reinforcement – A240C Ø 6 mm rebar located in the area without shear force [9].

Concrete class of tested beams was C32/40.

According to the testing program, three specimens were strengthened by reinforcing FRCM system with 70 mm wide strips (Fig. 1).

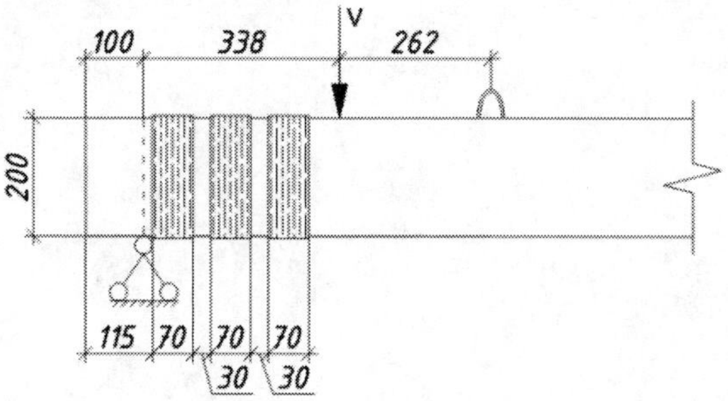

Fig. 1. Scheme of FRCM stripes placement on the RC beams.

Main PBO fibres properties of FRCM stripes are shown in the table 1.

Table 1. Physical and mechanical characteristics of composite fabric stripes RuredilXMeshGold.

№	Characteristics	Value
1	Density, g/cm^3	1,56
2	Tensile strength, MPa	5800
3	Modulus of elasticity, MPa	270000
4	Ultimate deformation, %	2,15
5	Equivalent dry fabric thickness in the direction of the warp, mm	0,0445
6	Equivalent dry fabric thickness in the direction of the weft, mm	0,0115
7	Ultimate tensile stress of the warp per unit of width, kN/m	264,0
8	Ultimate tensile stress of the weft per unit of width, kN/m	66,5

Beams marking were adopted as follows: BO – beam ordinary, BSC – beam strengthened by composite material; the first digit – serial number, the second digit –

sample number and the third digit – the section number. Index 0 – 0.5 means the loading level under which the strengthening was performed and depends on shear strength (V_{Ed}) of non-strengthened control sample. For example, BSC 1.2.2-0.3 means that the tested sample from the first series of the second beam of the second section, with initial load $0.3 \cdot V_{Ed}$.

3 Results of experimental research

3.1 The overview of shear failure of beams

For the RC beam, the limit value of concrete compressed deformation above the inclined crack was taken as the criterion of shear strength (ULS) [**Błąd! Nie można odnaleźć źródła odwołania.**]. Achievement of limit deformations in transverse reinforcement can not be taken as a criterion for the exhaustion of bearing capacity, because there is a cut of the rebar, which leads to the onset of the yield strength before approaching the limit tensile deformations of the reinforcement.

The bearing capacity of reinforced concrete beams in shear was for BO 1.1 - V_{Ed}=95 kN, BSC 1.1-0 – V_{Ed}=137.5 kN, BSC 1.2-0.3 – V_{Ed}=121.5 kN, and for BSC 1.3-0.5 - V_{Ed}=115 kN [**Błąd! Nie można odnaleźć źródła odwołania.**].

All samples failed in shear. The RC beams after testing are shown in Fig. 2 - 5.

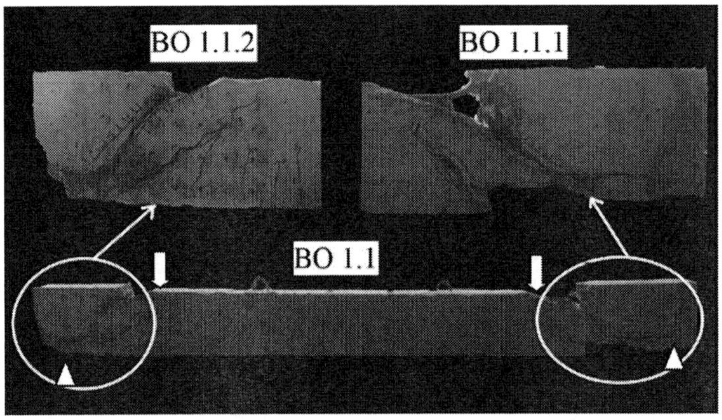

Fig. 2. The tested sample BO 1.1.

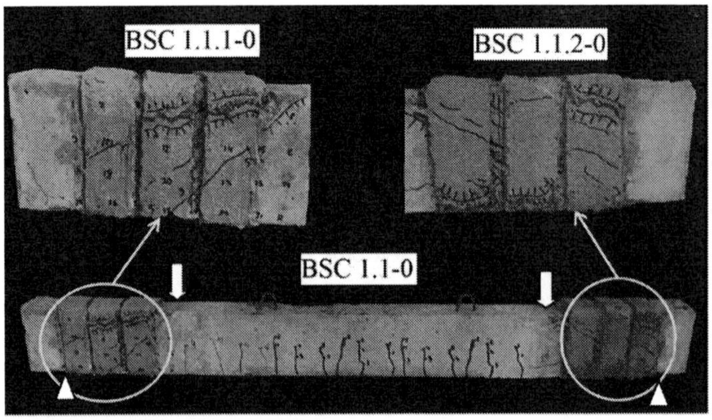

Fig. 3. The tested sample BSC 1.1-0.

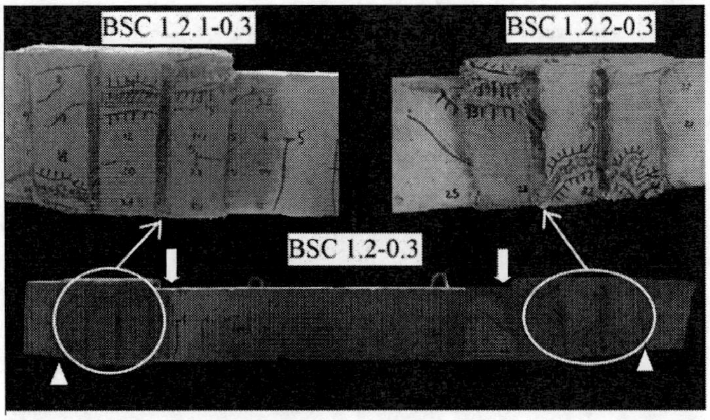

Fig. 4. The tested sample BSC 1.2-0.3.

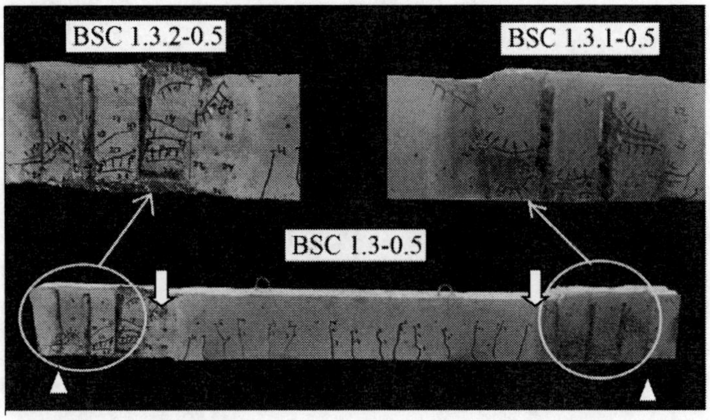

Fig. 5. The tested sample BSC 1.2-0.5.

Exhaustion of the bearing capacity in shear occurred at the time of the destruction of the compressed concrete zone with a sharp extension of the FRCM tape and damage to the protective layer of the FRCM system in the zone of inclined cracks.

3.3 Fracture toughness in testing samples

For detailed information about fracture toughness, we measured the width of the crack opening, depth of distribution and location. Width of crack were measured by microscope MBP-3, with accuracy 0.05 mm.

Table 2. The serviceability limit state (SLS) of testing beams on the shear strength.

Types of the beam	**Types of the beam's cross section**	**Inclined crack width, mm**	**Experimental shear strength, V_{Ed}^{exp}, kN**	**Average experimental shear strength, V_{Ed}^{exp}, kN**	**Increasing of shear strength, $V_{Ed}^{exp}/V_{Ed}^{BO\,1.1}$**
BO 1.1.	BO 1.1.1		80	80	-
	BO 1.1.2		80		
BSC 1.1-0	BSC 1.1.1-0		120	110	1.38
	BSC 1.1.2-0	0.4	100		
BSC 1.2-0.3	BSC 1.2.1-0.3		100	100	1.25
	BSC 1.2.2-0.3		100		
BSC 1.3-0.5	BSC 1.3.1-0.5		90	90	1.13

Limit width of crack w_{max}=0.4mm [7, 10] opened on the level V_{Ed}=80kN for sample BO 1.1. BSC 1.1-0 increased serviceability 1.38 times, BSC 1.2-0.3 – 1.25 times, and BSC 1.3-0.5 -1.13 times. Rapidly decreas of serviceability for BSC 1.3-0.5 caused by strengthening after opening of the first inclined crack.

The boundary values of the crack opening width are fixed at 78 - 84% of the load carrying capacity of the samples (Table 3).

Using the limit crack opening width as criteria of exhaustion of the serviceability (SLS) we determined that the beam still has 16 - 22% reserve of bearing capacity, before the danger of shear failure of the beam. Bearing capacity reserve for all samples are in one range and are close to the value of control beam.

Table 3. The comparison of testing beams in shear.

Types of the beam	**Types of the beam's cross section**	**Inclined crack width, mm**	**Average experimental shear strength, V_{Ed}^{exp}, kN**	**Average experimental shear strength by SLS, V_{Ed}^{SLS}, kN**	**Increasing of shear strength, $V_{Ed}^{SLS}/V_{Ed}^{exp}$**
BO 1.1.	BO 1.1.1	0.4	95	80	0.84
	BO 1.1.2				
BSC 1.1-0	BSC 1.1.1-0		137.5	110	0.80
	BSC 1.1.2-0				
BSC 1.2-0.3	BSC 1.2.1-0.3		121.5	100	0.82
	BSC 1.2.2-0.3				
BSC 1.3-0.5	BSC 1.3.1-0.5		115	90	0.78
	BSC 1.3.2-0.5				

Values of maximum crack opening are described in the table 4.

Table 4. Maximum inclined crack opening width of testing beams.

Types of the beam	**Types of the beam's cross section**	**Shear strength, V_{Ed}, kN**	**Inclined crack width (in concrete), mm**	**Average inclined crack width, mm**	**Inclined crack width (in FRCM), mm**	**Average inclined crack width, mm**
BO 1.1.	BO 1.1.1	90	0.80	0.85	-	-
	BO 1.1.2	90	0.90		-	
BSC 1.1-0	BSC 1.1.1-0	130	0.50	0.55	0.25	0.30
	BSC 1.1.2-0	140	0.60		0.35	
BSC 1.2-0.3	BSC 1.2.1-0.3	120	0.60	0.55	0.20	0.17
	BSC 1.2.2-0.3	115	0.50		0.15	
BSC 1.3-0.5	BSC 1.3.1-0.5	110	0.70	0.67	0.15	0.15
	BSC 1.3.2-0.5	110	0.65		0.15	

Maximum crack opening in FRCM strengthening system are 2 times less than crack width in the concrete for beam BSC 1.1-0. In this sample maximum level of strengthening, increased serviceability were achived, so the maximum width of inclined crack in FRCM system and concrete is at the shear distance. For another two beams, width of crack opening is on the same level, lower than for BSC 1.1-0.

However, maximum width of crack opening does not exceed limit value. In contradiction to crack in concrete, there are more cracks in the FRCM system.

4 Conclusion

Based on the data above, the following conclusions can be drawn:

1. The beam still has 16 - 22% reserve of bearing capacity, before the danger of shear failure of the beam.
2. Maximum crack opening in FRCM strengthening system are 2 times and more less than crack width in the concrete.
3. Bearing capacity reserve for all samples are in one range and are close to the value of control beam.
4. Maximum width of crack opening in the FRCM system does not exceed limit value.

References

1. T. Greenough, M. Nehdi, *Shear Behavior of Fiber-Reinforced Self-Consolidating Concrete Slender Beams* (ACI Mat. J., 2008)
2. A. Dasgupta, *Retrofitting of Concrete Structure with Fiber Reinforced Polymer* (IJIRST, 2018)
3. R. Azam., K. Soudki, J.S. West, M. Noël, *Strengthening of shear-critical RC beams: Alternatives to externally bonded CFRP sheets* (Const. and Buld. Mat. 2017)
4. H. Al-Ghanim, A. Al-Asil, M. Abdel-Jaber, M. Alqam, *Shear and Flexural Behavior of Reinforced Concrete Deep Beams Strengthened with CFRP Composites* (Mod. Appl. Sc., 2017)
5. S. Shahbazpanahi, *Mechanical analysis of a shear-cracked RC beam* (Acta Sc. Tech., 2017)
6. *National Standard of Ukraine, Concrete and reinforced concrete structures* (DBN B.2.6 - 98: 2009, 84, 2011)
7. *Eurocode 2: Design of concrete structures – Part 1-1: General rules for buildings* (EN 1992-1-1:2004 (E), 225, 2004)
8. P. Vegera, R. Khmil, Z. Blikharskyy, SCTP, **823**, 38 (2015)
9. P. Vegera, R. Vashkevych, R. Khmil, Z. Blikharskyy, SSP, **12**(2), 39 (2017).
10. *National Standard of Ukraine, Concrete and reinforced concrete structures* (DBN B.2.6 - 98: 2009, 84, 2011)

The analysis of friction properties after laser texturizing of the silicon carbide surface layer

Norbert Radek[1,*], *Agnieszka* Szczotok[3], and *Renata* Dwornicka[3]

[1]Kielce University of Technology, Faculty of Mechatronics and Mechanical Engineering, Al. 1000-lecia Panstwa Polskiego 7, 25-314 Kielce, Poland
[2]Silesian University of Technology, Institute of Materials Science, str. Krasinskiego 8, 40-019 Katowice, Poland
[3]Cracow University of Technology, Faculty of Mechanical Engineering, Al. Jana Pawła II 37, 31-864 Cracow, Poland

Abstract. The modification of the surface properties is a desired technological procedure. One of the possible method is the electro-spark deposition (ESD). Unfortunately, ESD process produces a surface with high roughness. The laser beam machining (LBM) has been involved to lower roughness of the coating made by ESD. The elements coated by ESD have been tested to determine tribological properties and they were compared before and after LBM. To achieve high reliability of the results, the test has been conducted in accordance with design of experiments methodology and the results which were obtained have been processed by a statistical analysis. The description of such an experiment performed for a silicon carbide SiC coating, the obtained results and the conclusions are included in this paper.

1 Introduction

During tribological investigations, it was found that employing heterogeneous surfaces models into boundary interaction of solid surfaces could make significant improvement [1]. Surfaces described as heterogeneous consist of areas, which are different one from another in geometrical, physicomechanical or physicochemical properties. The heterogeneity of surfaces is frequently due to the application of more than one technology, and can be constituted by [2, 3]:

- shaped surface features such as grooves, pits or channels resulting from milling, eroding, etching, laser-beam forming, etc.,
- areas with different physicochemical and physicomechanical properties, e.g. areas with varying hardness and mechanical strength accomplished by local surfacing or selective surface hardening (e.g. electron-beam machining, laser-beam forming or thermo-chemical treatment),
- areas with diversified surface microgeometry, e.g. areas eroded at the points of focus (laser treatment or electro-spark deposition), or areas with formed surface

* Corresponding author: norrad@tu.kielce.pl

microgeometry, for instance, in terms of desired microroughness directivity or load capacity (LBM and ESD technologies).

Heterogeneous surfaces can be obtained by different methods. The laser treatment of electro-spark deposited coatings being one of them.

The first publication on surface texturing appeared in Germany in 1993 [4]. It discussed the use of an excimer laser to texture elements of a magnetic memory disk drive with the aim of reducing friction at the start. Further experiments in this area involved texturing surfaces of punches applied to plastic forming. It was found that the process caused a 169% increase in the punch service life.

The current studies focus on the influence of texturing on the performance of various friction systems in internal combustion engines, e.g. precision bearing systems. Texturing is used to improve heat removal, vaporization, wettability, biological functions, absorptivity, etc.

In the considered experiment, two parameters controlling LBM process were investigated: the cavity diameter and the level of blackening. The average friction coefficient, being the investigated outcome, was determined on the tribological tester. The experiment was conducted according to response surface methodology (RSM) [5, 6], which is a part of the design of experiments (DoE) methodology [7, 8]. The appropriate RSM model was identified and accompanying plots were made.

2 Texturing methodology

Laser surface texturing is one of the most common and promising methods of surface roughening. Categorized as a metal removal process, laser texturing is usually performed at a power density of $10^6 \div 10^9$ W/cm^2. At present, it accounts for about 2% of all laser-based material processing processes used in the world. In laser surface texturing, a pulsed laser beam is focused on a material to melt a hole. The hole depth is dependent mainly on the power density and the pulse duration. The drilling debris is removed from a hole being drilled using compressed air or another inert gas.

The tests were conducted for SiC rings (certificate of material excellence) with the following dimensions: an outer diameter d_o of 35.3 mm, an inner diameter d_i of 25.1 mm, and a height h of 7 mm. The texturing was performed usingan ESI 5200 Nd:YAG laser (pulse mode). A view of the ESI 5200 laser μvia drill used in the tests is presented in Fig. 1. The wavelength λ= 355 nm (the laser uses radiation at the third harmonic frequency).

Fig. 1. A view of the ESI 5200 laser μvia drill.

The parameters of the laser surface texturing process were determined basing on the experimental results: the laser spot diameter d = 0.78÷150 μm; the laser power

P = 0.37÷0.4 W; the laser beam velocity v = 15.7÷23.56 mm/s; the distance from the focus Δf = 0 mm; the repetition frequency f = 6400 Hz.

The process of texturing was performed in two stages (two steps). In the first step, holes with a pre-determined diameter were drilled along a spiral path. In the second step, the drilling debris was removed from each pore, with the number and frequency of pulses being strictly defined.

3 Materials and methods

The samples of material were processed by a laser erosion and then analyzed visually and geometrically. Surfaces texturized by a laser beam were investigated to achieve the knowledge of technological parameters at different loads of a friction pair. The research was conducted as a designed experiment [9-13] to maximize an information amount about performance parameters at the fixed number of experiment.

Rings (Fig. 2) with external diameter of 35.3 mm, internal diameter of 25.1 mm and thickness of 7 mm served as the tested samples (discs). The material of discs was sintered SiC. These were original, commercially available rings for front sealing. The texture was made on rings in accordance to the design of experiment.

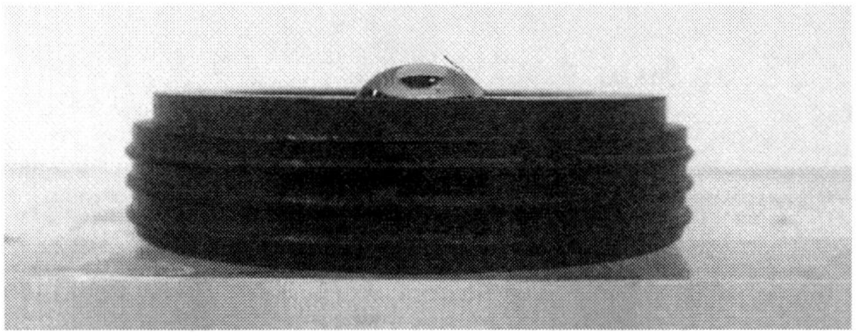

Fig. 2. The investigated ring (during a wettability test).

The effects of a laser induced erosion were investigated visually by scanning microscope Joel JSM-5400. The images was obtained at the magnitude 100x and 500x. Geometrical properties were investigated according to code ISO 25178 using the profilometer Hobson Form Talysurf 120L.

The tribological tester T-01M was used for measurements. The measured outcome was a coefficient of a friction, while the controlled factors were an average diameter of cavities and level of blackening. A typical stem of a tester was replicated with a specifically cut bearing ball (diameter of 6.3 mm). Such a circle-shaped cut had a diameter of 4.5 mm. Such a friction pair had a distributed contact with the possibility of self-adjusting. A full range of tester parameters was used sequentially: rotation varied from 100 rpm to 700 rpm; a normal load of the friction pair varied from 4.9 N to 39.2 N.

4 Experimental Design

The investigated LBM process was recognized as an object with two controlled factors and one measured quantitative outcome. The average diameter of cavities and the level of a blackening were assumed as input factors. The average coefficient of friction was assumed as an outcome. Due to the fact that both input factors were quantitative the response surface methodology appeared to be the most appropriate.

The average coefficient of friction being the outcome had the range of variation limited to only positive values. Such limitation requires a special treatment during regression analysis because potential negative predictions have no physical sense, so the dataset of measured values was transformed into logarithms of these values [14].

The response surface methodology was adopted to identify an appropriate model: the full-quadratic with additional two-way third-order interactions combined with the central composite design for two input factors. The center treatment was replicated five times to estimate a pure error (Tab.1).

The statistical analysis followed a typical approach: the analysis of effects, the analysis of variance (ANOVA), the analysis of standardized effects, the possible correction of a model terms set, a test of normality for residues and, finally, prognostic plots: the contour and 3D. The calculations have been provided in Statsoft STATISTICA software system.

Table 1. Approved design of the experiment.

Treatment Number	Diameter of cavity, A, μm	Level of blackening, B, %
1	78	18
2	134	18
3	78	42
4	134	42
5	70	27
6	150	27
7	102	15
8	102	50
9...13	102	27

5 Results and discussion

As an effect of the energy impact the grid of cavities in the shape of a crater (Fig. 3) have appeared. After treatment by means of lapping and superfinishing, flat and hard areas were obtained to transfer normal loads. Neighboring cavities generate hydrodynamic forces during liquid lubrication. Such surfaces may be used among others in sliding friction pairs [7, 8]. The raw data are presented in 3D scatter plot (Fig. 4). The distribution of datapoints suggests the possibility of the existence of a strong non-linearity.

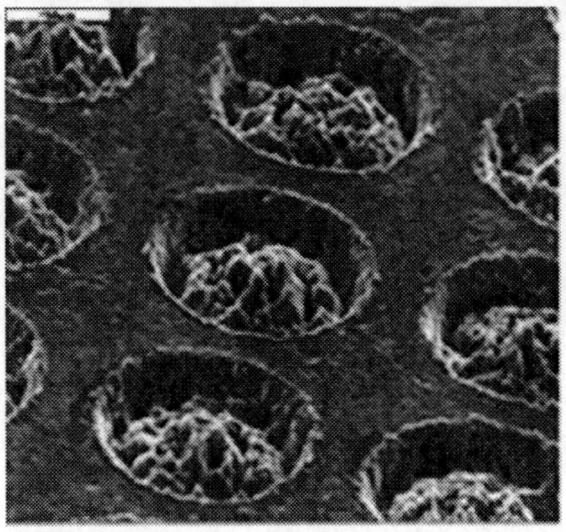

Fig. 3. Texturized surface (magnification of x500).

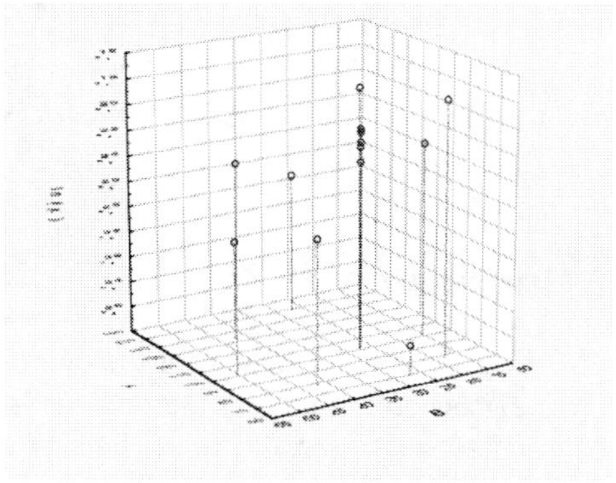

Fig. 4. The 3D scatter plot of raw data.

Statistical analysis of the obtained data revealed that classic polynomial models did not fit in any way. The deeper analysis of the 3D scatter (Fig. 4) plot suggested that outcome value associated with treatment 5 is a source of the problem. It probably was the outlier. After removing this value from the dataset, the linear model fitted well. Finally, the obtained linear model is as following:

$$\ln(\mu) = -1.03 - 7.03 \cdot 10^{-3} \cdot A - 1.79 \cdot 10^{-2} \cdot B \quad (1)$$

or in *explicite* form:

$$\mu = \exp(-1.03 - 7.03 \cdot 10^{-3} \cdot A - 1.79 \cdot 10^{-2} \cdot B) \quad (2)$$

The descriptive statistics of the fitted model are presented in Table 2.

Table 2. Approved design of the experiment.

Term	Coefficient	SE	t	p-Value	Confidence interval	
					-95%	+95%
const	-1.03	0.18	-5.79	0.004	-1.52	-0.54
A	-0.0070	0.0014	-4.96	0.008	-0.0110	-0.0031
B	-0.0180	0.0029	-6.21	0.003	-0.0260	-0.0099

The residuals were tested for a conformity of the normality by Shapiro-Wilk test and p-Value = 0.72 was obtained. This means that the model is consistent with the assumption of errors normality. The model was visualized by a contour plot (Fig. 5) and 3D response surface (Fig. 6).

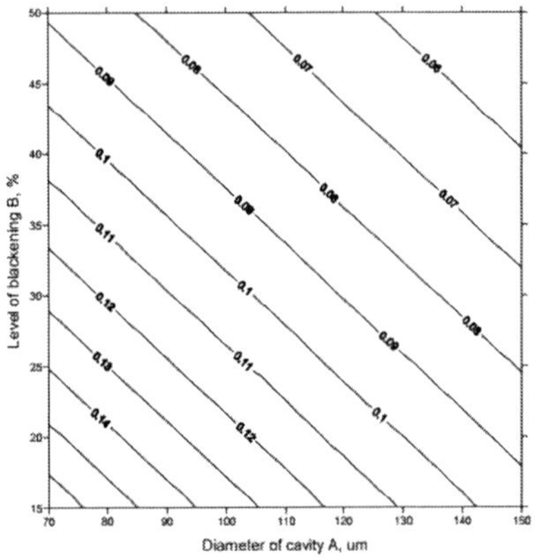

Fig. 5. Contour map of the identified model.

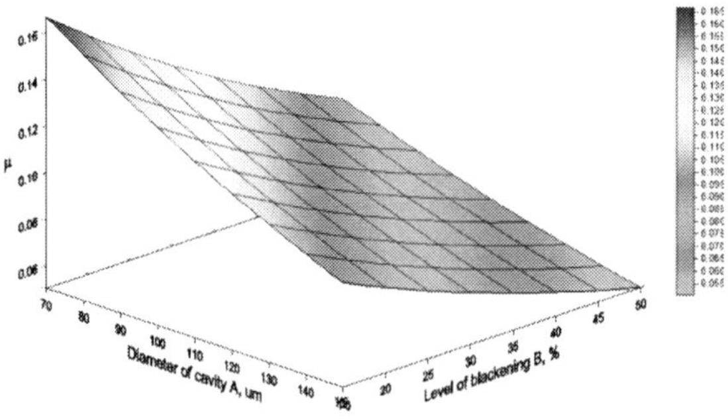

Fig. 6. Response surface plot for the identified model.

6 Conclusion

The identified model allows for the combination of such values of a cavity diameter and the level of blackening so that the designer's needs concerning the coefficient of friction may be satisfied. However, due to the lack of a physically based predicting formula, the obtained model may be utilized only locally and any extrapolation is not allowed. In further investigations, it seems it to be worth to employ the techniques of microstructures analysis [15-28], especially with the multivariate analysis [29-31]. Problems with the stability and repeatability of the technology process, encountered during the work, have shown the need to apply a quality management system [32-35] in the future research.

References

1. B. Antoszewski, *Influence of laser and plasma modification on surface properties of sliding and rubbing components* (Kielce University of Technology, Kielce, 1999)
2. G. Gyk, I. Etsion, Wear, **216**, 792 (2006)
3. Y. Wan, X. Dang-Sheng, J. of Mater. Proc. Tech., **197**, 96 (2008)
4. J. Arnold, G. Muller, H. Schneider, H.K. Muller, H. Hugel, Laser und Optoelektronik, **25**, 12 (1993)
5. G.E.P. Box, N.R. Draper, *Empirical Model-Building and Response Surface* (John Wiley & Sons, Hoboken, 1987)
6. R.H. Myers, D.C. Montgomery, *Response Surface Methodology. Process and Product Optimization Using Designed Experiments* (John Wiley & Sons, Hoboken, 1995)
7. D.C. Montgomery, *Design and Analysis of Experiments* (John Wiley & Sons, Inc., Hoboken, 2008)
8. O. Kempthorne, K. Hinkelmann, *Design and Analysis of Experiments.Vol.1. Introduction to experimental design* (John Wiley & Sons, Inc., Hoboken, 1994)
9. B. Antoszewski, Prz. Elektrotechniczny, **87**(7), 6 (2011)
10. J. Pietraszek, *Proc. 6th Int. Conf. Neural Net. Soft Comput.*, (Zakopane, Poland, 250, 2003)
11. J. Pietraszek, A. Gądek-Moszczak, T. Torunski, Adv. Mat. Res.-Switz., **874**, 139 (2014)
12. J. Pietraszek, E. Skrzypczak-Pietraszek, Adv. Mat. Res.-Switz., **874**, 151 (2014)
13. J. Pietraszek, A. Goroshko, Adv. Mat. Res.-Switz., **874**, 145 (2014)
14. J. Pietraszek, N. Radek, K. Bartkowiak, Solid State Phenom., **197**, 198 (2013)
15. T. Pieczonka, J. Kazior, A. Tiziani, A. Molinari, J. Mater. Process. Technol., **64**, 327 (1997)
16. M. Ulewicz, U. Lesinska, M. Bochenska, Physicochem. Probl. Mi., **44**, 245 (2010)
17. M. Ulewicz, E. Radzyminska-Lenarcik, Physicochem. Probl. Mi., **46**, 119 (2011)
18. T. Pieczonka, J. Kazior, A. Szewczyk-Nykiel, M. Hebda, M. Nykiel, Powder Metall., **55**, 354 (2012)
19. R. Ulewicz, P. Szataniak, F. Novy, *Proc. 23rd International Conference on Metallurgy and Materials* (Brno, Czech Republic, 784, 2014)
20. P. Szataniak, F. Novy, R. Ulewicz, *Proc. 23rd International Conference on Metallurgy and Materials* (Brno, Czech Republic, 778, 2014)
21. T. Lipinski, A. Wach, *Proc. 23rd International Conference on Metallurgy and Materials* (Brno, Czech Republic, 738, 2014)
22. D. Klimecka-Tatar, S. Borkowski, P. Sygut, Arch. Metall. Mater., **60**, 735 (2015)
23. D. Klimecka-Tatar, G. Pawlowska, M. Sozanska, Arch. Metall. Mater., **60**, 153 (2015)
24. R. Ulewicz, F.R. Novy, J. Balk. Tribol. Assoc., **22**, 1147 (2016)
25. J. Korzekwa, A. Gadek-Moszczak, M. Bara, Prakt. Metallogr.-Pract. Metallogr., **53**, 36 (2016)

26. R. Dwornicka, N. Radek, M. Krawczyk, P. Osocha, J. Pobedza, *Proc. 26th Int. Conf. on Metall. and Mater. – METAL* (Brno, Czech Republic, 1252, 2017)
27. A. Gadek-Moszczak, Image Anal. Stereol., **36**, 151 (2017)
28. A. Gadek-Moszczak, P. Matusiewicz, Image Anal. Stereol., **36**, 207 (2017)
29. E. Skrzypczak-Pietraszek, J. Pietraszek, Acta Biol. Crac. Ser. Bot., **51**, 62 (2009)
30. E. Skrzypczak-Pietraszek, I. Kwiecien, A. Goldyn, J. Pietraszek, Phytochem. Lett., **20**, 443 (2017)
31. A. Gadek-Moszczak, J. Pietraszek, B. Jasiewicz, S. Sikorska, L. Wojnar, *Proc. 6th Int. Conf. on Comp. Collective Intell.* (ICCCI, Seoul, South Korea, 127-136, 2014)
32. R. Ulewicz, *Proc. Int. Conf. Path Forward for Wood Products: A Global Perspective* (Baton Rouge, LA, 51, 2016)
33. A. Pacana, R. Ulewicz, Polish J. Environ. Studies, **16**, 165 (2017)
34. Kozien, E., Kozien, A., *Proc. 23th Int. Sci. Conf. on Economic and Social Dev. (ESD)* (Madrid, Spain, 38, 2017)
35. E. Kozień, *Proc. 31st Int. Sci. Conf. on Economic and Social Dev. – Legal Challenges of Modern World (ESD)* (Split, Croatia, 152, 2018)

The effect of the addition of polypropylene fibres on improvement on concrete quality

Alina Pietrzak[1*], and *Małgorzata* Ulewicz[1]
[1]Czestochowa University of Technology, Faculty of Civil Engineering, Poland

Abstract. Concrete serves as one of the most important construction materials, which thanks to its resistance and durability, gives total freedom in using it in building industry, and by extension in defining urban areas. Despite numerous advantages (ease of handling, high compressive strength, low production cost, heat resistance) it also has many disadvantages, including low tensile strength as well as sensitivity to destructive action of chemical agents. A constant growth in expectations concerning increasing the quality of concrete led to its polymer modification, whose aim was to eliminate above-mentioned disadvantages. The scope of the present research involved designing and making of concrete mixtures modified by 19, 30, 40 and 60 mm long polypropylene fibres. For the experiment the authors used: Portland cement CEM I 42.5R, sand, gravel aggregate of the 2-8 and 8-16 fractions, water, Master Pozzolith STD plasticizer and polypropylene fibres of various length. The authors made 4 series of concrete samples modified by fibres, for which the following factors have been assigned: compression strength after 7, 28 and 56 days of maturing as well as absorptivity, volumetric density and frost resistance after 100 cycles of freezing and thawing.

1 Introduction

Concrete and steel are the most commonly used construction materials, which form the image of the contemporary architecture. A constant growth of the quality of these products is caused by both an increase in users' expectations and an increase in demands determined by certain regulations [1-3]. Polymers, as modifiers, have been used to improve the concrete quality for over fifty years [4-6]. The group of polymer concretes include: resin concrete (PC), polymer-cement concrete (PCC), polymer impregnated concrete (PIC) [6-9].

Concrete modified by polypropylene fibres is used for making industrial flooring, road and airport surfaces, and prefabricated thin-walled elements. The main advantage of using such fibres is to prevent the occurrence of anti-shrinkage cracks in 'young' maturing concrete. However, having reached the resistance designed and the modulus of elasticity such fibres cease to work. Then, the stresses are transferred by concrete itself or main anchoring bars [9, 10-12]. The aim of the paper was to determine the influence of the length of the polypropylene fibre on the basic properties of concrete.

* Corresponding author: apietrzak@bud.pcz.pl

2 Materials and the research methodology

For the experiment the authors used: Portland cement CEM I 42.5R, sand, gravel aggregate of the 2-8 and 8-16 mm fractions, water, Master Pozzolith STD plasticizer and polypropylene fibres of various length (Table 1).

Table 1. Properties of polypropylene fibres used in the study.

Fibre designation	Properties		
	Shape	**Colour**	**Fibre length, mm**
A1	bundle	white	19 ± 1.5
A2			30 ± 1.5
A3			40 ± 1.5
A4			60 ± 1.5

5 series of samples were made. Control concrete sample with the w/c=0.55 ratio with the addition of plasticizer in the amount of 0.3% of the cement bulk – (SK) series. In the subsequent attempts the control concrete was modified with polypropylene fibres (A1-A4 series) in the amount of 0.9 kg/m^3. Samples for measuring compressive strength were made in forms according to requirements of the Norm PN-EN 12390-1. For all series of concrete samples there were twelve 15x15x15 cm cubic samples and twelve 10x10x10 cm samples formed according to Norm PN-EN 12390-2. The experiment of durability of compressive strength was conducted after 7, 28 and 56 days of sample maturing under laboratory conditions according to Norm PN-EN 206:2014 Concrete – "Specification, performance, production and conformity". The experiment was carried out by mean of Toni Technik type 2030 testing machine according to the Norm.

3 Research results

With the experimentally assumed water/cement ratio the concrete mixture was designed in the way to gain appropriate consistency and workability. For the control concrete series the S3 consistency was assumed (slump within 100-150 mm). The research results on the compression strength, absorptivity and concrete density for particular series are presented in Table 2. The average compressive strength of the control concrete series (1K) measures after 7 days equalled f_{cm} = 38.2 MPa. Adding polypropylene fibres caused a slight increase in average compressive strength for all modified series. After 7 days of maturing, the biggest, 4.2% increase of compressive strength in relation to the control concrete series was observed in A4 series. The smallest, 0.2% increase in compressive strength was examined in A1 series. The average compressive strength of the control concrete series (1K), measures after 28 days f_{cm} = 46.7 MPa. In both concrete series (A1 and A2) the average compressive strength was on the similar level as in the control series and it equalled 46.8 MPa. An increase of the compressive strength in relation to the 1K series was noticed in A3 and A4 concrete series (0.9% and 1.8%, respectively).

The average compressive strength of the control concrete series (1K), measures after 56 days of maturing f_{cm} = 46.7 MPa. In all concrete series (A1 and A4) modified by fibres the average compressive strength was noticed. The increase equalled 0.13%, 0.8%, 0.3% and 4.1%, respectively in relation to the 1K series.

On the basis of the methodology included in Norm PN-88/B-06250 Plain Concrete, an experiment of absorbability of particular series was conducted and the findings are presented in Table 2. According to the above-mentioned Norm, the absorbability of concrete exposed to environmental influences should not be bigger than 5%. However, for concrete protected from direct weather conditions should not be bigger than 9%. Both the control concrete and the polypropylene fibres modified concrete reached the absorbability within 5.5-5.7%.

Table 2. Results of compression strength, absorptivity and concrete density for particular series.

Concrete series	Average compression strength, f_{cm} (MPa)			Absorptivity, n_w (% masy)	Volumetric density, ρ (kg/m^3)
	after 7 days of maturing	after 28 days of maturing	after 56 days of maturing		
1K	38.2	46.7	51.4	5.5	2283
A1	38.2	46.8	51.4	5.7	2247
A2	39.0	46.8	51.8	5.7	2238
A3	38.5	47.5	51.5	5.7	2277
A4	39.7	47.9	53.5	5.5	2287

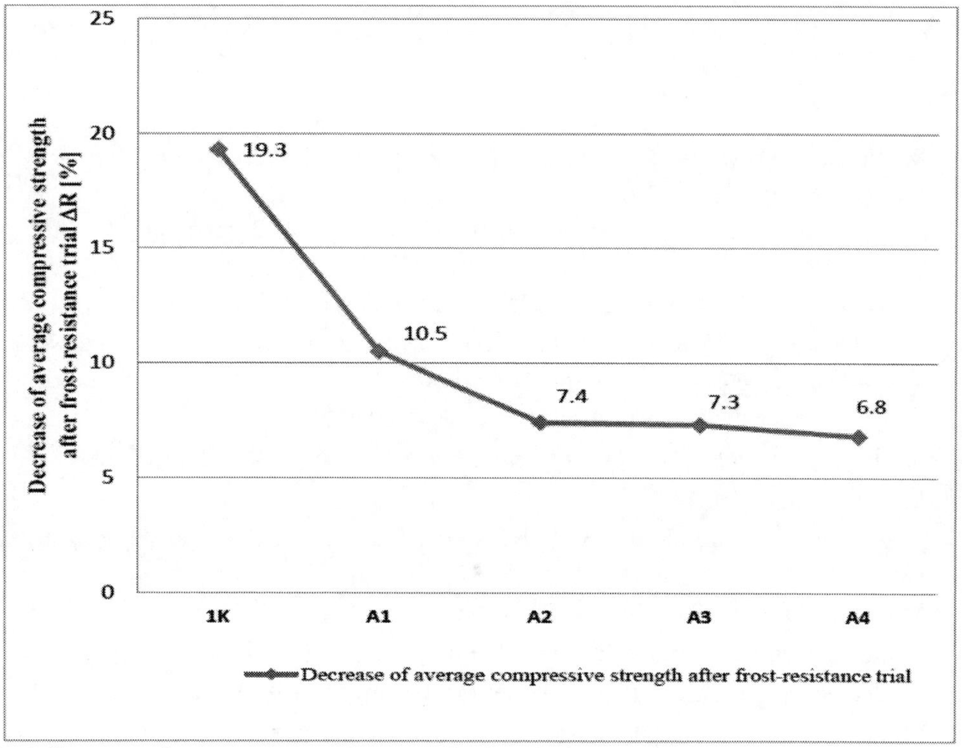

Fig. 1. Decrease of average compressive strength of particular concrete series after frost-resistance trial.

Using direct method (PN-88/B-06250) the authors conducted a study of concrete frost-resistance for a F100 resistance class. The results are presented on Figure 1. The analytical determination is taken as positive if after carrying out of *n* cycles of freezing and thawing, required by certain frost resistance class: the average decrease of compressive strength does not exceed 20%, the average loss in mass does not exceed 5% and none of the examined samples breaks or scratches. The decrease of resistance for alternate freezing and thawing of the control series equalled 19.3%. As presented in Figure 1, the A1=A4 concrete series modified by polypropylene fibres showed much smaller decrease of resistance for alternate freezing and thawing in relation to the control series. The smallest decrease of strength after concrete frost-resistance trial was noticed in A4 series and it equalled 6.8%. The loss of mass for all concrete series was in the range of between 0.15-0.31%.

3 Conclusions

Concrete series modified by 19, 30, 40 and 60 mm long polypropylene fibres obtained comparable drainage durability in relation to the control series. Along with the growth of the fibre length in the range of between 19-60 mm, a slight increase in compressive strength of concrete, equalling 1.2%, was noticed after 28 days (and 1.45 after 56 days). Series of concrete modified by polypropylene fibres demonstrated better resistance to freezing and thawing, as they obtained a smaller decrease in compressive strength in relation to the control series. The smallest decrease of compressive strength, equalling 6.8%, after frost-resistance trials demonstrated concrete samples containing fibres of 60 mm length.

Acknowledgment
The authors thank Wigolen S.A. from Czestochowa (Poland) for providing research material.

References

1. M. Nowicka-Skowron, R. Ulewicz, *24th International Conference on Metallurgy and Materials*(1707-1712, 2015)
2. A. Shahnawaz, H.S. Sharma, International Journal of Engineering Science and Computing, **7** (2017)
3. S. Kakooei, M. Hazizan Akil, M. Jamshidi, J. Rouhi, Constr Build Mater, **27**, 1 (2012)
4. M. Ulewicz, J. Selejdak, S. Borkowski, M. Jagusiak-Kocik, *22nd International Conference on Metallurgy and Materials* (1926-1931, 2013)
5. T. Aly, J.G. Sanjayan, F. Collins, Mater Struc, **41** (2008)
6. S. Chandra, *8th International Congress on Polymers in Concrete* (3-11, 1995)
7. L. Czarnecki, CWB, **2** (2010)
8. K.S. Yeon, L. Youn-Su,n B. Lee, C. Kim, *10th International Congress on Polymers in Concrete* (2001).
9. M.A. Galicki, BTA, **2** (2009)
10. T. Zuk, J. Pietraszek, M. Zenkiewicz, Polimery, **61**, 519 (2016)
11. M. Zenkiewicz, T. Zuk, J. Pietraszek, Przem. Chem., **95**, 1687 (2016)
12. M. Zenkiewicz, T. Zuk, J. Pietraszek, P. Rytlewski, K. Moraczewski, M. Stepczynska, Polimery, **61**, 835 (2016)

Outsorcing quality control in the automotive industry

Robert Ulewicz[1,*]

[1]Czestochowa University of Technology, Production Engineering and Safety Department, Armii Krajowej 19B, PL 42-201 Czestochwa, Poland

Abstract. Providing high quality of products not covered by defective products is an important part of the supply chain between the manufacturer of the automotive parts and the final assembly company. The paper presents the results of research related to determining the effectiveness of quality control in a company producing plastic components for automotive industry at separate levels. Upon detection of nonconforming products at his plant, recepient according to the contract asked the supplier to launch the Controlled Shipping Level 1 procedures. As a result of the lack of significant improvement, there was run in the supplier's plant the Controlled Shipping Level 2 procedure realized on an outsourcing basis by specialized external company. Presented results of the research have made it possible to determine the effectiveness of carried out control activities. Carried out research, also confirmed by expert interviews, show that automotive companies aim to make their business more flexible so that they respond faster and more easily to internal and mainly external factors. What in the opinion of production and quality managers provides outsorcing of quality control.

1 Introduction

Market conditions, new development strategies, focusing on processes with the highest added value, encourages many companies not only from automotive industry to choose outsourcing solutions [1]. Outsourcing can occur anywhere in the entire supply chain as well as in manufacturing processes, after-sales and service operations [2, 3]. In spite of the benefits associated with outsourcing flexibility there appear issues related to response time and costs associated with outsourcing activiestes control [4, 5]. The question arises as to what with the costs of control if the control, such as final or interoperational control, is the subject to voluntary outsourcing or imposed by the supplier under a signed contract? In such case, it is very important to eliminate the opportunistic approach of all process participants (supply chain participants) to the outsourcing process itself [6, 7]. From a productivity perspective, outsourcing can result in loss of information, asymmetry of information, and lack of knowledge on commissioned processes [4, 8], which in turn prevents the client from gaining proper insight into the quality issues that can arise both from outsourcing processes as well from automated supply chains and production processes

* Corresponding author: robert.ulewicz@wz.pcz.pl

(technological) [9-12]. The issue of outsourcing of quality control in the automotive industry looks a bit different. In this industry we are more and more often dealing with a case of quality control on the basis of voluntary outsourcing or imposed outsourcing by the supplier. The control in this case is understood as measuring, examining or checking one or more control characteristic of the product as well as comparing the obtained results with the technical specifications [19, 20]. Performing a control activity may involve a product or a process. In the analyzed case we are dealing with product control in the form of plastic mudguard. Control operations can be performed at a fixed location on a workstation or in a separate place. A permanent place of control at the end of the process is established when it is advisable to include quality control as a separate operation, e.g. in outsourcing control [21, 22]. Products quality control is performed after key operations where it is absolutely necessary to meet specific parameters or there is a high risk of forming incompatibilities after operations where the product and exactly its parameters are input base for next operation or final product (assembly operations), before bottlenecks [14, 15]. Industrial practice shows that in case of control of outsourcing, the place of carring out control usually is the end of the process.

2 Control outsourcing

Very high quality requirements are placed before any supplier of components. It's easy not to meet these requirements, and this generates considerable costs. A sudden increase in orders, process instability or operator error are sufficient that the manufacturing process become more complex. In situation where customer receives incompatible delivery, from which parts and components cannot be used in the further process of assembly, manufacturing facility has to take account with the risk of stopping the assembly line because of the lack of certainty as to the quality of components. In this case, more and more often companies themselves or by decision of the recipient employ external companies involved in quality control, among others: selection, repair and sorting of parts, components and finished products for assembly. Client-assembling plant in a crisis situation expects from the supplier immediate response and to solve the problem. Then the supplier has the opportunity to resorting to such outsourcing company, which will arrange for conducting selection in order to secure continuity of production to its customer.

Service of quality control of a third party is performed at the basis of the drawn up by the parties concerned so called specification of control. After completing the task entrusted the supplier and his customer receive a detailed report on the results of carried out control, and defective parts are separated and secured so that they do not enter the production line [16, 17, 27].

In case of detecting at the client non-compliant parts the supplier is obliged to immediately implement the Controlled Shipping procedure (sometimes also called Restricted Shipping).Controlled Shipping process assumes the introduction of an additional 100% inspection of section at the end of the production process or outside the production line, in order to search for parts of nOK, according to the accepted instruction. Controlled Shipping provides an immediate opportunity to protect the customer from receiving further non-compliant parts, and to the supplier gives time to find the root cause and to implement corrective actions [23-26]. Control in the process of Controlled Shipping can be visual or using specific control and measurement tools or gauges. Each time control of Controlled Shipping verifies 100% of parts and is confirmed by labeling parts, most commonly are used green labels. At the time of the imposition by the customer on the supplier the procedures of Controlled Shipping, the client defines the criteria for output.
Controlled Shipping process is divided into two variants:

Controlled Shipping Level 1: employees, suppliers carry out controls in order to prevent the customer from receiving non-compliant products. Control is carried out according to the instruction accepted by the customer.

- Controlled Shipping Level 2: it is introduced when despite introduced CSL 1 control and verification of part by the supplier, the customer still finds nOK parts that means that the control of CSL 1 is conducted not in accordance with the instruction or carelessly.
- CSL 2 assumes that the control is performed out of the line by an external company. Cost of CSL 2 service is very large and most frequently entirely covered by the supplier. It happens that the client provides to a supplier a choice of several companies that are certified by him to provide CSL 2 service on his behalf. Sometimes it happens that CSL 2control is carried out in the area of the customer's warehouse.

There is a third option, which is an extension of CSL 2.

- Controlled Shipping Level 3: activities become necessary when the supplier, faced with anomalies picked up during a CSL 2, shows that they do not have the technical and organizational skills to resolve the anomalies. In addition to the checks included in aCLS 2, the third party company checks application of suitable corrective and improvement actions.

In case of introduction of the CSL 2 control, the CSL 1 control is carried out in parallel. The client defines input criteria from the process of Controlled Shipping Level 2, in a similar manner as in the case of CSL1 but with the remark that the client often additionally defines the duration of the control - e.g.: 20 days of control without defect. This means that when within 20 consecutive days of CSL 2 control an external company will not find any part of nOK this control may be stopped. In case where in 19 day of control the external company performing control finds incompatible part and non-compliance is confirmed by a customer, counter of selection days without defect resets and the control is carried out for 20 consecutive days with no defect. If the procedure CSL 2 or CSL 3 lasts for a long period of time this may be information for the customer that the supplier is not able to improve its own production process or make proper quality control in the process. In this case, the customer most frequently decides to implement the New Business Hold procedure, and in a final step there is a transfer of part production to another supplier.

3 Case study

For the analysis and evaluation of benefits from the use of external quality control of the product there was selected medium-sized company producing products made of thermoplastic materials, among other things from polyamide and ABS. The element under control is the wheel arch of the car (Figure 1). The customer requirement is to ensure the minimum process capability at the level of Cpk of 1.33.

An important feature for the detail outside the geometrical parameters is uniformity of color and continuity of the surface. Detecting defect at the client launched at supplier, on command of the customer, the implementation the CSL 2 procedure. Functioning procedure CSL 1 proved to be insufficient. Indicated by the customer outsourcing company realizes control of so called third party. Practice shows that external companies in most cases, use visual control. Figure 1 shows variants of implementation of the procedure of Controlled Shipping. The situation before the implementation of the CSL procedure was as follows: injection molding machine operator after the opening of the matrix and pulling out molded piece made inspection directly on the workstation. In case of finding non-compliance of the product operator places incompatible molded piece in the separator of non-compliances. Products that meet the requirements are placed in special shipping boxes. We are dealing here with the case of alternative visual control and issuance of a two-stage evaluation of the compatible product "OK" and incompatible product "nOK". The controller (in this case, the operator is also the controller) can make two types of errors: defective product can pass up or reject good product, these errors are called: errors of the first type and errors of the second type. The second type of error found on production line in question does not detect incompatibilities and is called the supplier's risk, because the risk of handing over a defective product to the supplier is increasing drastically. This was the case in the analyzed case. The first type of error means that the control detects a defect that does not exist. This category is called the risk of the producer, as the producer runs the risk of rejection of good product. This generates costs, but from the point of risk of shipping of non-compliant product to receive this type of error is not so important. Table 1 shows a fragment of summary of the analysis of the effectiveness of visual control by the operator of the injection molding machine, by the plant employees on the position after the injection molding machine (CSL 1) and outsourcing control implemented by a third party (CSL 2) performed after the procedure CSL 1. In this case we are dealing with double 100 % quality control carried out by employees of the company and outside company.

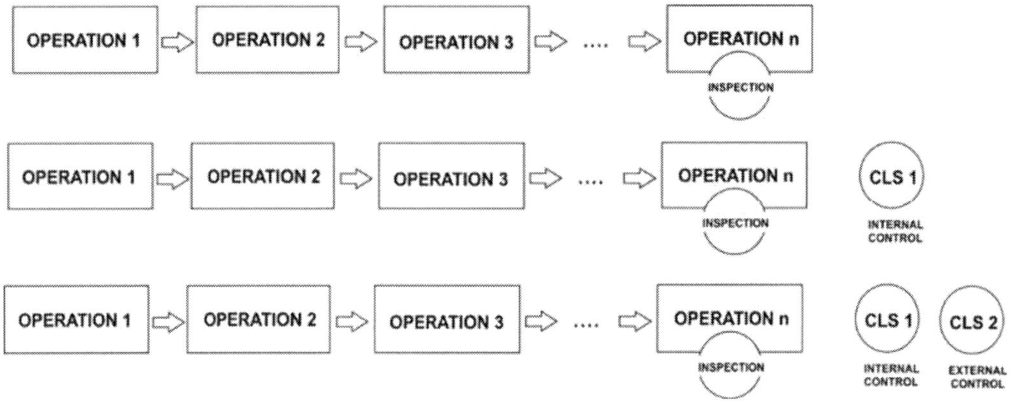

Fig. 1. Variants of implementation of the procedure of Controlled Shipping at the supplier of components made of thermoplastic materials.

Table 1. Fragment of analysis of the effectiveness of control implemented by the operator of the injection molding machine, internal controller - CSL 1 procedure and external controller - CSL 2 procedure.

Product No	**Control Operator of the Injection Molding Machine**	**Internal Controller CSL1 Procedure**	**External Controller CSL 2 Procedure**
1	OK	OK	nOK
2	nOK	nOK	nOK
3	OK	OK	OK
....	....	....	....
249	OK	OK	nOK
250	OK	OK	OK

4 Conclusion

The case analysis and the results obtained (Table 1) showed very high efficiency of used the CSL 2 control procedure, in particular in comparison with the control performed by the operator which is ineffective and worsens with the number of hours worked by the operator. At the end of the shift this type of control can cause up to 20% mistakenly identified "nOK" as "OK". At 250 carried out inspections, in case of visual control, by controllers according to CSL 1 and CSL 2 procedure, there was found 98.4% of the assessment repeatability (4 products were classified to another category). In case of comparing the results of control by operator and control realized on the basis of outsourcing (CSL 2), the repeatability of the evaluation was only 89%. The benefits of using quality control on the basis of outsourcing are effective. In practice, there can be found more and more often solutions involving the resignation of quality control realized by employees of own company for outsourcing without the CSL 1.

Carried out expert interviews have revealed one more problem related to the dramatic lack of hands to work and very high turnover of employees. Especially if we are talking about voluntary rotation, which is the result of own, individual decisions of employees who have decided to change jobs. This is very difficult to control because employers do not know when they can expect such a decision and it can involve very heavy losses for the company, e.g. in the context of building a quality culture. At present, the turnover of employees in the analyzed company exceeded the level of 15% (average rotation in Poland for the first quarter of 2017 was 10.8%).

The costs resulting from the voluntary staff rotation, including training and course costs, more increasingly convince companies from the automotive industry to use the external quality control. Using an external control company gives you a guarantee of control quality through the high competences levels of external controllers.

It is also important that companies specializing in quality control are actively involved in the process of developing suppliers and mobilizing them to improve the quality of their production, which translates explicitly into delivery of complying with the quality specification parts to the production line. The introduction of systematic quality monitoring allows protection against the consequences of human and material errors.

References

1. W.L., Tate, L.M., Ellram, L., Bals, E. Hartmann, International Journal Production Economics, **120(2),** 512-524 (2009)
2. T.R. Holcomb, M.A. Hitt, Journal Operations Management, **25(2)**, 464-481 (2007)
3. S. Wuyts, A. Rindfleisch, A. Citrin, Outsourcing customer support: the role of provider customer focus. Journal Operations Management, **35**, 40-55 (2015)
4. L.M. Ellram, W.L.Tate, C. Billington, Journal Operations Management, **26(2)**, 148-163 (2008)
5. J.M., Hallikas, I. Karvonen, U. Pulkkinen, VM. Virolainen, M. Tuominen, International Journal Production Econmics, **90(1)**, 47-58 (2004)
6. S.M. Handley, W.C. Benton Jr., Journal Operations Management, **31(3)** 109-128 (2013)
7. H. Kenneth, J.B. Wathne, Journal of Marketing, **64(4)**, 36-51 (2000)
8. S.M. Handley, J.V. Gray, Journal Operations Management, **22(6)** 1540-1556 (2013)
9. S. Borkowski, K. Knop, Production Engineering Archives, **1**, 25-28 (2013)
10. H.Golas, A. Mazur, P. Piasek, P. Czajkowski. Problemy Jakości, **2**, 10-14, (2017)

11. K. Mielczarek, K. Knop, Production Engineering Archives, **13**, 16-19 (2016)
12. K. Knop, K. Mielczarek, Technical Transactions Mechanics, **3**, 67- 72 (2016)
13. J. Selejdak, T. Corejova, R. Ulewicz, *Total Quality Management* (Wz-PCz, 2016)
14. C.G. Drury, M.A. Sinclair, Human Factors, **25**, 391–399 (1983)
15. A. Kujawińska, K. Vogt, Management and Production Engineering Review, **6(2)**, 25-31 (2015)
16. A. Gębalska-Kwiecień, *Proceedings of the 6th International Conference on Operations Research and Enterprise Systems,* 265-271 (2017)
17. J. Nowakowska-Grunt, M. Mazur, *METAL 2016 - 25th Anniversary International Conference on Metallurgy and Materials, Conference Proceedings,* 1956-1961 (2016)
18. E. Moulding: *5S: A Visual Control System for the Workplace* (Author House, Central Milton Keynes 2010)
19. S.M. Handley, J.V. Gray, Journal Operations Management, **22(6)**, 1540-1556 (2013)
20. Y-D. Jo, K-S. Park, Journal of Loss Prevention, **16**, 313-321 (2003)
21. New Trend in Global Production System
22. B.O. Örgüna, Procedia - Social and Behavioral Sciences, **181**,140-147 (2015)
23. J. Pietraszek, A. Gądek-Moszczak, T. Torunski, Adv. Mat. Res.-Switz., **874**, 139 (2014)
24. M. Zenkiewicz, T. Zuk, J. Pietraszek, P. Rytlewski, K. Moraczewski, M. Stepczynska, Polimery, **61**, 835 (2016)
25. J. Pietraszek, E. Skrzypczak-Pietraszek, Avd. Mat. Res.-Switz., **874**, 151 (2014)
26. T. Zuk, J. Pietraszek, M. Zenkiewicz, Polimery, **61**, 519 (2016)
27. M. Ingaldi, *METAL 2017: 26th International Conference on Metallurgy and Materials, Conference Proceedings, (Brno, Czech Rep.* 2017)

Control of incompatibilities in the production of helmets for mining

Piotr Sygut[1,*]

[1]Department of Production Engineering and Safety, Faculty of Management, Czestochowa University of Technology, ul. Armii Krajowej 19B, 42-200 Czestochowa, Poland

Abstract. The paper presents the results of tests carried out in one of the national enterprises producing personal protection equipment. Characteristics of individual protection measures are presented, including those found in the construction and equipment of a protective helmet for the mining industry. In addition, proposals for improving the production process were presented in order to eliminate or limit the number of incompatibilities occurring in a given process.

1 Introduction

There are many workplaces in production companies, where work can have unpleasant consequences in relation to the health and even life of the employee [1]. Therefore, in order to reduce or completely eliminate the risk of a hazard, all possible technical and organizational activities should be carried out. In the event of failure of that action and not receiving the expected results, individual protection measures that are the sole and final protection of the employee must be taken. It follows from the general health and safety regulations that the elements of individual protection should be used as a last resort and treated as the last link of protection.

An item of personal protection means an equipment or device that is intended to be worn or used by an explorer. They are used to protect against one or more elements that threaten the safety and life of the user. A protective agent can be [2, 3]:

- a set of a certain number of devices or types of protective accessories, e.g. a respiratory protection device connected to a protective helmet,
- a protective device that is connected with non-protective equipment, e.g. a protective helmet equipped with a mining lamp,
- replaceable components or parts of individual protection elements, eg filters for half-masks,
- systems that are used together with personal protective equipment to connect them to external devices. These devices are not intended to be held or worn, e.g. air filters, which are used for the hose apparatus as respiratory protection.

The employer is responsible for the used personal protective equipment. It should provide employees, free of charge, carefully selected for hazards, protective elements and provide a method for their storage, maintenance, cleaning, disinfection and possible repairs.

* Corresponding author: piotr.sygut@wz.pcz.pl

In addition, he should organize training in health and safety and the use of personal protective equipment [3].

To ensure proper functions, personal protective equipment should meet a number of specific requirements, including those that relate to aspects related to their design and manufacture. Confirmation that the product meets these requirements is the regulation of the Minister of Economy. On specific requirements for personal protective equipment. The basic requirements, the procedure and conditions for the assessment of conformity, as well as the design and method of CE marking are elaborated in this regulation [3].

Moreover, the product should guarantee proper protection of the user:
- be properly selected for the threat;
- not increase the risk;
- be selected in accordance with the prevailing conditions at a given workplace;
- be ideally suited to the user after the required adjustment.

In the countries belonging to the European Union, the criteria, scope, and compliance with the provisions of SOI are set out in Directive 89/686 / EEC, which introduced the divisions of these elements. It refers to the risks that occur in a given workstation, dividing the risk into three categories [4]:

Category I - includes measures that protect against diseases or injuries with a low degree of harmfulness, do not cause long-term discomfort and absence of an employee at work. A declaration of conformity of the European Union is required.

Category III is a means that protects against such threats as death or injuries that can cause severe and unversible injuries. It includes respiratory protection equipment, fall protection equipment and protection against electric shock.

Category II covers those personal protective equipment that do not belong to category I or III, they are: hearing protection, goggles, industrial helmets, safety goggles, protective clothing against possible mechanical hazards, shoes, gloves [4-6].

Before starting to choose protective measures, first identify the hazards that occur in the work environment and assess the occupational risk. If necessary, the emission of these harmful or dangerous factors is measured, and the results obtained are compared with the admissible values. If we take into account the manner in which harmful and dangerous factors affect the human body, the following sets of personal protective equipment are distinguished:
- protective clothing,
- measures to protect the upper limbs,
- measures to protect the lower limbs,
- face and eye protection,
- hearing protection,
- head protection measures,
- respiratory protection elements,
- protection against falling from a height.

There are many positions in various branches of industry, i.e. forestry, mining, construction, energy, which are at risk of head injuries. The most serious of them are mechanical injuries, which may be the result of an impact on dangerous stationary elements or by falling elements occurring at a given workplace. Activities performed in various positions have different specifications and it is not always possible to eliminate the threats arising from them. In such circumstances, the employer is obliged to provide the employee with an appropriate protective helmet, appropriate to the type of hazards [7].

Helmets, in addition, can protect the head against electric shock: electric shock, liquid metal splashes, transverse compressive forces, high temperatures. They can also be used as a base for the installation of eye shields, hearing protectors, respirators, neck guards, and ancillary equipment [7].

2 Characteristics of the analyzed subject of research

The study carried out research on the production of a protective helmet for the mining industry. It is a product of a Polish company producing the highest class protective equipment for civilians, as well as equipment and protective equipment for the army, police and etc.

The helmet has a constructionally reinforced shell, adjustable truss and many other necessary elements that allow its universal use, eg: chinstrap, visor protecting the face, potnik, rain gutter, ventilation holes, fixtures for fixing the lamp, cord and hearing protectors. The shell of the helmet is made of ABS. The manufacturer guarantees the operational parameters of the helmet within 5 years.

Materials used for the production of the helmet guarantee its electroinsulation up to 440 V and the ability to use at temperatures from -20 to +50 ° C. Due to its antielectrostatic properties, the helmet is approved for use in potentially explosive areas: 0, 1, 2, 20, 21, 22. In particular, it can be safely used in the presence of combustible media with a minimum ignition energy> 0.1 mJ, including - methane and coal dust, especially - in climatic conditions prevailing in underground mining plants (excavations classified as methane explosion hazard)", "b" and "c" and for the coal explosion hazard classes "A" and "B"). It is predicted for use mainly in mining, metallurgy, power engineering, construction and in those services where there is a risk of explosion. The helmet meets the requirements of the Polish standard PN-EN 397:1997/A1:2002 together with additional requirements, i.e. resistance to very low temperature (-20 ° C) and electrical insulation. The helmet is characterized by a small weight - below 350 g with full additional equipment for mining.

3 Analysis of the incompatibility of the subject of research

In the analyzed production company, research was carried out on the observation of the production process of the protective helmet. Nine discrepancies were detected during the weekly observations. The discrepancies are presented in Table 1 together with the frequency of occurrence and their percentage section. Based on this data, the Pareto-Lorenzo chart was made (Fig. 1) [8, 13].

Analyzing the received data in Table 1 and Fig. 1, it can be seen that the first two discrepancies (scratches, inaccurate shell casting and cracks and deformations on the eye-face shield) account for over 60% of all non-conformities observed in this process. Incompatibility scratches, inaccurate casting of the shell does not endanger the safety of the helmet operation by users, while for the new product it is undesirable due to the understatement of the aesthetic value of the product. In the case of the second in number of occurrences of non-compliance - the scratches and deformations on the eye-face cover are also the understatement of the aesthetic value of the product, but also some scratches and distortions caused visual field distortions and distorted vision for the user of the helmet.

For these discrepancies, guidelines for improving the production process have been developed. Fig. 2 presents the Ishikawa diagram [9-12] for the problem of cracks, inaccurate casting of the shell. This diagram was implemented on the basis of brainstorming carried out among the employees of the analyzed plant, which take an active part in the preparation and implementation of the helmet's production.

Table 1. Detected incompatibility during the period considered**Błąd! Nieprawidłowe łącze.**

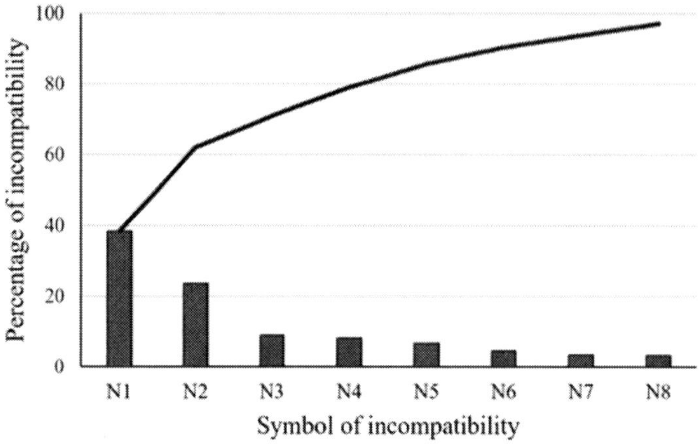

Fig. 1. Pareto-Lorenz chart for incompatibility in the production process of a protective helmet for the mining industry

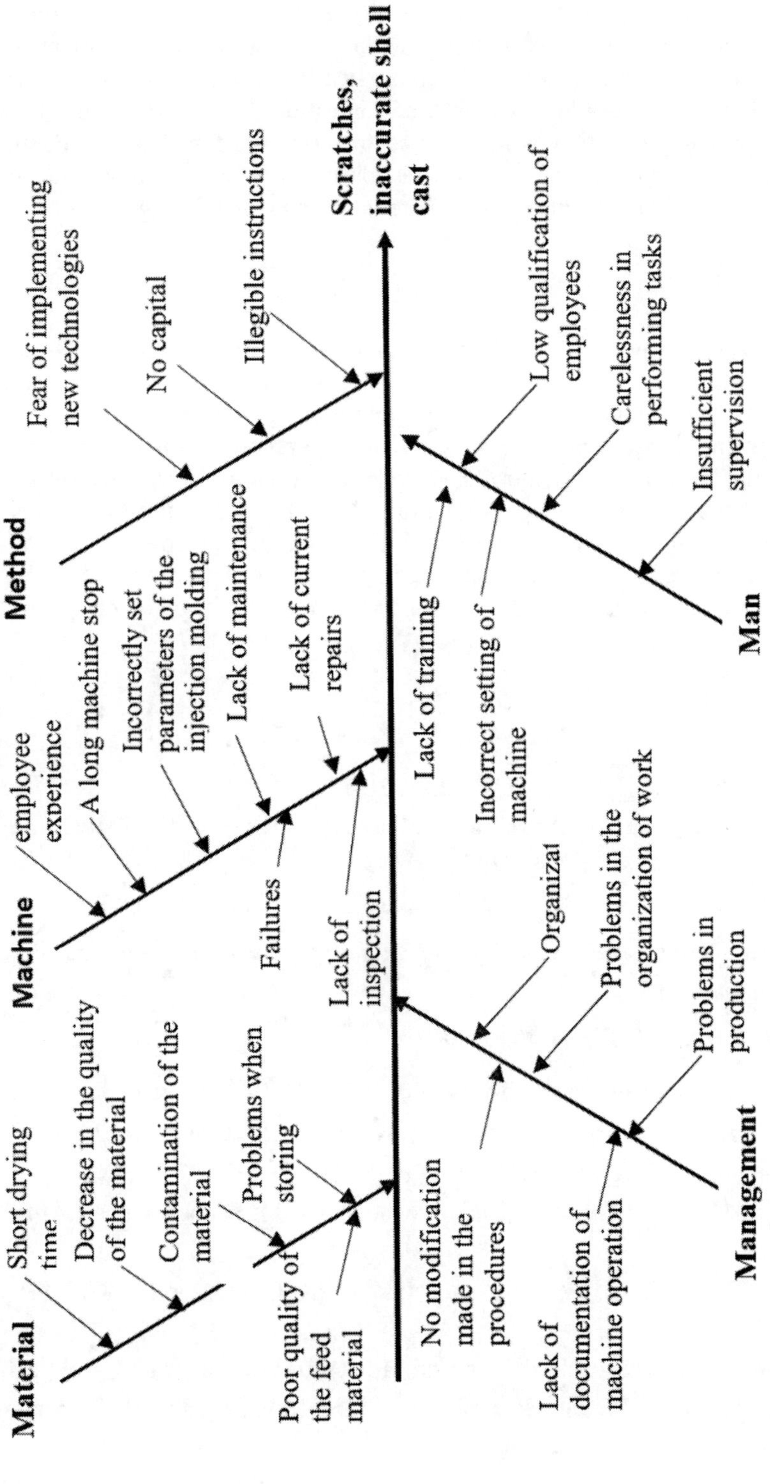

Fig. 2. Ishikawa diagram for incompatibility - Scratches, inaccurate shell cast

Analyzing the data shown in Fig. 2, it can be concluded that the non-compliance data is in most cases the resultant of human error (machine operator / warehouseman). This is due to the fact that the lack of sufficient human attention to the implemented process and procedures causes a malfunction of the machine. For example, lack of attention when retooling and maintaining the injection molding machine causes errors in the correct setting of the production process. Failure to observe the mold heating time causes incorrect casting or missing material in the cast product. The lack of control on the part of the management causes relaxation among production workers. The material itself had a low degree of disagreement. There were sporadic problems with improper material holding and drying. There were occasional problems with improper material holding and drying.

4 Conclusion

In order to minimize or completely eliminate a given nonconformity, corrective actions should be implemented in the entire enterprise with the associated control (process control, employee self-control, employee control). This requires greater involvement of the company's management division and people supervising the production line. It is necessary to carry out training for employees, introduce workplace training, introduce more clear instructions at the station, introduce employees' motivation. The works must be carried out at every level in the hierarchy of the analyzed enterprise. Because the company has a large production potential, the effects of the proposed improvement actions should be visible immediately after their introduction (impact on the improvement of production quality and its efficiency).

References

1. A. A. Godwin, T. R. Eger, WJoPA&H, **47,** 23 (2014)
2. C.J. Behr, A. Kumar, and G.P. Hancke, *Proceedings of the IEEE International Conference on Industrial Technology* (2016)
3. A. Pościk, Dobór środków ochrony indywidulanej,(CIOP, Warszawa, 2000)
4. E. Balańska, M. Flotyn, *Środki ochrony indywidualnej. Informacje dla pracodawcy,* (Urz. UIE, Warszawa, 2001)
5. Baszczyński, K. 2018, IJoOSE, **24**, 171 (2018)
6. M.A. Mououdi, J. Akbari and M. Mohammadi Khoshoei, EiD (2018)
7. K. Baszczyński, *Ochrona przed upadkiem z wysokości*, (CIOP, Warszawa, 2001)
8. P. Sygut, Production Engineering Archives **6** (1), 14-16 (2015)
9. B. Chokkalingam, V. Raja, J. Anburaj, R. Immanual, and M. Dhineshkumar, AoFE **17**, 174 (2017)
10. P. Milosavljevic, D. Pavlovic, M. Rajic, A. Pavlovic and C. Fragassa, IJoCEEaLL, **28,** 24 (2018)
11. J. Pietraszek, A. Gądek-Moszczak, T. Torunski, *Adv. Mat. Res.-Switz.* **874**, 139 (2014)
12. M. Zenkiewicz, T. Zuk, J. Pietraszek, P. Rytlewski, K. Moraczewski, M. Stepczynska, *Polimery,* **61**, 835 (2016)
13. R. Ulewicz, J. Selejdak, S. Borkowski, M. Jagusiak-Kocik, *Metal 2013: 22nd International Conference on Metallurgy and Materials,* 1926-193 (2013).

Quick Response Quality Control as an innovative approach in the conditions of rail transport

Eva Nedeliaková[1]*, *Vladimíra* Štefancová[1], and *Adrián* Kuka[1]

[1]Faculty of Operation and Economics of Transport and Communications, University of Žilina, Univerzitná 8215/1, 010 26 Žilina

Abstract. Customer satisfaction with the provided transport services is closely related to the ability of management to respond flexibly to process changes and maintain the required quality. QRQC is also considered as a method for troubleshooting and it can be applicable in many different areas along rail transport. The innovative qualitative approach QRQC highlights the principle of thorough analysis of the causes of problems. After a problem situation occurs in a traffic operation, it is necessary to proceed according to the specified sequence in order to reduce the undesirable influence in the shortest possible time. The article deals with the possibility of applying this method to rail transport processes.

1 Introduction

The achievement of true success lies in the ability to predict future trends and embrace them successfully rather than letting them adversely affect business. This fact is completely correct also in the case of rail transport undertakings. Adapting to change is particularly difficult for transport undertaking providing services with its long historic basics, assets which last for decades, the dominance of other means of transport which tend to react slowly and consequently lose good position in the transport market. A railway undertaking is affected by risks of internal environment, external, similarly interactive [1].

Development in the transport market includes the rapid growth of technology and globalisation, necessity to eliminate CO_2 emissions, pollution, and energy consumption, and the growing gap between various means of transport. There is also a whole range of social and economic issues. Rail transport has to face lots of issues, congestion is a problem in all countries, but there is a lack of infrastructure in developing countries while in developed countries it is very old.

Sustainable innovation connected with new quality management approaches can help to improve the quality of rail freight services. According to Mr Masaki Ogata, a president of the UITP: "we need to know the genuine needs of our customers, which will create the seeds for innovation." The knowledge economy describes a new stage in the development of society, the essence of which is sustainable economic growth based on information, knowledge and innovation [2].

Adapting to change is not an option but a requirement if rail wants to survive and prosper. Worldwide, there are several famous quality management approaches and according to the transport market´s requirements it is highly important to apply them on new approaches. Market influences emerged to challenge the status quo, including various approaches by Juran, Deming or other famous members of total quality control. It is necessary to interconnect the cost of quality monitoring with orientation on customer, with the process orientation in the area of quality [3].

There are typical visible characteristics for rail market development in connection with quality management such as liberalisation of transport market and specific conditions in product providing connecting customer needs. The specific purpose of transportation is therefore to fulfill a demand for mobility both in passenger and freight transport [4].

2 Needful innovative quality management approach in rail freight transport

Rail freight transport carriers endeavour to create such conditions for its customers as to maintain the existing transports and to gain new transport orders. For example, the state rail cargo carrier in Slovakia transport volumes have been largely dependent on the situation and development of metallurgical industry in Slovakia and the surrounding countries. The transport volume shares attributable to metallurgy accounts for as much as 66% of the total volume of transported goods. There are also great changes in the business relations of the steel plants or their production when the flows of transport of goods are redicted or the transported commodity is changed, which is consequently associated with varying requirements for the use of locomotives or wagon types for transhipment [5].

Rail freight transport creates difficult system where lots of impacts exist, but new challenge for it is characterised by seeking a progressive route for trespassing inferior services. Quality of services in rail freight transport is possible to follow within the frame of whole transportation chain or in division on its single stages. A current issue is to identify the quality not only prior or during actual transport but also after the transportation. At that time the customer often requires additional services, eventually, if the customer is not content with the transportation, he raises claim. Rail freight undertakings have to find relevant response to the problems and protect their customers [6].

3 Application of Quick Response Quality Control in rail freight transport

This philosophy emphasizes the team approach and consists of a quick response to accrued problem and the need to protect the customer and processes before staying in the undesirable situation. This method is universal and each railway undertaking can slightly modify it for particular transport processes. Simultaneously, it is necessary to constantly monitor the processes and solve particular tasks based on current trends and thus to operationally manage the internal processes in railway transport. The QRQC is based on activities such as the initial detection of the problem, the communication and the problem allocation at all levels within the rail transport processes from first contact employees until to high-level managers. After analysing and determining the solution to the problem, comes the verification itself, which determines whether the chosen solution was correct and the desired result was achieved.

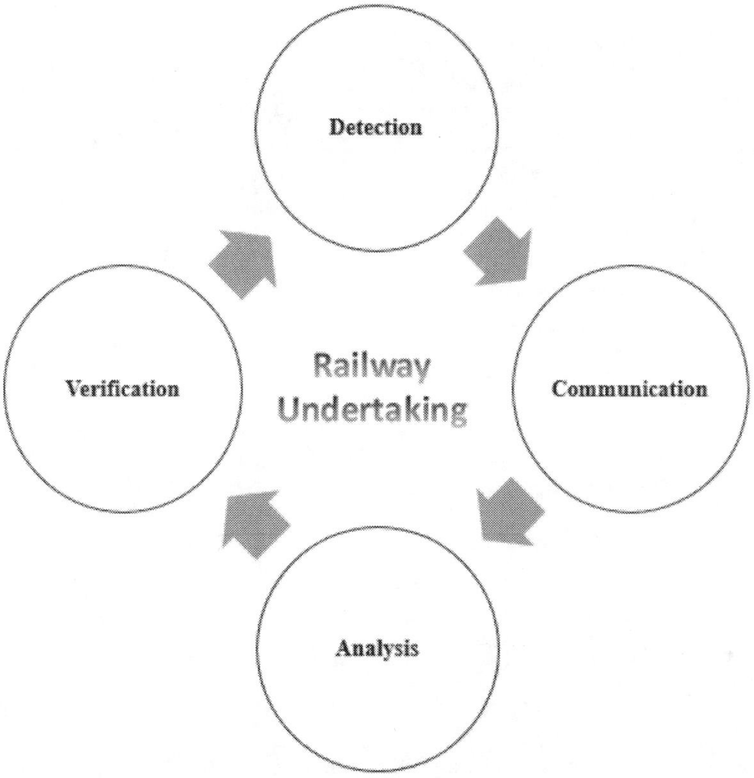

Fig. 1. QRQC in the conditions of a railway undertaking.

It allows clearly to capture the significant factors that cause problems in different areas such as logistics, transport operation and transport services with related quality. In terms of rail transport, this mentioned quality approach has been applied by University of Žilina in cooperation with the state railway company Cargo Slovakia, a.s. (ZSSK CARGO). This research was carried out in regards to application of this method in the case of a certified "logistic trains" product. This type of product is suitable for customers who utilize the significant logistics method just in time, these trains are given a better attention and in addition are considered to be priority trains. Rail company ZSSK CARGO has implemented the quality management system for these trains and also holds ISO 9001:2009 quality certificate. Research has shown the following advantages:

- an analysis that has to be carried out in the event of a problem situation in the process,
- a pre-arranged and functioning team meeting regularly on this product,
- capturing the problem in the process at the right time and immediately reporting the issue, then creating the subsequent documentation,
- introducing a permanent approach to solving the problem through a hardworking team,
- the ability to prevent problems by focusing on internal processes.

Head of department in rail company has to understand the principles of QRQC in order to be able to eliminate the growing problem and understand the main difficulties within transport processes [7]. The sequence of steps is illustrated by the following figure.

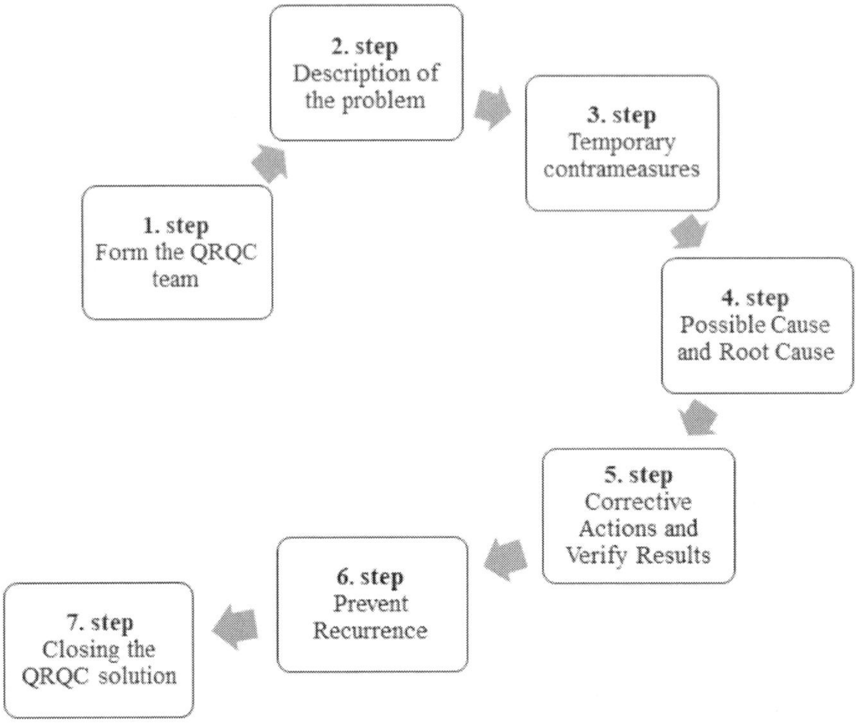

Fig. 2. QRQC in the conditions of a railway undertaking.

The accurate creation of the team is the basis for the successful functioning of the processes whereby individual team members should have the necessary knowledge of the issue [8]. The department manager or coordinator of transport process is considered to be the owner of the process where the problem has occurred. In solving the problem in rail transport according to QRQC, the team of experts in that field is first formed, then corrective measures are proposed which prevent recurrence. In order to eliminate the deficiencies in the process, the efforts are stressed on defining the origin that firstly caused them. The 5W1H method can be used precisely to identify the problem encountered in logistics trains operation. This is also related to determining and asking for specific questions when a problem arises.

- What is a logistical train problem?
- Which employee or customer has found this issue?
- When was the problem with logistics trains detected?
- Where was the problem identified?
- How was the problem recorded (form, complaint)?
- How many logistics trains have this problem?

The next step represents collecting as much information as possible as described in the idea "Well-described problem is a problem that is already solved halfway". The definition and implementation of corrective actions takes place within 24 hours on the basis of a quick and simple analysis to stop the spread of the problem into other areas. Subsequently, it is necessary to analyse and look for the factors and causes of the problem and to find out why the problem has not been detected in time. The controller in area of quality management should be involved to define quality objectives in cooperation with quality managers, so that goals can be quantified and tools developed to measure the quality [9]. A control point is determined for each factor, and so it is possible to compare the current state to its standard with the prescribed value. Within the railway undertaking interface, this is a percentage assessment using the change index over the previous period [10]. Finally, it is important to remove the root cause of the resource in order to resolve the problem and prevent its recurrence in the future. The railway undertaking must choose the right position to solve the problems [11]. If the problem is viewed as a possibility of growth, rail company will solve the problems more easily and at the same time it will move forward and will become more competitive on the rail market. After identifying and describing the problem, the railway undertaking has to take immediate corrective action. Preparation activities should contain also monitoring schemes of critical situation thus eliminating the flaws of particular processes [12]. These are recorded in the QRQC form for each problem whereby the responsible team and the finished date have to be assigned to particular problem situation.

Fig. 3. 5W1H (Description of the problem)

A lot of tools are used for a thorough analysis, one of the most used being 5W1H, which allows the problem to be broken down and determine a root cause. [13] Based on careful analysis, it is necessary to take corrective action to avoid recurrence of the problem and customers are satisfied with the logistic train product. In support of intermodal transport, logistics trains are an important element and therefore it is essential to meet customer expectations. Therefore, it is essential that speed and guaranteed quality are respected.

4 Conclusion

Through this method, various transport problems within rail operation can be clearly solved already in a relatively short time. Long-term arrangements will ensure that the root eliminated, thus ensuring that the problem is not repeated for the future. QRQC enables for certified products in rail transport to guarantee not only quality but also delivery times. It is also related to choosing the right wagon for transport, meeting the expected transport speed, and even ensuring train movement tracking. Once a long-term solution has been implemented, validation will be carried out on the basis of measurable results of the process and its improvement will be assessed.

Acknowledgement

This research was financially supported by the project KEGA 010ŽU-4/2017 New methods of teaching quality management in the study program Railway transport with a focus on optimization of extraordinary events in terms of customer orientation.

References

1. E. Nedeliaková, M. Panák, *Methodology for quality assessment within transportation chain*, CITPM, 309-314 (2016)
2. E. Hvizdová, V. Mokrisová, J. Polačko, *Changes in research and development after crisis in selected countries*, Economic Annals-XXI, 31-34 (2016)
3. A. Šatanová, M. Sedliačiková, *Model for Controlling the Total Costs of Quality*, WCBEM, 2-6 (2015)
4. A. Mačiulis, A.V. Vasiliauskas, G. Jakubauskas, *The impact of transport on the competitiveness of national economy*, 93-99 (2009)
5. J. Daniška, *Support for new transport orders and more favourable conditions for single wagonload*, Business (2015)
6. E. Nedeliaková, M. Panák, *New Trends in Process-Oriented Quality Management*, BEM, 172-179 (2015)
7. H. Aoudia, Q. Testa, J. Aschenbroich, *Perfect QRQC Quick Response Quality Control: The basics*, 172 (2016)
8. P. Polák, J. Žitňanský, M. Bujna, M. Prístavka, *Rapid response methodology for quality management*, Engineering magazine (2014)
9. A. Šatanová, *Quality management and quality controlling in Slovakia*, Perspectives, (2016)
10. J. Majerčák, P. Majerčák, *Logistics assessment of functional performance of material flows in railway transport*, Logi (2017)
11. Z. Dvořák, M. Čižlák, B. Leitner, R. Soušek, E. Sventeková, *Managing of risks in railway transport* (2010)
12. E. Nedeliaková, J. Sekulová, I. Nedeliak, *A New Approach to the Identification of Rail Risk at Level Crossing*, Transbaltica, 40-47 (2015)
13. P. Hudec, *Methodology of rapid reaction for quality management*, Environmental management (2010)

Significance of the quality of short - link chains for work environment

Katarzyna Midor[1,*]

[1]Silesian University of Technology, Faculty of Organization and Management, ul. Roosevelta 26, 41-800 Zabrze, Poland

Abstract. In the article basic elements influencing the quality and safety of the chain are discussed. The conducted analysis clearly indicates that price ought not to be the only determinant to take into account when choosing a supplier – instead, the users of these products should first of all consider certificates and approvals provided with the product by its manufacturer.

1 Introduction

Work environment has a big impact on the health and quality of the employee's life as it consists of a number of material and psychosocial factors that the employee deals with when performing or preparing for his/her work. Material factors of work environment include among others: the premises of an enterprise or institution, buildings, rooms, machinery, tools, equipment, but also physical and chemical factors (e.g. microclimate, lighting, vibrations, noise, mechanical factors as well as chemical substances and dust) and biological factors. Psychosocial (non-material) factors of work environment are among others: psychological work requirements, organisational factors, participation of employees, human relations, professional career development and corporate culture [1].

The paper deals with one of work environment elements influencing the employee's safety, namely hoisting equipment in which the chain is an important operating element.

Products such as technical chains, which are used to carry bigger loads, have become irreplaceable in many industrial and economic branches. They are mainly applied as transport chains, which can have short or long links. Short-link chains are used in a number of mechanical devices, mainly as drive chains cooperating with star wheels and auxiliary slings in tractors. Short-link chains are general purpose chains and they are most often applied in the mining industry [2].

The paper focuses on short-link chains, which are mainly used in hoists and winches. These two devices are designed to vertically lift a cargo by means of a rope or chain, using a gripping element, usually a hook, the difference between the two being the way of mounting the load-bearing structure. Depending on the manner of construction, the said devices can be used in various areas, such as: industry, building, automotive branch or forest industry [2]. According to the type of drive, hoists can be divided into: manual, electric and pneumatic, whereas according to the type of strand applied, they fall into two

* Corresponding author: katarzyna.midor@polsl.pl

categories: chain or rope ones. To lift a cargo, it is necessary to select an appropriate chain for the hoist/chain winch, depending on the construction of the hoist/winch, cargo weight and environment in which it is used. Depending on the type of hoist/winch drive, PN-EN 818-7 standard lists the following types of chain [3]:

T - hoist/winch with a manual drive or jacks with a low-speed motor drive, which do not work in abrasive conditions.

DAT - hoist/winch with a motor drive, which reach high speeds and high work intensity, in places requiring high abrasive resistance due to increased chain life.

DT - hoist/winch with a motor drive, used in abrasive work conditions.

Short-link chains for lifting cargo can be applied in many industrial and economic branches, among others in underground headings of mines, in areas endangered with coal dust or methane explosion.

On the Polish market there are a few producers of these chains, among others: FASING S.A., Pewag Polska Sp. z o.o., Grupa RUD, THIELE, Retezarna a.s.

In the article the authors base on chains manufactured by Retezarna a.s., whose main sales representative in Poland is PPHU "Anima" from Rybnik.

2 Requirements to be met by short - link chains

Handling equipment used in the mining industry includes among others hoists with a chain strand, which enable installation of machinery subassemblies and transport of materials. Due to hard operating conditions in limited spaces of excavations and assembly chambers, hoisting equipment for the mining industry, apart from appropriate lifting capacity, must be characterize by high lifting speed, low kerb weight, small dimensions and work safety [2].

An important operational element in these devices is the chain, which to a large extent determines the safety of the hoist operator. To be safe for the user, the chain must meet a number of technical requirements contained in such standards as:

- PN-G-46732 Mining chain hoists – Calibrated short-link chains,
- PN-EN 818-7:2002+A1:2008 Short-link chains for lifting purposes – Safety – Part 7 – Fine tolerance hoist chain– Grade T (Types T, DAT and DT),
- Safety requirements taking into account the provisions of "Geological and Mining Law" Act dated 9th June 2011 (the Journal of Laws of 2017 item 2126, with subsequent changes),
- Regulation of the Minister of Economy of 28th June 2002 concerning occupational safety and health, operation and specialist fire-fighting protection in underground mining plants (the Journal of Laws of 2002; No. 139, item 1169),
- The act on general safety of products dated 12th December 2003 (the Journal of Laws of 31st December 2003),
- PN-EN 10025 Hot-rolled products made of constructional steel,
- PN-EN ISO 643 Steel - Micrographic determination of grain size,
- PN-EN ISO 6507-1 Metals – Vickers hardness test – Part 1 – Testing method,
- Directive 2006/42/CE of the European Parliament and of the Council of 17th May 2006 on machinery, amending directive 95/16/CE

Moreover, each producer ensures the fulfilment of its technical conditions for the manufactured chain.

A basic notion characterizing the chain is so-called chain size. This value is defined as the product of the diameter of rod d that the link is made of and the biggest internal dimension p, referred to as the scale of link or chain. Moreover, the link is also characterized by such values as a – internal length of the link measured in the link inside diameter, *b* – external width of the link and *L11p* – internal length of the link with 11 links [4]. The above mentioned dimensions of the link have been presented in Fig. 1.

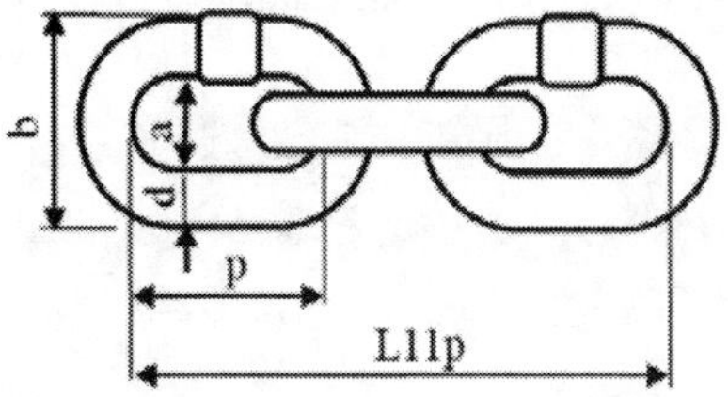

Fig. 1. Dimensions of the link/chain. Source: [4].

In its offer Retezarna a.s. has standard chain sizes, which are presented in Table 1. Apart from standard sizes, the company can also produce customised chains, suited to the customer's needs [4].

Table 1. Standard sizes of chains produced by Retezarna a.s. [4].

Size of chain *d x p*	*d*		*p*	*a* no less than	*b* no more than	Mass of 1 m of chain	Nomina l length *L = 11 x p*	Toleran ce	Max. diamete r of weld
mm						kg. ca	mm		
4 x 12	4	+/-0.2	12	4.8	13.6	0.35	132	0.6	4.3
5 x 15	5	+/-0.2	15	6.0	17.0	0.54	165	0.8	5.4
6 x 18	6	+/-0.2	18	7.2	20.4	0.8	198	1.0	6.5
7 x 21	7	+/-0.3	21	8.4	23.8	1.1	231	1.1	7.6
8 x 24	8	+/-0.4	24	9.6	27.2	1.4	264	1.3	8.6
9 x 27	9	+/-0.4	27	10.8	30.6	1.8	297	1.4	9.7
10 x 30	10	+/-0.4	30	12.0	34.0	2.2	330	1.6	10.8
11 x 31	11	+/-0.4	31	13.2	37.4	2.7	363	1.7	11.9
12 x 36	12	+/-0.5	36	14.4	47.6	3.1	396	1.9	13.0
13 x 39	13	+/-0.5	39	15.6	44.2	3.7	429	2.1	14.0
14 x 42	14	+/-0.6	42	16.8	47.6	4.3	462	2.2	15.1
16 x 45	16	+/-0.6	45	19.2	54.4	5.6	528	2.5	17.3
18 x 54	18	+/-0.9	54	21.6	61.2	7.0	594	2.9	19.4
20 x 60	20	+/-1.0	60	24.0	68.0	8.7	660	3.2	21.6
22 x 66	22	+/-1.1	66	26.4	74.8	1.5	726	3.5	23.8

Welded chains for Class T hoists and chain winches (T, DAT, DT), must be characterized by high quality, excellent functional advantages and long life. To meet these requirements, they have to be made with the utmost care so as to ensure user's safety [5,6].

Parameters which decide about the quality and safety when using the chains are among others mechanical properties and loads. In the production process the manufacturers of chains for winches/hoists use steel characterised by mechanical properties compliant with PN-EN 10025 standard [7-9], so that the final product can meet all the requirements imposed by rigorous European standards, in line with PN-EN 818-7 and PN-G-46732 standards. The above standards also oblige the producer to subject the chain to the calibration process. Another important element influencing the chain user's safety is the quality of workmanship, with particular focus on dimension deviations, which have been contained in Table 1. Moreover, PN-EN 818-7 standard requires that the producer should apply a complex programme of product quality testing in order to ensure operational safety while using the chain. Particular emphasis has been placed on the use of appropriate grades of steel as materials that the elements of a T class short chain are made of [10-13]. Each delivery to the customer must be marked with the producer's logo and provide information on the chain width, production serial number, chain length and the number of pieces in the series. At the ends of the chain, with a spacing of 1m, the following information must be placed: producer's logo, year of production, number which stands for month of production and chain class (T, DAT or DT- for 818-7 standard); 5,6 or 8 for PN-G-46732 standard – Fig. 2.

Fig. 2. Mark placed at the chain ends, with a spacing of 1 m.

Chains produced at Retezarna a.s. fulfil the requirements imposed by standards for particular products. To emphasise high quality and safe use, chains manufactured by this company are marked with H45 (Producer's identifier) granted by the certifying authority DGUV Deutsche Gesetzliche Unfallversicherung, which guarantees the product's compliance with PN-EN 818-7 standard. Fig. 4 shows a quality mark on the chain link.

Fig. 4. Quality mark H 45 on the chain link.

In Poland supervision over the production and quality of the discussed products is exercised by KOMAG Institute of Mining Technology.

3 Requirements to be met by suppliers

Requirements to be met by the supplier providing the customer with the product are contained in PN-EN ISO/IEC 17050 standard, which imposes on the supplier the duty to provide a certificate of compliance together with the product [14]. PPUH Anima, being one of the major distributors of products manufactured by Retezarna a.s. in Poland, with each delivery provides a certificate of conformity, which contains such information as: name of supplier, name of product, declaration of the product's compliance with the requirements imposed by legal acts and standards, such as: Geological and Mining Law as well as PN-EN 818-7 and PN-G-46732 standards, technical specification extract, certificates,

instructions for use, reports on tests provided with the product as well as a statement on accepting responsibility for the quality and safety of the offered product.
With every delivery of the chain, the buyer receives an audit certificate 3.1. In the case discussed in the article this is a certificate (approval) according to ČSN EN 10 204-3.1 standard, which contains producer's data, product identification data, information on the quantity, dimensions and test results.

4 Requirements to be met by recipients

An important element influencing safe use of the chain applied in hoists/winches is also the procedure to be followed by the user himself. For this reason Anima company, together with the delivered product, provides the operation and maintenance manual so as to ensure safe and long operation of the chain. The elements that influence proper operation include observing the load capacity for a particular grade of the chain as well as using a suitable chain, adjusted to the type of winch drive. Attention should also be paid to environmental conditions in which the device works, i.e. temperature (over 200°C the chain must be removed from service) and acidic environment, having an adverse impact on the product.

Due to safety reasons, it is also necessary to properly use the chain and observe the OSH regulations, among others such as [15]:

- No chain overloading.
- The chain must be placed on and removed from chain wheels in a straight position, without twists.
- The load must be well fastened on the hook to prevent it from sliding off.
- The chain mustn't be jerked sharply.
- The chain mustn't be used when if is deformed or torn.
- The chain mustn't be used for suspending by reeving or using a loop.
- The chain must be stored in a dry dust-free location.

An important element is also an inspection of the chain before its first use, on-going monitoring and checking the technical condition with as often as it is necessary in particular working conditions.

The chain should be treated as an element of machinery. In particular, it is recommended that the chain is not dragged on the ground or exposed to weathering. It is forbidden to overload the chain and carry out unprofessional repairs.

5 Summary

Development of mechanization in various branches of industry, including mining, necessitates the use of auxiliary equipment to support assembly, servicing or transport works. Handling of increasingly bigger weights requires the application of equipment characterized by better technical parameters. Such devices, belonging to so-called "moving equipment", are chain hoists and winches. Apart from technical and constructional parameters of these devices, work and safety are influenced by the chain by means of which goods are being handled. Therefore, from the point of view of work safety, the quality of the chain is extremely important. The article draws attention only to basic elements and determinants influencing the chain quality. This analysis clearly indicates that when selecting a chain, one should take into account the mechanical properties of the material that the chains are made of as well as the quality of their manufacture. The quality of the material used and the chain workmanship is confirmed by the producer's certificate, provided with the product.

Acknowledgements

This article has been prepared under principal research conducted at the Institute of Production Engineering of the Silesian University of Technology. Research reference number: BK: 13/030/BK_18/0039.

References

1. B. Ciecińska, W. Homik, MSPE, **2(22)**, 89 (2016)
2. D. Cebula, M. Kalita, Maszyny Górnicze, **1**, 22 (2014).
3. PN-EN 818-7:2002+A1:2008
4. PN-G-46732
5. M. Zasadzień, ICORES, SciTePress, **314** (2017)
6. J. Brodny, K. Stecuła, M. Tutak, SGEM, STEF92, **2**, 65 (2016)
7. PN-EN 10025
8. M. Kalita, Maszyny Górnicze, **3**, 21 (2013)
9. A. Dudek, B. Lisiecka, TANGER Ltd, **1045** (2016)
10. M. Nowicka-Skowron, R. Ulewicz, METAL 2015: 24th International Conference on Metallurgy And Materials, 1707-1712 (2015)
11. P. Klouda, V. Moni, J Blata, L. Donát, F. Helebrant, MSPE, **1(25)**, 68 (2017)
12. P. Szataniak, F. Novy, R. Ulewicz, METAL 2014: 23rd International Conference on Metallurgy and Materials, 778-783 (2014)
13. T. Lipinski, 16th International Scientific Conference: Engineering For Rural Development, 1069-1074 (2017)
14. PN-EN ISO/IEC 17050
15. PPUH „Anima" Krzysztof Śleziak, Instructions for use and maintenance, (in polish) Instrukcja użytkowania i konserwacji

The analysis of non-conformances as the way of assessment of quality of aluminium finishing strips in the selected company

Edyta Kardas[1*]

[1]Czestochowa University of Technology, Faculty of Production Engineering and Materials Technology, Department of Production Management and Logistics, al. Armii Krajowej 19, 42-201 Czestochowa, Poland

Abstract. The main purpose of the paper is the analysis of non-conformances of aluminium finishing strips in one of Polish companies. This kind of analysis may be one of the ways to assess the quality of products. In the first part of the paper qualitative analysis of non-conformances of strips was made using the Pareto chart, where 6 types of non-conformances were taken into account. In the second part of the paper the causes and effects of non-conformances of strips were assessed using FMEA method. This method showed that the biggest problem concerned the anodised coating of strips. In the last part of the paper the in-depth analysis of the causes of appearance of defects if anodising process using Ishikawa chart was made. All causes were discussed. The analysis takes into account the results collected in the company producing various types of strips. The analysis covers the period of two calendar years.

1 Introduction

Quality is one of the most important issues in the contemporary world. It companies people in every sphere of their lives. It is important in the activity of production company and accompanies during each stage of the creation of product or service. Customer satisfaction, and thus the success of the company, depends mainly on the level of product quality [1].
Quality of products depends of many different factors, among others: the quality of choosing materials from which the product is made; the level of team of employees, workstations, machinery and equipment and production technology [2].
In order to maintain the appropriate level of production quality, its continuous observation, analysis of the collected results as well as the implementation of corrective and improvement measures should be made. Various methods and tools that can be used for this are helpful. They help to identify problems and their causes, and look for solutions that improve quality. They cause a clear division of responsibility for the quality and adapt the system to better manage organizational form. The quality information can be used not only by the control staff, but by all organizational units, thanks to which the impact on the quality of the product or service occurs at every stage of their performance [3-6].

* Corresponding author: ekonstan@wip.pcz.pl

The literature presents a wide variety of quality management instruments, which can be used in various organizations. Their basic classification is shown in Table 1.

Table 1. Basic classification of quality management instruments [7].

Principles	Methods	Tools	Techniques
Team work KAIZEN POKA-YOKE Zero defects Quality management principles 14 Deming's principles Quality management principles used in the study and management of the object	FMEA QFD SPC DOE 8D Report 5S	Six sigma 5 Why Ishikawa chart Pareto chart Flow diagram Shewhart's control cards Histogram Brainstorm New quality tools	Measurement Record Organoleptic assessment Research sheets

Many publications present their use. They discuss use of such instruments in various organizations [8-11]. They also show examples of the use of specific instruments, such as e.g. Six Sigma [12-13]. It should be noted that their use is dictated by many diverse premises, among others: product life cycle stage, the type of problem being solved, size and type of organization, business profile, previous crew experience [8-15].
The main purpose of the paper is the analysis of non-conformances of aluminium finishing strips in one of Polish companies. This kind of analysis may be one of the ways to assess the quality of products. It usually concerns the analysis of the quality of finished products, but it can also be used during the production process to assess products in progress.

2 Analysis of results

The testing material taken to analysis are aluminium finishing strips with A02 profile. These are 120 cm long finishing strips, coloured in silver. Finishing strips, having the category of coating thickness AA15 (15μm), are made of anodised aluminium. They are used for aesthetic, as well as functional merging of floors, so that they create a chic and uniform whole. In most cases, they combine wooden floors, floor panels, parquet floors, with carpets or ceramic tiles as well as decorative stone [16].
In the paper the analysis of non-conformances of aluminium finishing strips was made. The analysis was divided into three parts: qualitative analysis of non-conformances of strips using Pareto chart, FMEA analysis of the problems of anodizing process and Ishikawa chart of causes of non-conformances. The analysis covers the period of two calendar year. Research was carried out in the Polish company that deals with the mechanical and surface treatment of aluminum, including the production of finishing strips.
In the company, apart from statistical tools, no quality instrument have been used earlier. Therefore, the use of three basic instruments was proposed: Pareto chart, FMEA and Ishikawa chart. This choice was dictated by two factors: economic - these methods were relatively cheap to use, their use in the company does not entail high costs associated with their use, and effective - these methods are quick and easy to use, and at the same time bring good results.

2.1 Pareto chart of the causes of non-conformances of aluminium finishing strips

The qualitative analysis of the causes of non-conformances of aluminium finishing strips was carried out. During the analysed period (two calendar year) six basic causes of non-conformances occurred: 1 - too thick coating, 2 - stains, 3 - pollution on the anodised

surface, 4 - fingerprints on the anodised surface, 5 - damages on the anodised layer, 6 - damages during transport [16].
Based on the number of non-conformances Pareto chart was made. This analysis sorted the causes of non-conformances from the most to the last frequent ones. The results of the analysis are presented in Fig. 1.

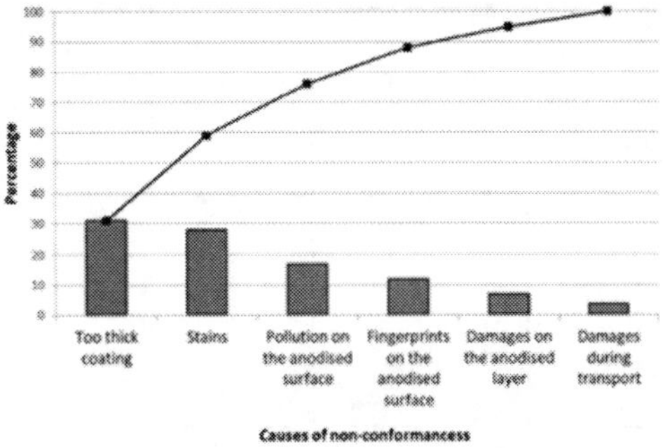

Fig. 1. Pareto chart of causes of non-conformances [Own study base on 16-17].

The results of the analysis presented in Fig. 1 indicate that the main causes of non-conformances are thickness of coating and stains on the profile. Non-conforming products included in these two groups are nearly 60% of total number. Causes of their appearance are the weakest points in the process of anodising finishing profiles.

2.2 FMEA analysis of the causes of non-conformances of aluminium finishing strips

The causes of the non-conformances are closely related to the problems occurring during the production process. Due to the series of performed operations and the specific properties of aluminium problems generated during profile anodizing process were subjected to a detailed examination. As a result, the most important factors were identified and based on these results the FMEA analysis were made. The results of the FMEA analysis are presented in Table 2.

Table 2. FMEA analysis of defects in profile anodizing process [Own study based on: 16-17].

No.	Problem	Effect	Cause	Frequency of occurrence	Possibility of detection	Importance of defects	Priority number	Corrective actions
1	inadequate coating thickness	too thick coating	too long anodizing time in sulfuric acid	5	3	7	105	adjust the anodizing time to the customer's guidelines
2	no repeatability of process parameters	stains	bathing pollution	4	1	8	32	frequent exchange of scrubbers and changes in process parameters
3	no repeatability of process parameters	pollution on the anodized surface	too short rinsing time	3	3	7	63	increase the rinsing time and changes in the process parameters

4	no repeatability of process parameters	fingerprints on the anodized surface	for a short sealing time, a low amount alfiseal 939	2	5	8	80	maintain the sealing time in accordance with the 3min to 1μm principle. Increase the content of alfiseal 939
5	mechanical damage	damages on the anodized layer	incorrect suspension or removal of an item from the pendant	2	4	7	56	modify pendants
6	inappropriate material protection during transport	damages during transport	low quality of protective materials	1	3	4	12	activities on the part of the supplier of the raw material

Based on the FMEA analysis the main problem of anodizing process was diagnosed. Inadequate coating thickness was caused by the wrong anodizing time in sulfuric acid. To prevent the effect of this week point of the production process, the verification of anodizing process was carried out. The implementation of necessary changes in the process parameters was the most important corrective action. In order to maintain the proper anodizing time, additional preventive actions were implemented, such as employee trainings and additional technological process control.

2.3 Ishikawa chart of causes of non-conformances

After performing a detailed identification of non-conformances the analysis of causes of their appearance using Ishikawa chart was made. This graph helps to identify all factors that affect the finishing strips during anodising process. It allowed to gather and organize information in specific groups of causes (Fig. 2). Based on the results the following improving method of anodizing process of finishing strips should be implemented:

1. According to the group of **Machine**:
- Technical conditions - assessment of the technical condition of machinery and equipment on time may have a significant impact on the production volume,
- Producer - his goal should be to improve the quality and productivity of machines,
- Construction-technology - construction, and at the same time the technology used by the device manufacturer is of particular importance to the level of products.
2. According to the group of **Measurement**:
- Accuracy of the meter – the meter should indicate value of 0.01μm to examine accurately the thickness of the profile,
- Measurement method – it is important that the strips from each tested series wad tested in five places at equal distance from each other,
- Imprecision of the meter – inaccurate readings may disturb the actual value of data.
3. According to the group of **Method**:
- Anodizing program – the use of the appropriate anodising program is one of the basic factors determining the obtaining high quality of products,
- The course of anodizing process – performing the anodising process in accordance with the selected anodizing program,
- Way of hanging profiles – fasten the strips to the carriers in optimal way and using appropriate pendants, so that each element is coloured properly.

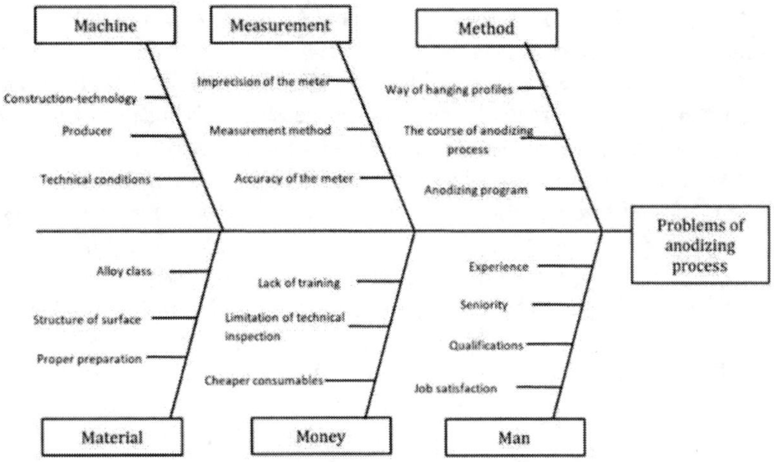

Fig. 2. Ishikawa chart of causes of problems of anodising process of finishing strips [Own study based on: 16-17].

4. According to the group of **Material**:
- Alloy class – it is important that used material is in the appropriate class of alloys, does not have a large amount of admixtures, what can significantly affect the aesthetics of the profile or prevent the process from being carried out.
- Structure of surface - the profile cannot have various types of blemishes (scratches, dents), this will worsen the appearance and aesthetics of the strips.
- Proper preparation - processes such as sanding or brushing must be performed by experienced workers, because it affects the final effect and the quality of the strips.

5. According to the group of **Money**:
- Lack of training - it is important to conduct periodic training to improve the employee's qualifications and professional skills,
- Limitation of technical inspection - a malfunctioning machine system can significantly reduce the production volume,
- Cheaper consumables - faster wear parts lead to frequent failures.

6. According to the group of **Man**:
- Experience – lack of experience results in improper performance of certain activities entrusted by the manager,
- Seniority – gaining knowledge and experience through many years of work teaches people to carry out tasks in proper way,
- Qualifications – lack of knowledge skills in handling equipment to perform tasks has great effect on the effects of work.
- Job satisfactions – satisfied employees show greater initiative, reliability and positive climate work, while lack of it may result in lowering organizational efficiency.

3 Conclusion

In order to maintain the high efficiency of the anodizing process of aluminium finishing strips, it is necessary to carry out the process in the shortest possible time. However, high level of quality of the finished product must be maintained. The qualitative analysis of aluminium finishing strips showed that:

1. Thickness of coating and stains on the profile are main causes of non-conformances that occurs the most frequent.
2. The main problem of anodizing process was inadequate coating thickness. It was caused by the wrong anodizing time in sulfuric acid. Necessary changes in the process

parameters, employee trainings and additional technological process control were the most important corrective action.
3. The most important causes of non-conformances in the production process are: ways of hanging profiles, the course of anodizing process, anodizing program, experience of workers, their qualifications, seniority and job satisfaction.
4. The following actions may be proposed to eliminate con-conformances: purchase of measuring and research equipment, implementation of the registry and equipment cards, proper supervision of these documents, the increased technological control of the process, additional laboratory tests, frequent trainings to improve the qualifications, skills and knowledge of people responsible for the process.
5. Other methods and quality management tools can be proposed for the assessment of quality of finishing strips in this company in the future. Their application can also bring a lot of interesting results and can be very effective. However, it should be assessed whether the costs of their use will not exceed the benefits they will bring.

References

1. A. Hamrol, W. Mantura, *Quality analysis. Theory and practice* (PWN, Warsaw 2009)
2. E. Kardas, METAL 2017 – Proceedings, 2188 (2017)
3. S.T. Dziuba, M. Ingaldi, M. Kadlubek, 16th International Scientific Conference Proceedings, PTS I-V, 425-432 (2016)
4. M. Ingaldi, S.T. Dziuba, M Kadlubek, 16th International Scientific Conference Proceedings, PTS I-V, 750-758 (2016)
5. P. Sygut, Production Engineering Archives **1/1**, 16-18 (2013)
6. P. Szklarzyk, D. Klimecka-Tatar, P. Sygut, T. Lipiński, Production Engineering Archives **4/3**, 11-13 (2014)
7. A. Mazur, H. Gołaś, *Principles, methods and techniques used in quality management* (Publishing House of Politechnika Poznańska, Poznań, 2010)
8. F. Bartes, M. Hodinka, J. Dufka, Proceedings of International Conference on Business and Competitiveness of Companies, 11-16 (2011)
9. D. Y. Negroni, L. G. Trabasso, Proceedings of the 17th International Conference On Engineering Design, Vol 7, Design For X, Design To X, 127-136 (2009)
10. L. K. Tsironis, Benchmarking-An International Journal, **25/1**, 47-75 (2018)
11. A. Pabedinskaite, R. Vitkauskas, Proceedings of 6th International Scientific Conference Business And Management Vols I And II, 905-912 (2010)
12. R. Banuelas, J. Antony, M. Brace, Qual. Reliab. Engng., **21**, 553–570 (2005)
13. R. Joghee, International Journal Of Quality & Reliability Management, **34/1**, 2-17 (2017)
14. M. Muchova, Management 2012: Research In Management And Business In The Light Of Practical Needs, Proceedings, 266-231 (2012)
15. P. C. Valentina, G. Sorescu, Management Of Technological Changes, **1**, 137-140 (2009)
16. Information materials from Company
17. D. Lawinski, *Control and analysis of the quality of products in a selected production company* (Engineering work under the direction of Edyta Kardas, Czestochowa University of Technology, Czestochowa, 2016)

Using 5W-1H and 4M Methods to Analyse and Solve the Problem with the Visual Inspection Process – case study

Krzysztof Knop[1,*], and *Krzysztof* Mielczarek[1]

[1]Technical University of Czestochowa, Faculty of Management, Department of Production Engineering and Safety, 42-200 Czestochowa, Poland

Abstract. The article presents a case study on the use of specially prepared 5W-1H and 4M sheets for the analysis of the problem during the visual inspection process of the electric device, in order to solve it. The identified problem was related to inconsistent assessments during the visual (alternative) inspection of chamber gaps of the electric switch. The research methodology was presented the same as results confirming the effectiveness of the problem analysis in the area of quality control by using these two methods of Lean and WCM concepts. The article aimed to show that a skilful and pragmatic approach to the problem supported by appropriate tools can contribute to its effective solution.

1 Introduction

Visual inspection is the most common variant of alternative control of the product [1], where the product is assessed on the basis of measurable or non-measurable features and its result is classification of product to one of two (OK, nOK) or several quality categories [2]. The main purpose of the visual inspection is to ensure that the product will be free from nonconformities and defects when it is forwarded to the next steps of the process or to the user [3]. Visual inspection is particularly important in the case of processes whose repeatability and reproducibility are limited, and the process results differ and require an individual approach when assessing the quality of their performance [4]. The condition for the effectiveness of visual inspection is the knowledge of errors that can potentially be found at a given inspection station and a clear definition of the inspection criteria (how many and what types of nonconformities cause the product to be considered as nonconforming).

Visual inspection, in particular in the organoleptic version, is an error-prone inspection method due to the high proportion of the human factor [3-5]. Errors in visual inspection are unfortunately unavoidable. In the visual assessment, the controller may commit two types of errors, i.e. accepts an nonconforming product (IInd type error) or rejects a conforming product (Ist type error). The reasons for these errors can be very different, dependent or independent of the controller itself (related to, for example, work organization, work time,

* Corresponding author: krzysztof.knop@wz.pcz.pl

workstation organization, time of inspection, etc.) [4-8], but they can be grouped into 4 or 5M categories (i.e. Material, Machine, Method, Man, Management), which come from the Ishikawa diagram [9]. The consequences of errors in visual inspection are costs, and in the worst case the decline of the company's reputation in the eyes of customers and decrease in sales (in the case while nonconforming product getting into the customer – as a result of the IInd type of error) [10-11]. Analysis of losses and errors should be based on the cost criterion [10]. In effects visual control process, like any other process, should be subject to random changes (random variation) [13]. The large variability of internal and external factors is a serious problem hampering production management [14].

Every error or other problem occurring during or after visual inspection processes should be thoroughly analysed. To solve the problem permanently, one should understand in depth the situations and reach the source causes of the problem. For this purpose, tools from the Lean and WCM concepts, i.e. 5W1H and 4M method, can be used.

2 Aim and scope

The aim of the article is to analyse and solve the problem related to low effectiveness of visual inspection of the product – an electrical switch, more precisely one of the characteristics of its element – slots in the chambers, by using specially prepared sheets supporting analysis with the use of 5W1H and 4M methods.

3 Methodology

The paper presents the practical use of two methods in the field of analysis and problem solving in visual inspection process, i.e. 5W1H and 4M methods.

The 5W1H method is a direct reference to the rules: "if you don't ask, you won't find out", "the problem well described is a half-solved problem". It is used in describing and analysing a given problem by answering 5 questions beginning with the W letter (What, Where, When, Who, Which) and 1 question beginning with the H letter (How). Due to the fact that all questions are open, i.e. none of them can be answered YES or NO, they do not allow to stick to one aspect of a given problem, but show different "sides of the coin". The 5W1H method alone will not solve the problem, but it creates the conditions for the proper identification of the problem under analysis [9].

The 4M is a method that allows to identify and group causes that impact to a specific effect. 4M categories (Material, Method, Machine, Man) are often used in the Cause-Effect Diagram created by Kaoru Ishikawa [9]. It is a good, intermediate tool of problem analysis.

Both, the 5W1H and 4M methods can be used independently of each other, as well as together in relation to a given problem. By the use of these methods it could be analyse any production, service or management problem. They are tools used in Lean and WCM concepts [15]. There are some examples of the use of these methods, based on special sheets, for the analysis of production problems (especially in quality, maintenance), but it was noticed lack of dedicated sheets for quality control processes. I decided to fill this gap. The inspiration to create such sheets was their successfully used versions in the FCA Canada car factory – Windsor Assembly in Ontario, Canada.

The analysed problem, which was decided to deal with the use of 5W1H and 4M method, was the problem of inconsistent assessments during visual inspection of the electrical switch. The analysed product – an electric switch is a product produced in a factory in Malopolska (one of the regions in Poland). Its task is (after adapting with the accessory) to disconnect the power supply in the event of an overload in the mains.

4 Analysis and results

First, after identifying the problem – inconsistent assessments during the visual inspection, it was decided to analyse it using the 5W1H method. The purpose of using the 5W1H method was to examine in detail the analysed problem, before the next stage of the analysis, i.e. looking for the cause or causes of the problem using the 4M method. The result of applying the 5W1H method to the analysed problem was shown in Table 1.

Table 1. 5W1H Analysis Sheet to Describe the Problem in the Quality Control Process in more detail.

Initial Description of the Problem: ***Inconsistent assessment of the size of the chambers gaps during visual inspection***		
5W1H	**Consider These Questions ("Q"):**	**Answer As Many "Q" As You Can!**
What	* **What** does the problem look like? * Is the problem related to Ist type or IInd type error? * During the production of which model / version of the product did the problem take place?	*The problem is the lack of coherence in the assessments of the appraisers performing the visual inspection of the gaps. It applies to both errors of type I and type II. It occurs during visual inspection of slots in the chambers of the electrical switch C1.*
When	* **When** did the problem occur? * When in the sequence of inspection: initial, first piece, self-control, inter-operative, final? * At what time and in what period?	*The problem arises at every stage of the control process in which the size of the gaps is controlled.*
Where	* **Where** did the problem occur? * Where was the problem detected? * In relation to which product's characteristic/nonconformity in particular the problem occurred?	*The problem arises at 5 control stations: quality controllers, leaders from the Compact line, leaders from the line of chambers, trainers from the Compact line, trainers from the line of chambers and setters from the chamber line. The problem was detected in the next stages of the control process, in the final control as well as by the client.*
Who	* **Who** does it affect? Everyone? * Is problem affected by human factor? * Is problem caused by operator error-QCD employee error-other error? * Is the problem related to specific controllers' skill? * Is the problem related to special abilities? * Could any specific behaviour cause the problem? * Do only some self-controllers (operators) had this problem? * Is the problem occurred only on some changes? * Do QC department controllers show a problem but self-controllers (operators) not or vice versa?	*All persons who have been in contact with the chamber tightness check are involved in the problem. It is the result of errors of operators, leaders and employees of the quality control department (QCD). It is associated with specific assessment abilities in the scope of reading the requirements given in the process standard. The ability to read and interpret these requirements affects on the effectiveness of the visual inspection. The problem related to the uncertainty of decisions regarding qualification of the product to qualitative categories (OK, nOK) is demonstrated by operators, leaders and employees of the QCD.*
Which	* **Which** trend or pattern does the problem have? * Does the problem happen by	*The problem appears chronically, it occurs practically on every shift (there were errors related to it). The frequency of this problem varies with each shift.*

	accident or does it have a tendency or is it related to something? * How often does the problem occur? * Every hour, every shift, every day, once a month? * Does the problem happen in any particular direction?	
How	* **How** is the state of the product changed from the optimal? * How many times does the problem occur? * What consequences (& quality, cost, safety) are related to the problem?	*The consequence of errors is that the conforming products are considered as nonconforming and vice versa, appearing false alarms and production is stopped , overregulation of the threader and ultimately high value of NQC.*
Revised Description of the Problem: ***Lack of consistency of assessments during visual inspection of the gap size of the C1 chambers, related to the Ist and IInd type of errors, at each stage of the inspection process, in which this characteristic is inspected, related to each appraisers, appearing chronically, generating high costs.***		

After thoroughly defining the examined problem, the root causes of its occurrence were identified and analyzed by the use of the 4M method (an approach to this analysis in the Windsor Assembly in Ontario, Canada was used). In the next step, based on the information from the 5W-1H method, the possible causes of problem in the analysed area have been defined. The most probable causes of the problem were marked with the red frame in "4M box" and transferred to the Ishikawa diagram (Fig. 1). The indicated causes were verified by the 3G type analysis. Next, the causes that had impact to the problem were marked with a red frame on the Ishikawa diagram, after their verification. Other causes have been deleted. The best corrective actions for root causes have been marked (Table 2).

Table 2. 4M Analysis Sheet to Describe the Problem in the Quality Control Process.

Man: *"Training, Experience, Qualification"* *"Knowledge of QC process, standards"* *"Following the Standards/SOP/OPL"*

Potential Root Causes	Possible Corrective Actions
Unfamiliar with Some Elements of the Job	Train of Unfamiliar Elements
	Review Standards/SOP/OPL
Elements of the Job Not Taught	Train the Elements of the Job
	Review Standards/SOP/OPL
Not Trained on Visual Aids/Standard Usage	Train on how to use the visual aids
	Review Standards/SOP/OPL
Not Following Standards/SOP/OPL (All Shifts)	Review Standards/SOP/OPL
	Post Standards/SOP/OPL in Job Station
	Review Flex Charts/Training Matrix
	Implement LPA for Critical Craftsmanship Elements
Is Irresponsible (you can not count on him)	Conversation with the employee
	Coaching
Inadequate Qualifications/Insufficient Experience	Review Flex Charts/Training Matrix
	Participation in Training
	Using other forms of improving qualifications (study, others)
Does not Maintain Good Interpersonal Relations	Conversation with the employee
	Coaching
	Team building (integration games, trip, discussions, exercises)
Poor personal mental/physical state	Review the medical examinations
	Referral for medical examination
	Analysis corelletion between wrong physical/mental state & work environment
	Ensuring safe and hygienic working conditions
Other cause:	

Method: *"Craftsmanship - The Knack"* *"Best Inspection Sequence or Standard"* *"Inspection Complexity", "Inspection Condition"*

Potential Root Causes	Possible Corrective Actions
Process Standard is Incorrect	Specify Best Inspection Sequence
	Specify Best Inspection Tool
	Specify Best Way to Use the Tool
	Update Standard/SOP/OPL
	Implement LPA for Critical Craftsmanship Elements
Inspection process is complex	Complete Mura Action to Determine Corrective Action
	Rebalance Job Elements
	Implement Downstream Validation
	Implement Mistake Proofing
Inspection method is not effective (does it ensure that only a good product is handed over to the next operation)	Evaluate the effectiveness of the inspection process (MSA approach) to Determine Corrective Action
Temperature and humidity are not appropriate	Measure the Level of Temperature and Humidity and Physical Ailments Associated with It to Determine Corrective Action
Lighting and ventilation are not adequate	Measure the Level of Lighting and Ventilation and Physical Ailments Associated with It to Determine Corrective Action
Work station is not ergonomic	Evaluate Physical Discomfort During the Inspection to Determine Corrective Action
Other cause:	

Material/ Nonconformities: *"Complexity"* *"Difficult to Detect"* *"Known Location"*

Potential Root Causes	Possible Corrective Actions
Errors related to the type of nonconformities detected	~~Train the Unfamiliar Type of Nonconformities~~
	Review Standards/SOP/OPL/QC instruction
	Post Standards/SOP/OPL/QC instruction in Job Station
Errors related to the number of detected nonconformities	Verify Operator's Inspection Station and Its Ergonomic Correlates to Effectiveness of Nonconformities Detection
	Train on how in effective way detect particular type of nonconformities
	Update Standard/SOP/OPL/QC instruction
Errors related to the unknown location of nonconformities	Train the Unfamiliar Location of Nonconformities
	Review Standards/SOP/OPL/QC instruction
	Post Standards/SOP/OPL/QC instruction in Job Station
Other cause:	

Management: *"Preasure"* *"Motivation"* *"Commitment"*

Potential Root Causes	Possible Corrective Actions
The pressure to results	Verify the bonus system
	Establish new/improved bonus system
Managers do not help in problems, they only judge	Trainings
	Change the rules of managers' assessment
Distrust of managers	Verify managers' participation in solving problems in gemba
	Implement Ohno circle/gemba gembutsu approach
Inspection station not at basic condition	Verify standard work correctly
	Implement 5S solution
	Evaluate Lighting, Noise level, and temperature of the area
Poor motivation of employees	Verify level of motivation
	Coaching
	Verify the financial rewards, bonuses system
	Verify non-wage and non-cash means of motivating
	Establish new/improved motivation system
Problems related to the production/inspection process are not visible to everyone (especially for managers)	Implement visual management solution (information board, Andon, visual aids)
Other cause:	

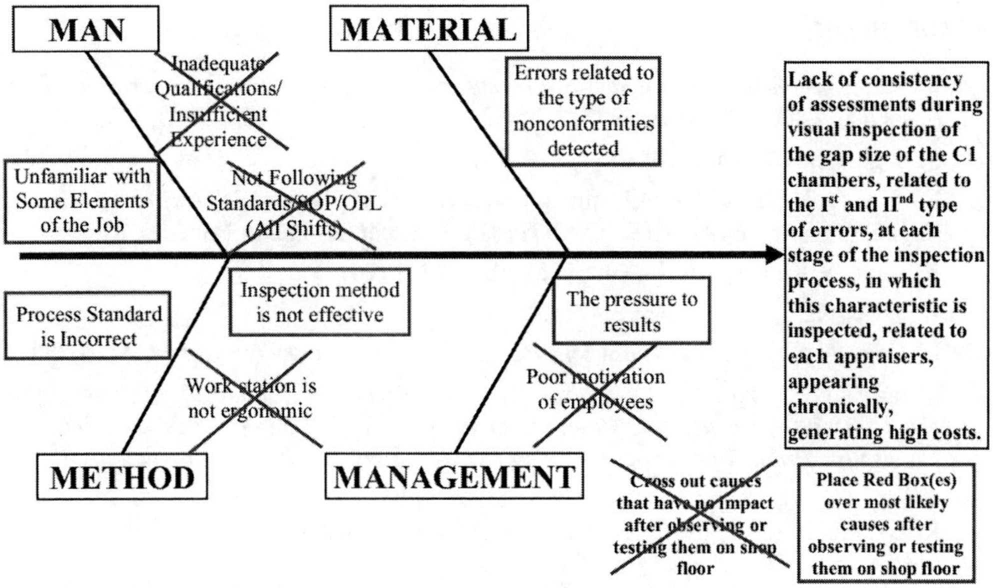

Fig. 1. Ishikawa diagram to analyse the problem with the visual inspection process.

4 Conclusions

The analysis carried out using the 5W1H and 4M methods showed that the problem with the lack of coherence of gaps assessments is on the side of imprecise documentation describing this activity. Hence, employees making such control interpreted the size of the gaps differently. During the investigation (as part of the MSA procedure) it was noticed that not all operators used feeler gauges in the case of doubts about the permissible size of gaps, and those who did it, used it in different ways. Some of them inserted a feeler gauge gently, and when the resistance appeared, they stopped further evaluation and others tried to move the gauge over the entire length of joining the housings. It was shown that the leaders of the

Compact line were more restrictive to the control, but this was the result of fear of rejection of dubious chambers on the line of main assembly. In turn, the leaders of the chamber lines approached the control more liberally, i.e. more chambers were not allowed for further production than it should. Interviews conducted among the appraisers showed that a large group of people could not precisely specify the distances in which the 0.05 mm feeler gauge should stop. In addition, the rejection of questionable chambers also resulted from the fear of rejecting them at the next stage of the control. Passing the defective product and detecting this fact in the next stage resulted in lowering the value of the quarterly bonus. It was easier for all appraiser groups to consider the dubious chamber as nonconforming and dismantle it, generating losses on the NQC.

In order to solve the problem, a visual pattern was created with marked zones in which a section of at least 2 mm in length with no gap was to be found and all people were trained out to use it. There have been changes in the standard documentation used during a visual inspection, a more clarifying and explaining "contentious issues" with the use of visual elements (images). Also, persons responsible for the replacement of used feeler gauges for new ones were established. The corrective actions introduced have brought the intended effect – the number of errors and the amount of related costs have been significantly reduced.

References

1. R. G. Rosandich, *Intelligent Visual Inspection. Using artificial neural networks*, (Chapman & Hall, London, 1997)
2. S. Borkowski, K. Knop, MPER **7**, 3 (2016)
3. K. Knop, M. Ingaldi, M. Smilek-Starczynowska, *Lecture Notes in Mechanical Engineering, Advances in Manufacturing* (Springer Inter. Pub., Cham, 2018)
4. K. Vogt, A. Kujawińska, Inżynieria Maszyn, **18**, 1 (2013)
5. A. Kujawińska, K. Vogt, MPER **6**, 2 (2015)
6. K. Vogt, A. Kujawińska, *Innowacje w zarządzaniu i inżynierii produkcji*, **2** (2014)
7. M. Diering, K. Dyczkowski, *International Conference on Industrial Engineering and Engineering, Assesing the Raters Agreement in the Diagnostic Catheter Tube Connector Production Process Using Novel Fuzzy Similarity Coefficient*, (IEEM, 2016)
8. R. Nakajima, K. Shida, T. Matsumoto, APMS 2013, Part I, IFIP AICT 414, 29-39 (2013)
9. B. Starzyńska, A. Hamrol, Tot. Qual. Man. and Bus. Exc. **24**, 5-6, 577-595 (2013)
10. A. Pacana, R. Ulewicz, Pol. J. of Manag. Stud. **16**, 1 (2017)
11. M. Nowicka-Skowron, R. Ulewicz, *METAL 2015, 24th International Conference on Metallurgy and Materials, Quality management in logistics processes in metal branch*, in (Tanger Ltd., Ostrava, 2015)
12. A. Maszke, Production Engineering Archives **17** (2017)
13. S. Borkowski, M. Ingaldi, *METAL 2015, 24th International Conference on Metallurgy and Materials, Brno, 2015)*
14. J. Trojanowska, K. Zywicki, E. Pajak, *Information Technologies in Environmental Engineering: new trends and challenges* (Springer Verlag, 2011)
15. F. De Felice, A. Petrillo, S. Monfreda, *Operations Management* (Intech, 2013)

Application of function of quality losses G. Taguchi for study of roller burnishing parameters

Andrzej Pacana[1,*]

[1]Rzeszow University of Technology, The Faculty of Mechanics and Technology, al. Powstancow Warszawy 8; 35-959 Rzeszow, Poland

Abstract. The results of roller burnishing parameters analysis with self-acting feed are presented in this paper. Method of parameter process design based on the quality losses G. Taguchi function was applied. The following controlled parameters were assumed: Ra – roughness parameter before roller burnishing process, w - interference between a roller and a burnished sample. k – multiplication factor of each sample point deformation, V – burnishing velocity. The following factors were assumed as process disturbances: treated material hardness and lubrication of burnished surface. Basing on experimental results it was found that strongest influence (in tested range of parameters variations) on roller burnished surface roughness has initial surface roughness and burnishing velocity. These factors also showed have the smallest resistance to process disturbances.

1 Theoretical introduction

Basing on the achievements of mathematical statistics and classical methods of experimental design, Taguchi developer his own methods of experimental design and results analysis [1-4]. Taguchi created among the others theory of quality losses function based on experimental design. Decrease in the number of experiment with respect to traditional experiment is the fundamental advantage of Taguchi's approach. However selection of suitable factors combination causes controversy.

The roller burnishing is one of the most often used methods of burnishing of cylindrical outer surfaces. Usually it is used for cold hardening. Smoothness burnishing is the other application. This kind of burnishing is applied when required shape and dimensional accuracy was obtained in previous treatment and very good smoothness of burnished surface is required. Smoothness roller burnishing is used to finish machine parts in which grinding pollutions in surface layer are unacceptable or for elements which can't be finished with too strong forces. For mentioned above elements ball burnishing or slide diamond burnishing are usually used. However these finishing methods are characterized by very small productivity and whenever possible the roller burnishing should be used because of its this higher productivity. In order to obtain higher shape-dimensional

* Corresponding author: app@prz.edu.pl

accuracy the roller burnishing should be applied with rigid pressing of tools (rollers) [5, 6, 16-18].

2 Description of the research problem

The information mentioned in the introduction and observed in last years technology the tendency to eliminate of abrasive machining methods for machine parts operating under sliding friction conditions (because of disadvantageous surface layer state after abrasive machining) caused that problems connected with design and manufacture of device for shaft roller burnishing are now present-day problems in Department of Manufacturing Processes and Production Engineering in Rzeszow University of Technology. This device makes smoothness shafts burnishing in automatic machining line possible, with the analysis of process parameters effects and parameters optimisation [7-15, 19, 20].

This paper presents attempt of determination of machining process selected parameters influence on achieved surface roughness of cylindrical outer elements after roller burnishing.

3 The analysis of rolling burnishing parameters with Taguchi method application

In order to analyse of the rolling burnishing process (described in references [5, 7] using Taguchi method the following control factors were selected:
1) Ra – surface roughness parameter before roller burnishing (μm),
2) w – depth of tools (rolls) interference into burnished surface (μm),
3) k – deformation multiplicity of each point of burnished surface,
4) V – burnishing velocity (rev/min)/

As a disturbances the following factors were selected:
1) HRC – hardness after quenching and tempering,
2) lubrication before roller burnishing (samples were lubricated and no lubricated). Cylindrical turned specimens of about 90 mm length were made from 1.503 steel. The number of repetitions was three. Diameter of specimen was measured in two perpendicular directions using micrometer and basing on the average value of both results the interference of tool was calculated.

Surface roughness was measured by Surtronic 3+ surface measurement gauge. High-pass filtering was done with 0,8 mm cut-off. The assessment length was 4 mm. Surface roughness R_a parameter of each specimen before and after roller burnishing was calculated in the same place as arithmetic average from 4 measurements in axial and circumferential shaft directions, in centre part of specimen.

The roller burnishing tests were realized on special centreless roller burnishing machine constructed and produced in Chair of Manufacturing Processes and Production Organization of Rzeszow University of Technology. This machine is characterized following features: is assigned to finish burnishing of shafts, turn axis of rollers is inclined in relation to treated shaft axis and pressing of rolls is realized by rigid method. The machine is detailed described in [5].

Controlled (input) parameters signatures were as follows: A = R_a, B = w, C = k, D = V and disturbance factors signatures were as follows: X = HRC and Y = lubrication. Theoretical upper values of parameters (sign. +1 or MAX) and lower values (-1 or MIN) are presented in Tab. 2. The effect of roller burnishing parameters ob surface roughness was determined basing on Taguchi method design of experiment. Table 2 shows plan and obtained results of experiment. Arithmetical mean value of the Ra parameter was denoted as Yi and calculated from 3 preceding columns, however each from these results is arithmetical mean from 20 measurements (5 cut-offs x 4 circumferential measurement).

Table 1. Maximum and minimum parameters values.

PARAMETER	R_a [μm]	w [μm]	k [-]	V [r.p.m.]	HRC	Lubrication
	A	B	C	D	X	Y
MAX (+1)	5.7	105	16	150	50	YES
MIN (-1)	1.5	11.5	4	50	32	NO

Table 2. The results of roughness measurement and their average values (Yi) and values of SN index.

					X	-1	+1	+1		
	A	B	C	D	Y	-1	-1	+1	Y_i	SN_i
1.	-1	-1	-1	-1		0.90 0.67 0.79 *0.787*	1.73 1.23 1.03 *1.330*	1.56 1.42 1.24 *1.407*	1.175	-1.632
2.	-1	-1	-1	+1		1.11 0.93 0.82 *0.953*	0.76 0.57 0.41 *0.580*	0.78 0.46 0.51 *0.583*	0.705	2.772
3.	+1	+1	-1	-1		2.64 2.54 2.40 *3.175*	1.05 1.08 0.99 *1.040*	0.95 1.11 0.94 *1.000*	1.738	-6.079
4.	+1	-1	+1	-1		5.39 5.42 5.40 *5.400*	2.05 2.09 1.98 *2.040*	1.95 2.13 2.00 2.027	3.156	-10.961
5.	-1	+1	+1	+1		0.28 0.25 0.27 *0.267*	0.37 0.51 0.53 *0.470*	0.47 0.61 0.51 *0.530*	0.422	7.189
6.	+1	+1	+1	+1		0.36 0.37 0.55 *0.427*	1.73 1.23 1.03 *1.330*	1.32 1.13 1.09 *1.180*	0.979	-0.471

The values of SN index for features: „the less finally value the better" were calculated from formula [1]:

$$SN_i = -10 \cdot \log\left(\frac{1}{n}\sum y_{ij}^2\right) \tag{1}$$

As the results of calculation of each parameters influence on surface roughness after roller burnishing, basing in presented in Tab. 1 experiment plan and formulae according to [1, 3], the following equations were obtained:

$$Y(A) = \frac{Y_3 + Y_4 + Y_6}{3} - \frac{Y_1 + Y_2 + Y_5}{3} = 1{,}958 - 0{,}767 = 1{,}191 \tag{2}$$

$$Y(B) = \frac{Y_3 + Y_5 + Y_6}{3} - \frac{Y_1 + Y_2 + Y_4}{3} = 1{,}046 - 1{,}679 = -0{,}633 \tag{3}$$

$$Y(C) = \frac{Y_4 + Y_5 + Y_6}{3} - \frac{Y_1 + Y_2 + Y_3}{3} = 1{,}519 - 1{,}206 = 0{,}313 \tag{4}$$

$$Y(D) = \frac{Y_2 + Y_5 + Y_6}{3} - \frac{Y_1 + Y_3 + Y_4}{3} = 0{,}702 - 2{,}023 = -1{,}321 \tag{5}$$

However the resistance of particular parameters on process disturbances for selected levels was as follows:

$$SN(A) = SN(+A) - SN(-A) = \frac{SN_3 + SN_4 + SN_6}{3} - \frac{SN_1 + SN_2 + SN_5}{3} = \\ = -17{,}511 - 8.330 = -25{,}841 \tag{6}$$

$$SN(B) = SN(+B) - SN(-B) = \frac{SN_3 + SN_5 + SN_6}{3} - \frac{SN_1 + SN_2 + SN_4}{3} = \\ = 0{,}639 - (-9{,}820) = 10.459 \tag{7}$$

$$SN(C) = SN(+C) - SN(-C) = \frac{SN_4 + SN_5 + SN_6}{3} - \frac{SN_1 + SN_2 + SN_3}{3} = \\ = -4{,}243 - (-4{,}938) = 0{,}695 \tag{8}$$

$$SN(D) = SN(+D) - SN(-D) = \frac{SN_2 + SN_5 + SN_6}{3} - \frac{SN_1 + SN_3 + SN_4}{3} = \\ = 9{,}490 - (-18{,}671) = -28.161 \tag{9}$$

 https://doi.org/10.1051/matecconf/201818303007

Figure 1 shows graphical presentation of the results.

a) b)

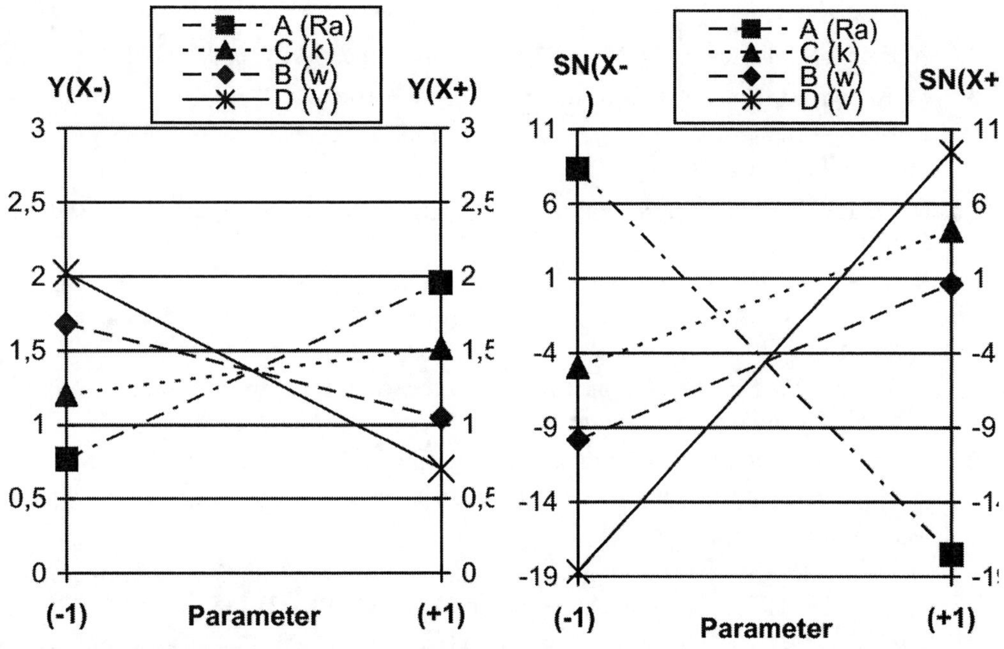

Fig. 1. The dependencies between surface roughness after roller burnishing and analysed parameters (a) and its resistance SN for process disturbances (b).

4 Conclusion

The following conclusions can be drawn:

- For tested change area of parameters burnishing velocity (V) and previous surface roughness (R_a) strongly effect surface roughness after roller burnishing. The smaller initial surface roughness the smaller roughness after roller burnishing. The higher burnishing velocity the smaller surface roughness height after roller burnishing. In this case, it is difficult to determine reason of this phenomenon
- The deeper depth of interference the smaller surface roughness height after roller burnishing.
- Substantial influence of deformation multiplicity on surface roughness height after roller burnishing was not observed.
- Previous surface roughness height and burnishing velocity show the least resistance for process disturbances

The conclusions obtained basing on carried out tests, relate only to presented technological conditions and cannot be the base general conclusions. But they can be the fundamentals for future tests of roller burnishing process taking into account the effect of other constructional-technological parameters (for example R, R/r) or burnished materials properties.

References

1. A. Hamrol, W. Mantura, *Zarządzanie jakością. Teoria i praktyka,* (PWN, Warszawa-Poznań, 2008)
2. A. Gazda, A. Pacana A., D. Malindžak, Przemysł Chemiczny, **6**, 980 (2013).
3. M. Korzyński, *Metodyka eksperymentu*, (WNT, Warszawa, 2006)
4. R. Ulewicz, *International Conference on the Path Forward for Wood Products: A Global Perspective* (Baton Rouge, 2016)
5. M. Korzyński, A. Pacana, Journal of Materials Processing Technology **210,** 1217 (2010).
6. S. Aysun, *Analysis and optimization of surface roughness in the ball burnishing process, using response surface methodology and desirability function, Advances in Engg.* (Software, 2011)
7. A. Pacana, W. Koszela, *Nagniatarka do bezkłowego krążkowania wałków, Materiały VII KNT Technologia obróbki przez nagniatanie* Bydgoszcz, 2000)
8. R. Wolniak, B. Skotnicka-Zasadzień, Metalurgija, **4**, 709 (2014)
9. H. Luo, J. Liu, L. Wang, Q. Zhong, The International Journal of Advanced Manufacturing Technology, **28**, 707 (2006)
10. N. Kumar, S. Anish, L. pal Singh, H. Tripathi, J. Liu, L. Wang, Q. Zhong, International Journal of Machining and Machinability of Materials, **18(1/2)**,185 (2016)
11. P.R. Babu, K. Ankamma, K., T.S. Prasad, A.V.S. Raju, N.E. Prasad, International Journal of Engineering Research and Applications, **2**, 2,1139 (2012)
12. N.H. Loh, S.C. Tam, Precision Engg., **4**, 215 (1988)
13. R. Ulewicz, J. Selejdak, S. Borkowski, M. Jagusiak-Kocik, *Metal 2013, 22nd International Conference on Metallurgy And Materials (*Tanger Ltd, 2013)
14. M. Nowicka-Skowron, R. Ulewicz, *Metal 2015, 24TH International Conference on Metallurgy And Materials*, (Tanger Ltd, 2015)
15. A. Pacana, Production Engineering Archives, **16**, 28-31,(2017)
16. J. Pietraszek, A. Gądek-Moszczak, T. Torunski, Adv. Mat. Res.-Switz. **874**, 139 (2014)
17. N. Radek, J. Pietraszek, B. Antoszewski, Adv. Mat. Res.-Switz. **874**, 29 (2014)
18. J. Pietraszek, A. Szczotok, N. Radek, Arch. Metall. Mater. **62**, 235 (2017)
19. A. Dudek, B. Lisiecka. R. Ulewicz, Arch. Metall. Mater. **62,** 281-287 (2017)
20. T. Lipinski, A. Wach, *23rd International Conference on Metallurgy And Materials, 23rd International Conference on Metallurgy And Materials* (2014)

Instruments of identification of hydraulic components potential failures

Joanna Fabiś-Domagała[1,*], *Grzegorz* Filo[2], *Hassan* Momeni[1], and *Mariusz* Domagała[1]

[1]Cracow University of Technology, Institute of Applied Informatics, Al J. Pawla II 37, 31-841 Cracow, Poland
[2]Western Norway University of Applied Sciences, Department of Mechanical and Marine Engineering, Inndalsveien 28, 5063 Bergen, Norway

Abstract. This paper presents an analysis of potential failures of hydraulic components on the example of a hydraulic gear pump with the use of instruments of quality management. The "five why" method was implemented to evaluate failures as well as the bases for the causes which have an influence on pumps failures. The causes are connected with operation condition of gear pumps and the quality of working fluid. In the next step, by using a systematic diagram, appropriate improvement actions for the identified causes were determined.

1 Introduction

Product defects are almost unavoidable and may occur during the entire life cycle. They may appear already at the design stage or later during operation. When a defect is detected, the most common approach is undertaking an improvement actions, removing incompatibilities and implementing appropriate corrective actions. However, such actions do not include processes of identification of causes of defects, which may lead to defects occurring again. There are various methods of the identification of possible defects, among which quality management methods, such as systematic diagram (decision tool) and the "five why" method might be used. These techniques help to acquire information about a possible (or occurred) defect in a short time and to undertake appropriate improvement actions. They can be used as an independent tool to identify causes and effects of defects or support other methods of quality management including FMEA, KAIZEN or the Ishikawa diagram.

2 The "five why" method principles

The "five why" method is one of the methods allowing the identification of the causes of the problem/defect. The method was developed by a Japanese industrialist and inventor Sakichi Toyoda and was used for the first tie in the automotive industry at Toyota. Currently, this method is known all over the world and can be used regardless of the type of business. The "five why" method is one to ask "why" five times, or, if needed, so many

* Corresponding author: fabis@mech.pk.edu.pl

times that the cause of the problem is clearly defined, understood and corrective and preventive actions are undertaken. However, quite often the answer to the fifth question determines the cause of the problem. Therefore, this method was named "five why" and it provides the answer to three basic questions [1]:

- why did the problem/defect occurr?
- why wasn't the problem/defect detected?
- how to solve it? - possible application of the systematic diagram.

The analysis of a cause-problem conducted by the "five way" method requires following a few principles:

- correct formulation of the problem/defect to formulate correct questions,
- preserve the sequence of cause and effect relationships,
- each answer is the basis for the next "why" question,
- "why" questions should be asked until the source of defect/problem is detected, the removal of which will eliminate the problem/defect completely so that a similar situation does not happen again in the future,
- "why" questions should be stopped when the answer to the following question are the same or leads to wrong conclusions.

Most often, the "five why" method is used when the problem/defect has already occurred, however, it can also be successfully applied to the analyses aimed at detecting potential problems/defects (those that may still occur). In such a case, preventive measures can be taken or measures that would minimize the problem/defect occurrence by preparing appropriate corrective actions in a proper place and time. The analysis of cause and effect relationships using the "five why" method is conducted in three stages: preparation, analysis and improvement. The first stage of the analysis which is preparation includes:

- Preliminary identification of a problem that has been identified in the object/process.
- Collecting information to create input data for further analysis of the problem, the greater number of collected data, the greater the chance that the cause of the problem will be properly determined.
- Selecting working team and a moderator – employees, experts or clients – anyone who can help determine the causes of the problem correctly. A moderator is a person who controls the whole process, reminds about the problem and monitors the time frame of the analysis.

The second stage of the analysis is a proper analysis, which consists of:

- Naming and describing the problem precisely and finally after the first meeting of the working team. A precisely formulated problem allows the team to focus on important issues and go straight to the root causes of the problem.
- Asking a few questions to help to solve the problem and find the basis for the cause.
- Providing answers - in accordance with the principle of "five why", each answer will be the basis for the following question and for the better understanding of described problem.

When "five why" is completed the final answer is verified whether it is logically related to the problem and if it is based on the principle of cause-and-effect relationships.

The third stage of the analysis is the stage of undertaking improvement actions. If the analysis has dealt with the existing problem the corrective actions are taken. If it is an analysis of potential problems, the preventive actions are formulated and are later monitored. Once the improvement actions are completed, their effectiveness is verified. If the problem still exists despite the implemented actions another "five why" analysis might be necessary to be carried out. The systematic diagram can be used to define improvement actions.

3 Systematic diagram

The systematic diagram is one of the newest quality improvement tools, which is also called a tree diagram, a tree type graph or a decision tree [2]. It is most often used in planning and managing processes in order to predict the consequences of the decisions taken. It can also be used while developing concepts and designing new products to identify possible defects or to improve actions for identified problems. It is therefore, a graphical presentation of necessary activities in a given process or factors affecting the occurrence of a given defect. Additionally, in a logical and chronological way, it describes causes or tasks due to the identified problem/purpose in accordance with the "from general to specific" principle. Fig.1 presents the method of proceeding when analysing potential defects, with the use of a systematic diagram.

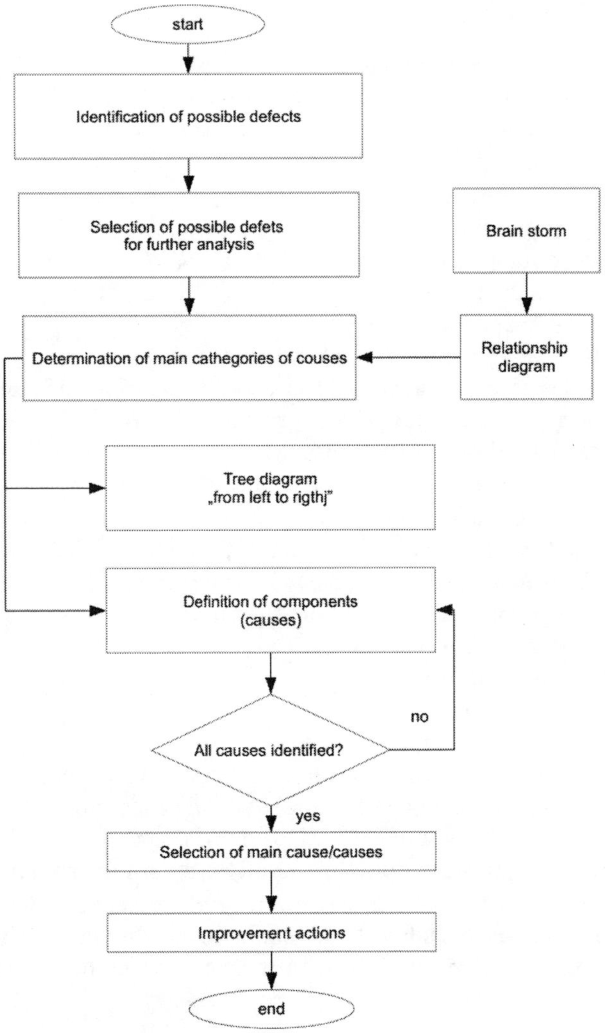

Fig. 1. Algorithm of a systematic diagram.

3 Analysis of possible defects of a hydraulic gear pump

One of the most popular hydraulic rotary pumps is a gear pump. The principle of its operation is based on moving the liquid from the suction port to the discharge port between the tooth spaces. The gears are meshing in such a way that they separate the suction area from the discharge area at the meshing point [3]. There are several types of gear pumps. They can be distinguished by the shape of the teeth or the number of gears. Hydraulic gear pumps consist of such components as: body, a drive shaft with a keyway, a drive gear, a driven gear, a cover, a bearing, sealing rings, screws, a spring washer and a snap ring. The bearing, drive shaft and gears are those components which are responsible for the proper operation and durability of the entire gear pump and, as a result, affect its efficiency and reliability. However, according to the research [4], about 60% of failures are caused by the gears. Owing to the use of the "five why" method and the systematic diagram, the cause-and-effect relationship was created for the problem of "failure of the gear pump" based on the analysis of the wheel gears.

3.1 The "five why" method for a gear pump

Research [5] shows that the wheel gears failure is mainly caused by operational conditions and quality of working fluid. Therefore, the "five why" method was used to identify the main failures in the area of "operation" and "working fluid" for the "gear pump failure" problem. The question-and-answer method provided, in five steps, the information about the source cause in the investigated area.

Problem/effect: gear pump failure (example for operation)

1. Why did the gear pump fail?
   Because the gear is damaged.
2. Why is the gear damaged?
   Because it has been worn out, cracked and corroded.
3. Why has it been worn out, and why did cracks and corrosion appear?
   Because the operating conditions of the gears are incorrect
4. Why are working conditions inappropriate?
   Because there were excessive vibrations and contact pressure on teeth.
5. Why did excessive vibrations and contact pressure have occur?
   Because there is a small contact area between the teeth and the gears are seized.

The reason for the source: inadequate clearance between the gear wheels

Problem/effect: gear pump failure (example for fluid)

1. Why did the gear pump fail?
   Because the gear is damaged.
2. Why is the gear damaged?
   Because it has been worn out, cracked and corroded.
3. Why has it been worn, and why did cracks and corrosion appear?
   Because of poor quality of a working fluid of or improper properties.
4. Why is the quality of the liquid poor or has improper properties?
   Because there are impurities in it or the liquid does not fulfil pump requirements
5. Why is the liquid contaminated or why has it been incorrectly selected?
   There were no filters responsible for the quality of the liquid, poor maintenance.

Reason: filters not used, working fluid with improper properties due to poor maintenance instructions.

3.1 The "five why" method for a gear pump

The next step of the analysis is creating improvement actions for the identified source reasons and the main problem obtained in the "five why" method. These improvement actions are presented in Fig. 2.

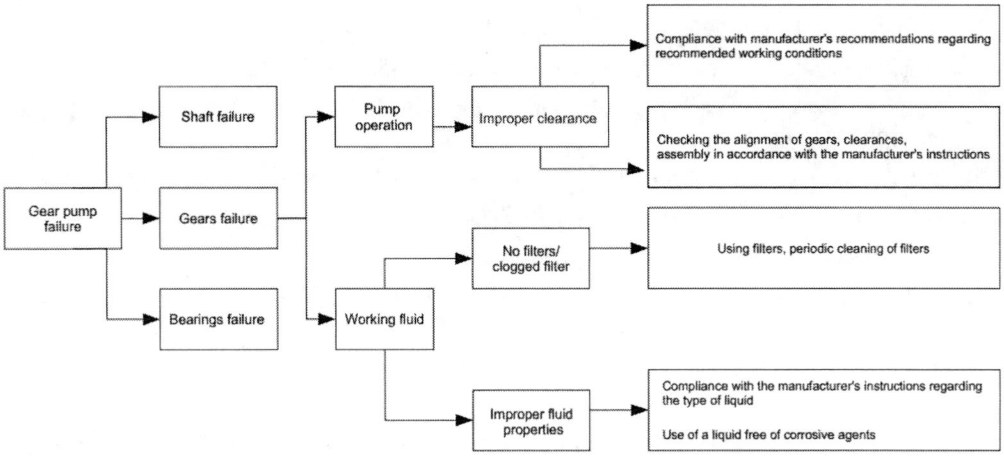

Fig. 2. A systematic diagram for the failure analysis and improvements for a gear pump.

4 Summary

The "five why" method supported by the systematic diagram appeared to be an effective tool for identifying potential problems/defects and related causes. It allows to identify areas and causes of failures and specify necessary improvement actions in a relatively short time. Additionally, the details of improvement actions might be obtained by using other tools of quality management, such as an event flow diagram, which determines the sequence of necessary actions. Going further, it should be noted that this method may be especially useful for hard-to-control processes like e.g. biotechnological [8, 9], where a biomass may have a wide variation, or machining [10, 11], where sources of variability are weakly identified and they may require a specific approach to the uncertainty assessment [12]. It may be also systematically applied in a laboratory scale investigations, if replications have wide variability and it is no idea, what source of variation is a cause e.g. materials science research [13-22], but the greatest impact is achieved at a huge scale enterprises like e.g. large accelerator building [23, 24] or power plants [25]. Such method may be also alternative approach to identify the process factors of the greatest interest in numerical experiments e.g. [26-29] instead of a typical Shainin's approach or factorial screening designs more appropriate for real world experiments with uncontrolled noise factors. Last but not least, it cannot be omitted the applicability of this method in the management area [30-34], where a "human factor" is the greatest noise source.

References

1. J. Liker, G. Trachilis, *Developing Leadership Skills 12: Root Cause – Using The 5 WHY'S* (Lean Leadership Inst., Winnipeg, 2014)
2. J. Żuchowski, E. Łagowski, *Tools and methods for quality improvements* (Wyd. Politechnika Radomska, Radom, 2004)
3. J. Stryczek, *Gear wheels of hydraulic machines* (Wyd. Politechniki Wrocławskiej, Wrocław, 2007)
4. J.D. Limmer, Wear, **241**, 26 (2000)
5. J. Fabis-Domagala, Tribiologia ,**2**, 33 (2017)
6. P. Pietkiewicz, Techn. Sci., **12**, 219 (2009)

7. S. Haidong, L. Zhiqiang, Q. Lele, Q. Liang, Mech. Syst. Signal Proc., **99**, 921 (2018)
8. E. Skrzypczak-Pietraszek, A. Hensel, Pharmazie, **55**, 768 (2000)
9. E. Skrzypczak-Pietraszek, I. Kwiecien, A. Goldyn, J. Pietraszek, Phytochem. Lett., **20**, 443 (2017)
10. J. Pietraszek, *Proc. 11th Int. Conf. on Artificial Intell. and Soft Comp. (ICAISC), Lect. Notes. Artif. Int.* 7267, 310 (Zakopane, 2012)
11. E. Lisowski, G. Filo, Flow Meas. Instrum., **53**, 269 (2017)
12. J. Pietraszek, M. Kolomycki, A. Szczotok, R. Dwornicka, Lect. Notes Artif. Int., **9875**, 260 (2016)
13. T. Pieczonka, J. Kazior, A. Tiziani, A. Molinari, J. Mater. Process. Technol., **64**, 327 (1997)
14. M. Ulewicz, U. Lesinska, M. Bochenska, Physicochem. Probl. Mi., **44**, 245 (2010)
15. D. Klimecka-Tatar, S. Borkowski, P. Sygut, Arch. Metall. Mater., **60**, 735 (2015)
16. M. Ulewicz, E. Radzyminska-Lenarcik, Physicochem. Probl. Mi., **46**, 119 (2011)
17. D. Klimecka-Tatar, G. Pawlowska, M. Sozanska, Arch. Metall. Mater., **60**, 153 (2015)
18. N. Radek, K. Bartkowiak, *Proc. Sixth Int. Wlt Conf. on Lasers in Manufacturing*, Physics Procedia, Vol. 12, 499 (Munich, 2011)
19. A. Szczotok, K. Rodak, *Proc. Conf. Technol. Properties of Modern Utilised Materials (TPMUM)* (Katowice, 2012)
20. R. Ulewicz, P. Szataniak, F. Novy, *Proc. 23rd International Conference on Metallurgy and Materials* (Brno, 2014)
21. P. Szataniak, F. Novy, R. Ulewicz, *Proc. 23rd International Conference on Metallurgy and Materials* (Brno, 2014)
22. R. Dwornicka, N. Radek, M. Krawczyk, P. Osocha, J. Pobedza, *Proc. 26th Int. Conf. on Metall. and Mater. – METAL* (Brno, 2017)
23. B. Singh et al., Eur. Phys. J. A, **52**, 325 (2016)
24. B. Singh et al., Nucl. Phys. A, **954**, 323 (2016)
25. R. Dwornicka, Adv. Mater. Res.-Switz., **874** 63 (2014)
26. T. Styrylska, J. Pietraszek, ZAMM-Z. Angew. Math. Mech., **72**, T537 (1992)
27. I. Dominik, J. Kwasniewski, K. Lalik, R. Dwornicka, *Proc. 32nd Chinese Control Conf.* (Xian, China, 2013)
28. J. Pietraszek, A. Gądek-Moszczak, T. Torunski, Adv. Mat. Res.-Switz., **874**, 139 (2014)
29. J. Pietraszek, E. Skrzypczak-Pietraszek, Adv. Mat. Res.-Switz., **874**, 151 (2014)
30. E. Kozien, *Proc. 20th Int. Sci. Conf. on Economic and Social Dev.* (Prague, 2017)
31. E. Kozien, A. Kozien, *Proc. 23th Int. Sci. Conf. on Economic and Social Dev.* (Madrid, 2017)
32. E. Kozien, M.S. Kozien, *Proc. 26th Int. Sci. Conf. on Economic and Social Dev. – Building Resilient Society* (Zagreb, 2017)
33. E. Kozien, A. Kozien, *Proc. 26th Int. Sci. Conf. on Economic and Social Dev. – Building Resilient Society (ESD)* (Zagreb, 2017)
34. E. Kozień, *Proc. 31st Int. Sci. Conf. on Economic and Social Dev. – Legal Challenges of Modern World* (Split, 2018)

The idea of fuzzy logic usage in a sheet-based FMEA analysis of mechanical systems

Grzegorz Filo[1,*], *Joanna* Fabiś-Domagała[1], *Mariusz* Domagała[1], *Edward* Lisowski[1], and *Hassan* Momeni[2]

[1]Cracow University of Technology, Faculty of Mechanical Engineering, Jana Pawła II 37, 31-864 Cracow, Poland
[2]Western Norway University of Applied Sciences, Inndalsveien 28, 5063 Bergen, Norway

Abstract. The main purpose of the work which was carried out and is presented in this paper was to examine the possibility of using fuzzy logic inference for conducting a risk analysis with the help of a sheet-based Failure Mode and Effects Analysis method (FMEA). At the beginning, the main features of the analysed method were presented, with particular emphasis put on the *Risk*, *Priority* and *Number* parameters. Then, a proposal has been made which suggests using Matlab Fuzzy Logic Toolbox package in order to convert the factors into the form of fuzzy sets and to define rules for fuzzy inference process has been made. Finally, the created fuzzy logic model was used to present an example analysis of a turbocharger failure in the fuzzified form.

1 Introduction

Today, in engineering design, a lot of emphasis in engineering design is put on the reliability of the product. One of the methods used for this purpose is Failure Mode and Effects Analysis (FMEA). It consists in an analysis of the causes and effects of defects, and became one of the fundamental methods used to improve the quality of products, processes and services. It was developed in the United States as a military procedure MIL-P 1629 [1]. Then, in the 1960s, it was used by NASA in the *Apollo Saturn* space flight project. The flight to the moon succeeded, therefore the method was used in other industries, especially in nuclear and aerospace industry. Since 1970 FMEA has become more and more popular in the automotive industry, e.g. Ford Motor Company used FMEA to analyse cars in terms of their safety for the user and the fulfilment of legal requirements. Suppliers of parts and subassemblies for the production of Ford cars were obliged to use the method in accordance with the ISO/TS 16949 standard previously known as QS-9000. Currently, FMEA has a very wide usage and can be applied in many fields, including medicine, metallurgy, food services, engineering, and even public administration. The analysis can be carried out in accordance with the EN 60812:2006 standard [2], and it has been included into the ISO standard [3]. The main goal is to consistently and systematically identify potential defects in products or processes and then eliminate or minimize the risks associated with them. The

* Corresponding author: filo@mech.pk.edu.pl

greatest benefits are brought when the method is used at the initial stage of concept and design, because in this case the costs of defect removal is the lowest.

Several variants of the FMEA method can be distinguished. However, due to the preparation process, the way of conducting and presenting the obtained results, it can be divided into a sheet-based analysis (classical) and a matrix-based analysis (FFDM) [4]. A state of the art summary of the classical FMEA method was published in the form of a review article by Spreafico et al. in 2017 [5]. The review consists in analysing a representative pool of over 200 scientific papers and over 100 patents in order to have an overview of the evolution of the method. It shows a large number of problems which have already been solved using FMEA. However, the interest in the method is still growing. One of the ways of its development is combining the classic FMEA analysis with the fuzzy set theory. A novel modelling approach based on the intuitionistic fuzzy approach was proposed by Tooranlo and Ayatollah in [6]. The approach offers advantages over earlier models since it accounts for degrees of uncertainty in relationships among various criteria or options, specifically when relations cannot be expressed in definite numbers, and provides a tool to evaluate failure modes in the case of insufficient data. Another methodology, which integrates the concepts of similarity value measure of fuzzy numbers and the theory of possibility was developed by Mandal and Maiti [7]. New proposals for the application of fuzzy logic met with great interest of scientific centres, which resulted in creation of many hybrid FMEA methods and models in the fuzzy environment, such as extended MULTIMOORA or AHP, applied e.g. by Fattahi [8] and Hu-Chen [9].

The developed methods have already been used in a number of practical applications. Considering the fields related to mechanical engineering, most of the articles are devoted to the analysis of drives and control systems. Zaifang [10] proposed a fuzzy RPN-based method that integrates the weighted least square method, the method of imprecision and partial ranking method to generate more accurate fuzzy RPNs and ensure robustness against the uncertainty. The method was demonstrated by the design example of a new horizontal directional drilling machine. Xu [11] presented a fuzzy logic based method for the FMEA analysis of a diesel engine turbocharger system, while Renjith [12] carried out fuzzy FMECA (failure mode effect and criticality analysis) of a liquid natural gas storage facility, and Yazdi [13] used a similar FDFMEA method (Fuzzy Developed FMEA) for an aircraft landing system. With regard to the presented state of art, this article concerns creation of a sheet-based FMEA model using fuzzy logic in Matlab environment. The model was used to demonstrate the analysis of an example turbocharger failure.

2 Fuzzy model of a sheet-based FMEA method

In general, risk assessment in the sheet-based FMEA method is carried out using a risk priority number (*RPN*), which can be estimated on the basis of three factors: *Occurrence* (*O*), *Severity* (*S*) and *Detection* (*D*). In the conventional approach, the value of the *RPN* is a simple product of all three factors: $RPN = O \cdot S \cdot D$. The value of each factor is an integer number of range 1 to 10. The analysis is usually carried out in three stages [14]: preparation of the analysis, conducting the analysis, as well as introduction and supervision of improvement activities. The diagram of a complete procedure is shown in Fig. 1.

The application of fuzzy logic requires transformation of integer numbers (crisp values) into degrees of fuzzy set memberships (fuzzy values). For this purpose, first fuzzy sets of all input parameters (*O*, *S*, *D*) as well as outputs (*RPN*) must be defined. The detailed process of a fuzzy logic model creation is presented e.g. by Piegat [15], while its practical application illustrated on the example of a sterilization unit risk analysis is shown by Dağsuyu [16]. Crisp values of the factors with verbal description and names adopted for defining fuzzy sets are given in Table 1.

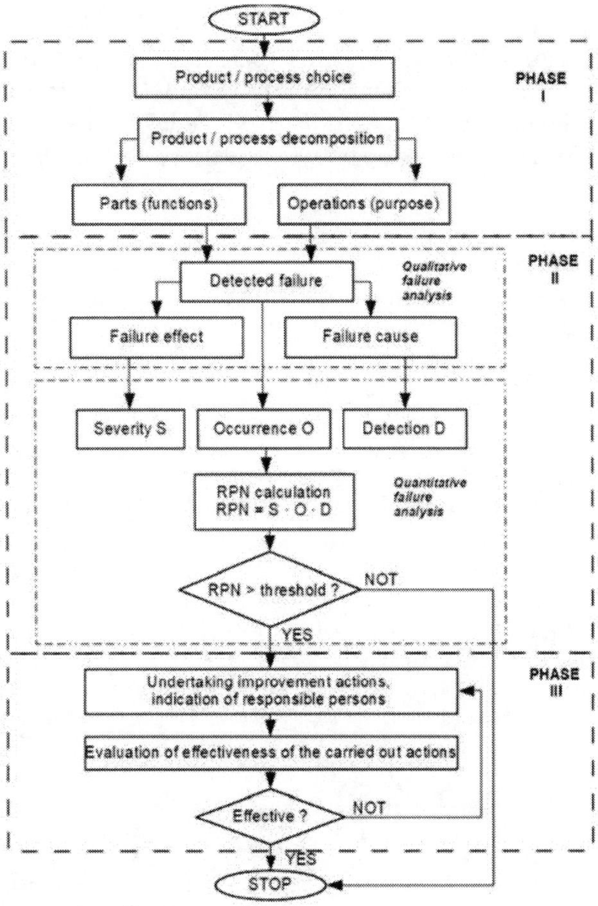

Fig. 1. Process of a conventional FMEA analysis.

Table 1. Crisp values and names of fuzzy sets defined for the *O*, *S*, *D* factors.

Crisp value	Factor					
	Occurrence O		*Severity S*		*Detection D*	
	verbal	fuzzy set	verbal	fuzzy set	verbal	fuzzy set
1	Absent	VL	Unnoticeable	VL	Very high	VH
2,3	Rare	L	Small meaning	L	High	H
4,5,6	Sometimes	M	Moderate	M	Medium	M
7,8	Often	H	Serious	H	Low	L
9,10	Very often	VH	Critical	VH	Very low	VL

According to Table 1, the following fuzzy sets were defined for each input factor: VL (very low), L (low), M (medium), H (high) and VH (very high). Such unification of the fuzzy set names significantly increases consistency of the model. Due to the fact, that the cores (subsets with the maximum value of the membership functions) have the form of intervals in most fuzzy sets (e.g. the core of the fuzzy set L of *Occurrence* factor is in the range from 2 to 3), instead of the commonly used linear-type membership functions [6, 7, 10, 14], the Gauss-type functions were applied in this model, similarly to [11]. The scheme of the model created in *Anfis* editor is presented in Fig. 2. The diagram of fuzzy sets defined for the *Occurrence O* input factor is shown in Fig. 3, while the diagram of *Detection D* is presented in Fig. 4. The *Severity* diagram is omitted, since it has the similar form to *Occurrence*. The *RPN* failure classification factor has the crisp value from the range of [1,

1000]. It was divided into seven fuzzy sets, using a similar membership function set as introduced in [12], respectively: VL (very low, core less than 10), ML (medium low, core between 20 and 30), LM (low medium, core 90-110), MM (medium medium, core 180-220), HM (high medium, core 330-370), MH (medium high, core 460-540) and VH (very high, core 750-1000). The diagram of fuzzy sets defined for the *RPN* factor is shown in Fig. 5.

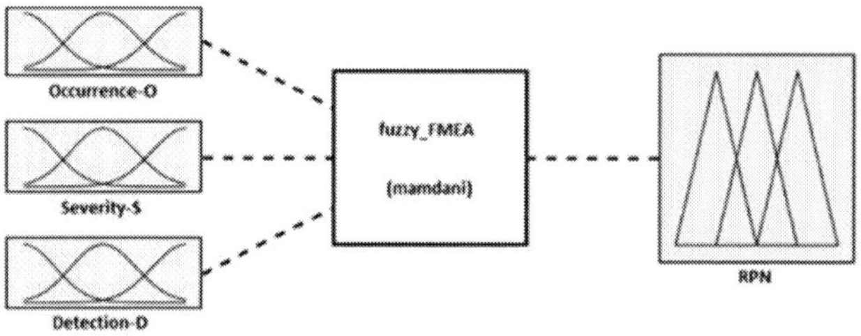

Fig. 2. Scheme of fuzzy FMEA model created with Matlab *Anfis* editor.

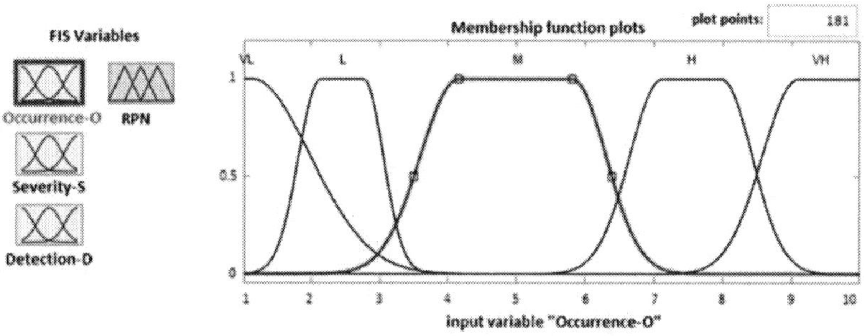

Fig. 3. Fuzzy sets defined for *Occurrence* factor using Gauss-type membership functions.

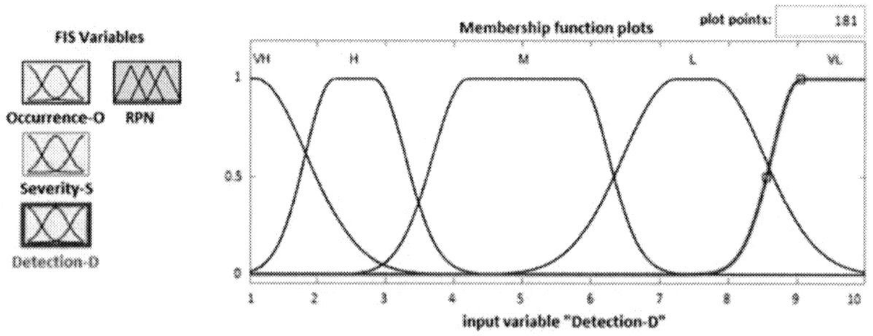

Fig. 4. Fuzzy sets defined for *Detection* factor using Gauss-type membership functions.

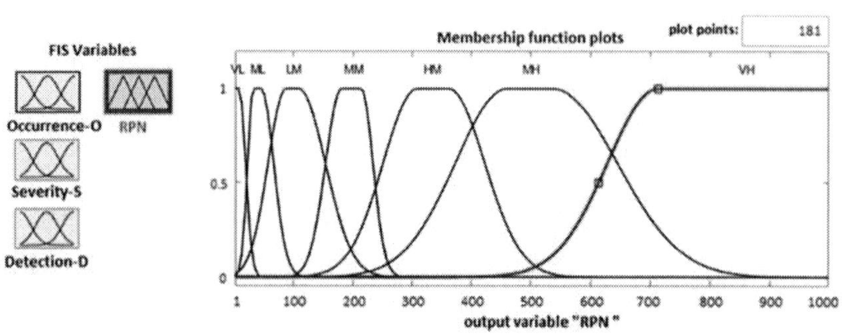

Fig. 5. Fuzzy sets defined for *RPN* factor using Gauss-type membership functions.

In the next step, fuzzy mathematical operators as well as the defuzzification method required by the inference process were determined. Based on the analysis of previous research, the following parameters have been selected: fuzzy product operator: *prod*, fuzzy sum operator: *probor*, the defuzzification method: *Centroid*. Finally, a database of rules which represent connections between the fuzzy sets of input factors and *RPNs* has been created. The database consists of more than 100 rules in the form of *if-then* statements.

3 Practical application of created fuzzy FMEA model

As a practical example will be presented application of the proposed fuzzy model in the analysis of a selected turbocharger defect, resulting in decrease in engine power [17]. The visible symptoms of the fault were noise and vibrations, hence the following factor values were assigned: $S = 7$, $O = 8$, $D = 4$. The fuzzy system gives in this case the *RPN* value of 500 (Fig. 6), which means the problem is considered to be serious (the most activated fuzzy set of RPN is VH).

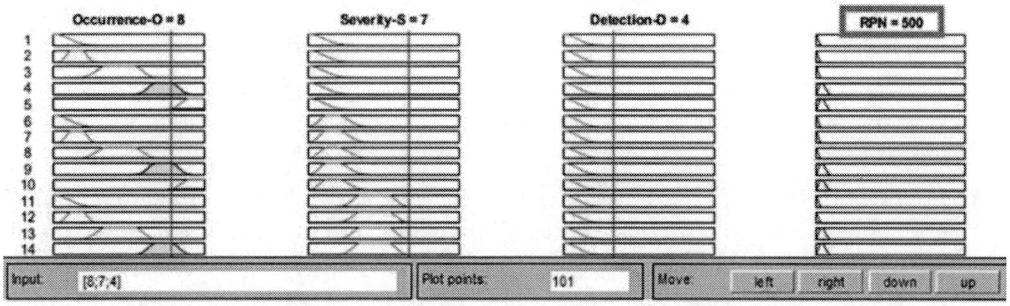

Fig. 6. Fuzzy FMEA system output for S = 7, O = 8, D = 4.

After the repair, which consisted of replacing the air filter, values of the *Severity* and *Occurrence* factors changed respectively to: $S = 5$, $O = 6$. The *RPN* generated by the fuzzy model in this case has been reduced to RPN = 190, which means the highest activation of the LM and MM fuzzy sets.

4 Conclusion

This paper has presented a proposal to use Matlab Fuzzy Logic Toolbox to carry out a sheet-based FMEA analysis. A fuzzy model was created in Anfis editor and then was used to analyse a selected turbocharger defect. The obtained *RPN* values are higher than those from the classical FMEA method [17]: RPN = 224, and RPN = 120, before and after the repair, respectively. However, in both cases the *RPN* value was approximately halved.

Regardless of the specific described application, similar approach may be used to manage the risk in interdisciplinary long-term research like e.g. biotechnological investigation [18, 19] or materials science industry-oriented investigations [20-32]. Similar problems are also met in strict management area of the industry and the academia e.g. [33-37]. Another related issues worth investigating more closely are studies on non-classically estimated fuzzy uncertainty e.g. finite element combined with time series [38, 39], a fixed-effect approach [40] or a bootstrap simulations [41, 42]. Further investigations intend fine-tuning of the created model parameters and using it to analyse more complex systems.

References

1. USA military standard, *Procedure for Performing a Failure Mode, Effects and Criticality Analysis* (MIL-P 1629, USA, 1949)
2. EN 60812:2006 standard, *Analysis techniques for system reliability. Procedure for failure mode and effects analysis (FMEA)*, (2006)
3. ISO 9001:2015 standard, *Quality Management Systems – Revision*, (2015)
4. J. Fabiś-Domagała, Tribologia, **2**, 33-38 (2017)
5. C. Spreafico, D. Russo, C. Rizzi, Comp. Science Rev., **25**, 19-28 (2017)
6. H. S. Tooranlo, A. Ayatollah, Appl. Soft Comput. **49**, 238-247 (2016)
7. S. Mandal, J. Maiti, Expert Syst. Appl. **41**, 3527-3537 (2014)
8. R. Fattahi, M. Khalilzadeh, Saf. Sci., **102**, 290-300 (2018)
9. L. Hu-Chen, F. Xiao-Jun, L. Ping, C. Yi-Zeng, Eng. Appl. Artif. Intell., **34**, 168-177 (2014)
10. Z. Zaifang, C. Xuening, Expert Syst. Appl., **38**, 206-214, (2011)
11. K. Xu, L.C. Tang, M. Xie, S.L. Ho, M.L. Zhu, Reliab. Eng. Syst. Saf. **75**, 17-29 (2002)
12. V.R. Renjith, M.J. Kalathil, P. H. Kumara, D. Madhavan, J. Loss Prev. Process Ind., (2018)
13. M. Yazdi, S. Daneshvar, H. Setareh, *An extension to Fuzzy Developed Failure Mode and Effects Analysis (FDFMEA) application for aircraft landing system* (Safety Science, 98, 2017)
14. A. Hamrol, *Management and quality engineering* (PWN, Warszawa, Poland, 2017)
15. A. Piegat, *Modeling and fuzzy control (*AOW EXIT, Warszawa, Poland, 1999)
16. C. Dağsuyu, E. Göçmen, M. Narl,A. Kokangül, Comput. Ind. Eng., **101**, 286-294 (2016)
17. J. Fabiś-Domagała, *Technical Trans.,* **111(4)**, 11-17 (2014)
18. E. Skrzypczak-Pietraszek, J. Pietraszek, Chem. Biodivers., **11**, 562 (2014)
19. E. Skrzypczak-Pietraszek, I. Kwiecien, A. Goldyn, J. Pietraszek, Phytochem. Lett., **20**, 443 (2017)
20. T. Pieczonka, J. Kazior, A. Tiziani, A. Molinari, J. Mater. Process. Technol., **64**, 327 (1997)
21. A. Dudek, C. Kolana, Solid State Phenom.**,** **165**, 25 (2010)
22. M. Ulewicz, U. Lesinska, M. Bochenska, Physicochem. Probl. Mi., **44** 245 (2010)
23. D. Klimecka-Tatar, S. Borkowski, P. Sygut, Arch. Metall. Mater., **60**, 735 (2015)
24. M. Ulewicz, E. Radzyminska-Lenarcik, Physicochem. Probl. Mi., **46**, 119 (2011)
25. D. Klimecka-Tatar, G. Pawlowska, M. Sozanska, Arch. Metall. Mater., **60** 153 (2015)
26. R. Ulewicz, P. Szataniak, F. Novy, *Proc. 23rd International Conference on Metallurgy and Materials* (Brno, 2014)
27. P. Szataniak, F. Novy, R. Ulewicz, *Proc. 23rd International Conference on Metallurgy and Materials* (Brno, Czech Republic, 2014)
28. A. Szczotok, Materialwiss. Werkstofftech., **46**, 320 (2015)
29. A. Szczotok, J. Nawrocki, J. Pietraszek, Arch. Metall. Mater., **62**, 587 (2017)
30. R. Dwornicka, N. Radek, M. Krawczyk, P. Osocha, J. Pobedza, *Proc. 26th Int. Conf. on Metall. and Mater. – METAL* (Brno, Czech Republic, 2017)

31. J. Pobedza, A. Sobczyk, Key. Eng. Mater., **542**, 143 (2013)
32. M. Niemczewska-Wojcik, W. Pielzoszewski, Arch. Civ. Mech. Eng., **17**, 344 (2017)
33. Kozien, E., *Proc. 20th Int. Sci. Conf. on Economic and Social Dev. (ESD)* (Prague, Czech Republic, 2017)
34. Kozien, E., Kozien, A., *Proc. 23th Int. Sci. Conf. on Economic and Social Dev. (ESD)* (Madrid, Spain, 2017)
35. Kozien, E., Kozien, M.S., *Proc. 26th Int. Sci. Conf. on Economic and Social Dev. – - Building Resilient Society (ESD)* (Zagreb, Croatia, 2017)
36. Kozien, E., Kozien, A., *Proc. 26th Int. Sci. Conf. on Economic and Social Dev. – - Building Resilient Society (ESD)* (Zagreb, Croatia, 2017)
37. E. Kozień, *Proc. 31st Int. Sci. Conf. on Economic and Social Dev. – Legal Challenges of Modern World (ESD)* (Split, Croatia, 2018)
38. I. Dominik, J. Kwasniewski, K. Lalik, R. Dwornicka, *Proc. 32nd Chinese Control Conf. (CCC)* (Xian, China, 2013)
39. E. Augustyn, M.S. Kozien, Acta Phys. Pol. A, **125**, A164 (2014)
40. J. Pietraszek, A. Gądek-Moszczak, T. Torunski, Adv. Mat. Res.-Switz., **874**, 139 (2014)
41. J. Pietraszek, A. Gadek-Moszczak, Solid State Phenom., **197**, 162 (2013)
42. J. Pietraszek, M. Kolomycki, A. Szczotok, R. Dwornicka, Lect. Notes. Artif. Int., **9875**, 260 (2014)

Improvement of quality by using an integrated management system of construction processes

Andrzej Sobczyk [1*], *Janusz* Pobędza[1], and *Małgorzata* Sobczyk[2]
[1]Cracow University of Technology, Mech. Eng. Department, Al. Jana Pawla II 37, 31-864 Poland
[2]M.Sc. Dip. Arch., BIM Specialist, London, UK

Abstract. The subject of the article is the concept of an innovative integrated construction management process connecting Building Information Modelling (BIM) and construction equipment with unique control systems. It would require developing an advanced software application allowing data exchange and keeping track of cost-consumption, efficiency, use of building materials, as well as many other essential parameters, which would facilitate the improvement of quality of the technological process of construction works.

1 Introduction

The basic features of the proposed technology would be the introduction of an innovative system of managing construction works with organizational support, highly automated construction equipment and an IT system based on the integration of technical, organizational and economic information. Due to the dispersed nature of activities, starting from designing tasks to selection of materials to technology and related equipment, an IT system will be developed to enable remote integration of process elements and data exchange in real time as well as to monitor the schedule and quality of works on site. The BIM process has been used for planning, designing, implementation and operation so as to develop or utilize materials after renovation or demolition of a worn-out building (Fig.1). So far this management process has not been widely used in construction of machines for positioning, monitoring or machine management [1, 5]. Therefore, the integration of the BIM process [2] will be an integral and necessary activity of this proposal.

In order to obtain high quality and efficiency of construction works in the same or even shorter time of the cycle, it would be essential to establish, as part of the development of this system, a research and development and training centre for the management staff, as well as for operators of the machines equipped with advanced information technologies and traffic control systems.

One of the examples of promising systems already in use worldwide is Machine Control System developed by TRIMBLE. Primary benefits of this solution include:

- Increased efficiency
- The accuracy of performing tasks in 3D environment and BIM process
- Savings in the use of building materials
- Savings in operation by reducing the number of hours to perform a specific task
- Increased reliability of machines
- Improvement of ergonomics of the machine operators' work
- The possibility of ongoing monitoring, management and possible adjustment of the technological process.

* Corresponding author: andrzej.sobczyk@mech.pk.edu.pl

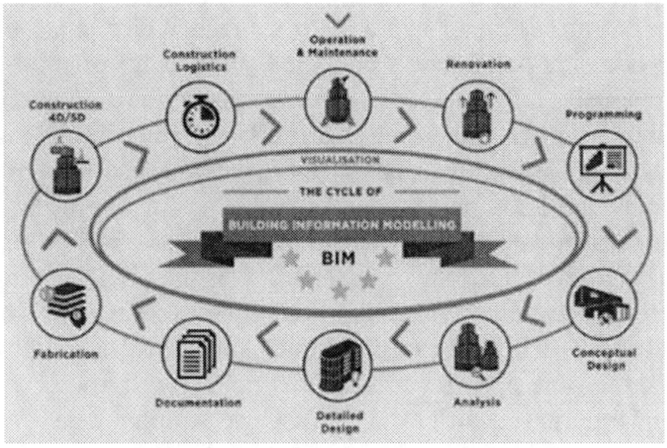

Fig. 1. BIM process structure used in civil engineering construction [2].

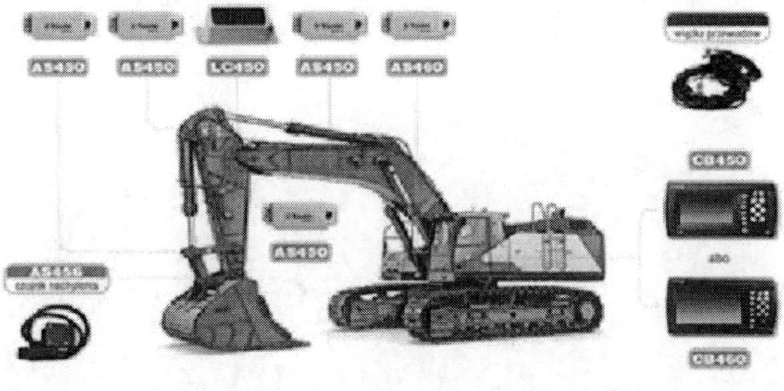

Fig. 2. An example of excavator with control-measuring equipment as part of work management system [3].

An additional reason for the application is the offer of SITECH Ltd, which is a licensed representative for Poland of the TRIMBLE EARTHWORKS 3D system, including its latest components, such as: Flexible Valve Driver with LED Status, The TD520 10" Touch Display, Joystick Remote Switch, GS520 Inertial Measurement Unit, The EC520 Electronic Controller, CAN interfaced I/Expansion module, The CI510. The AA510 is a CAN interfaced and configurable audible alarm. This module will also reduce the time required to perform individual construction works and minimize the costs of the business activity. Real estate site management will enable accurate planning of earthworks performed by excavators and other heavy mechanical equipment in real time. On-going planning of works on the construction site will allow for an increase of the efficiency of construction works and will contribute to the optimization of operations. While in use, the system will enable obtaining highly precise observational data for exchange.

2 Highly precise position data project

Currently, construction processes often rely on rented GPS receivers as base stations that give corrections after obtaining data in by radio transmission. The stations are also rented and then adjusted using Internet connection and radio wave combined. The services are provided only locally and the range of services included only cover only limited areas. The corrections are only corrections from a single reference station and are characterized by a decrease in accuracy, proportional to the distance from that station. The proposed solution to be implemented on the basis of innovative technology available on the international market will enable the provision of services in a much wider area - hundreds of square kilometres – by processing and sharing collected satellite information in real time. As it was mentioned previously, information processing was not carried out accurately by radio. Once received by the customer, the data was not updated in real time, as using the satellite signal was not possible to integrate. VRS3Net Plus innovative system software solutions will allow you to generate RTK corrections in real-time for any point inside the network area with the highest and equal accuracy, which has not been possible so far (the innovation on the international market). Until now, the accuracy of the point determination dropped in direct proportion to the distance from the reference station. For static measurements (determination of accuracy with increased accuracy), the user was required to have at least 3 receivers, at the same time observe at least two points with known coordinates and one point determined for a period of at least 1 hour. Observational data was transferred to the headquarters and it was subjected to analysis and technical processing. Only after these activities, processed data could be made available to recipients. Currently, the registered user of the system may have only one GNSS receiver and observe several minutes of data over the designated point and obtain a position of the point with a degree of precision not lower than 1 cm even in a few / several minutes after the measurement.

Fig. 3. Selected pictures illustrating application of Trimble equipment and software in cycles of field geo-work, design construction, earth moving and paving works and analysis of road construction processes [4].

The implementation of the project will allow users to provide satellite information on a regular basis, thanks to which they will be able to determine their position in the field clearly and without delay, which is an innovation on the international market. This information will be available to download from the website of the system from the tab reserved for registered customers. The data obtained through the reference station system will be passed on-line to users thanks to multiple systems sharing the produced results as well as online customer service. An innovative solution, in this case, will be the use of a multi-track system for sharing measurement data. Until now, users have been forced to be content with information obtained only via radio, which due to frequent disruptions and a limited range of waves significantly impeded the effective acquisition of coordinates of points in space. Thanks to the implementation of the proposed methodology, the basic data transmission system will be the Internet, whereas the radio system will be only complementary (in areas with limited GPRS coverage).

The investment plan assumed the purchase and implementation of a technological solution that has been used in the world for a period not longer than 3 years. In particular, the latest components of the Earthwork system in earthworks are used in the world for no longer than one year. Detailed justification for the purchase and use of equipment and technology could be made after receiving introductory data of the particular project.

The reason for the application is the lack of comprehensive solutions for the implementation of construction tasks in the country, especially in the field of road infrastructure. The proposed concept suggests implementing the most modern equipment available on the market, both machines and electro-informatics equipment for geodetic equipment connected systemically with precise machine movement control systems. Such an approach, with the help of an information system that is part of the project, integrates and allows effective design of works, selection of necessary materials and hardware resources, technologies, management and supervision. The system proposed for implementation will also ensure a high level of quality while reducing construction costs through the optimal use of materials and precise work of construction machines, as well as shortening the time of implementation and thus reducing costs. It should also be added that the implementation of this integrated on-line construction management system - in road engineering or similar infrastructure - will prevent many mistakes at the implementation stage. Additionally, the applied IT and control systems will increase employee safety, which will reduce failure rates and damage to machines thanks to modules controlling their work parameters, independent even from the operator, thus eliminating inevitable human errors in the realities of the construction site.

3 Conclusion

To sum up, it is worth emphasising that machines equipped with the Control System described above prove that the system may be considered to be highly innovative. This is an integrated system which has not been used so far in Poland in machine construction in terms of not only positioning but also monitoring and management of machines, using advanced software that allows to keep track of cost-consumption, efficiency, consumption and material supplies and other interesting parameters of the technological process of construction works.

This high precision positioning system may be also very interesting for other application like e.g. precise positioning of the ground plant location in a biotechnology investigations [6, 7]. Regardless, such system may be useful for creating of local mappings with support of a photogrammetry and an image analysis [8, 9], and an additional supporting system for an non-parametric uncertainty estimation [10-12]. The ability to obtain precise position records and appropriate control is very valuable during field testing in engineering research, both machine [13, 14] and materials science [15-23]. It is obvious that such RTK GPS utilizing system will have a great impact on a structure and capabilities of management systems in any domain where it will be applied [24-29].

References

1. A. Gawlik, A. Sobczyk, P. Walczak, Technical Trans., **113(5),** 13 (2016)
2. *BIM Object Cloud Solution*. http://www.bim.com
3. *Sitech Polska*. http://www.sitech.pl
4. Trimble Inc., 935 Stewart Drive, Sunnyvale, CA 94085
5. P. Kucybała, J. Pobędza, A. Sobczyk, *J. KONES* **21(2)**, 169 (2014)

6. E. Skrzypczak-Pietraszek, A. Hensel, Pharmazie, **55**, 768 (2000)
7. E. Skrzypczak-Pietraszek, I. Kwiecien, A. Goldyn, J. Pietraszek, Phytochem. Lett., **20**, 443 (2017)
8. A. Gadek-Moszczak, Image Anal. Stereol., **36**, 151 (2017)
9. A. Gadek-Moszczak, Image Anal. Stereol., **36**, 207 (2017)
10. J. Pietraszek, A. Gadek-Moszczak, Solid State Phenom., **197**, 162 (2013)
11. J. Pietraszek, A. Gądek-Moszczak, T. Torunski, Adv. Mat. Res.-Switz,. **874**, 139 (2014)
12. J. Pietraszek, E. Skrzypczak-Pietraszek, Adv. Mat. Res.-Switz. **874**, 151 (2014)
13. M.S. Kozien, J. Theor. Appl. Mech., **47**, 411 (2009)
14. J. Selejdak, R. Ulewicz, M. Ingaldi, *Proc. 23rd International Conference on Metallurgy and Materials*, (Brno, Czech Republic, 2014)
15. T. Pieczonka, J. Kazior, A. Tiziani, A. Molinari, *J. Mater. Process. Technol.,* **64**, 327 (1997)
16. M. Ulewicz, U. Lesinska, M. Bochenska, Physicochem. Probl. Mi., **44**, 245 (2010)
17. M. Ulewicz, E. Radzyminska-Lenarcik, Physicochem. Probl. Mi., **46**, 119 (2011)
18. N. Radek, J. Pietraszek, B. Antoszewski, Adv. Mat. Res.-Switz., **874**, 29 (2014)
19. T. Lipinski, A. Wach, *Proc. 23rd Int. Conf. on Metallurgy and Materials – METAL*, (Brno, 2014)
20. D. Klimecka-Tatar, S. Borkowski, P. Sygut, Arch. Metall. Mater., **60**, 735 (2015)
21. D. Klimecka-Tatar, G. Pawlowska, M. Sozanska, Arch. Metall. Mater., **60**, 153 (2015)
22. R. Dwornicka, N. Radek, M. Krawczyk, P. Osocha, J. Pobedza, *Proc. 26th Int. Conf. on Metallurgy. and Materials – METAL*, (Brno, Czech Republic, 2017)
23. A. Szczotok, J. Nawrocki, J. Pietraszek, Arch. Metall. Mater., **62**, 587 (2017)
24. Kozien, E., *Proc. 20th Int. Sci. Conf. on Economic and Social Dev. (ESD)*, (Prague, Czech Republic, 2017)
25. Kozien, E., Kozien, A., *Proc. 23th Int. Sci. Conf. on Economic and Social Dev. (ESD)*, (Madrid, Spain, 2017)
26. A. Pacana, R. Ulewicz, Polish J. Environ. Studies, **16**, 165 (2017)
27. Kozien, E., Kozien, M.S., *Proc. 26th Int. Sci. Conf. on Economic and Social Dev. – - Building Resilient Society (ESD),* (Zagreb, Croatia, 2017)
28. Kozien, E., Kozien, A., *Proc. 26th Int. Sci. Conf. on Economic and Social Dev. – - Building Resilient Society (ESD)*, (Zagreb, Croatia, 2017)
29. E. Kozień, *Proc. 31st Int. Sci. Conf. on Economic and Social Dev. – Legal Challenges of Modern World (ESD)*, (Split, Croatia, 2018)

Analysis of production incompatibilities and risk level in series production of assembly elements for the automotive industry

Magdalena Mazur[1,*]

[1]Częstochowa University of Technology, Department of Production Engineering and Safety, Armi Krajowej Str. 19B, Częstochowa, Poland

Abstract. The paper presents issues connected to risk analysis of production incompatibilities occurrence. Low-pressure cords thermoforming process analysis was presented, as well as analysis of types of incompatibilities during the phase of marking product (pad printing). The risk analysis is based on three factors: occurrence, detection, and severity of a specific incompatibility. 6 incompatibility types were defined and their various causes were determined. One critical incompatibility was determined, as well as 4 severe production problems. Presented analysis is an element of implementation of project of new products into the existing thermoforming processes.

1 Introduction

An important element of quality management and assessment is involving the management in the area of actions connected to detected production incompatibilities. An element of process supervision system is management involvement both during the phase of new production implementation and realisation maintenance and supervision. One of component of production preparation and supervision is production incompatibilities occurrence risk analysis. The analysis regards at first production planning, and then it is continued as supervision over production tasks fulfilment and achieved results monitoring. Utilising FMEA analyses is a system activity, the purpose of which is to ensure correct level of sureness that the product will meet assumed quality requirements [1].

One of the most important quality processes in the w automotive field companies is PPAP - Production Part Approval Process. By obtaining PPAP confirmation, the company simultaneously receives from the customer approval of competence of assembly elements provider. Processes connected to obtaining PPAP are focused on process approach, the basic element of which is focus on the customer and his expectations. PPAP requirements include the following documents: project provisions, materials composition, engineering changes documentation authorised by the customer, project FMEA, process flow chart, process FMEA, analysis of measurement systems defined by the control plan, measurement

* Corresponding author: magdalena.mazur@wz.pcz.pl

report, statistics control records, test results, appearance compatibility report, part samples, model sample [2, 3].

Assembly elements producer is obliged to present full PPAP documentation only if:

- a new part or product appears, which the customer has not been previously supplied with,
- inaccuracies in comparison to the previous PPAP need to be rectified,
- changes regarding materials or technical specification occurred in the area of provided product,
- changes in the product production process itself occurred – using new or modified tool or form,
- occurs a change of: supplier, control methods, production technology.

Presented actions are connected to the quality management in low-pressure cords production processes. The necessity of providing full PPAP documentation was connected to structural changes in a new model of the product. Research entity prepared full documentation and assessed risk connected to the planned production processes quality. PPAP processes are realised according to the internal guidelines of the company, ant the responsibility for specific implementation phases of PPAP requirements lies within responsibilities of the Head of the Process Engineering Department.

2 FMEA analysis – main assumptions

Presented FMEA analysis is an element of set of documentation prepared as preparation of new production on low-pressure cords production lines. According to the requirements, a basis for beginning a cooperation with a customer (automotive corporation) is ensuring proper supervision of production processes (PPAP). An important element of the documentation is risk analysis of potential production incompatibilities, which is conducted in both FMEA process analysis and FMEA product analysis.

The presented analysis of planned production processes uses basic risk factors based on severity, occurrence, and detection values. The values were evaluated on scale 1-10, according to the guidelines of an external customer [4-8]. Risk priority number in the FMEA analysis was assessed on the basis of the formula:

$$RPN = SEV\, x\, DET\, x\, OCC \tag{1}$$

where: RPN – risk priority number, SEV – severity, DET – detection, OCC – occurent.

For the needs of PPAP production parts approval process, quality tests were conducted mainly on characteristics regarded as important by the customer. Serial production implementation process analysis was conducted on the basis of low-pressure cord production. Design team is decomposing the system and preparing comparisons related to new production implementation: amounts of necessary materials, their quality, machine requirements (production devices and their wiring).

The analysis itself describes and classifies potential faults and their severity. The next phase is determining probability of fault occurrence and determining its severity for functioning of the product. the next phase is providing proper guidelines in order to decrease faults occurrence risk. Table 1 presents detailed FMEA analysis of product marking process. FMEA analyses results become a basis of creation of standard operating procedures for processes operators at specific work stations after undercure process.

Table 1. FMEA method in process of tampoprin-marking.

Requirement	Failure Mode	Failure Effect	SEV.	Potential Failure Cause	Preventive Action	OCC.	Detection Action	Control mean	DET.	RPN
Orientation of marking according to drawing tolerances	Incorrect orientation of marking	Risk of contact with adjacent elements	6	Use wrong tool (gabarit)- operator mistake	SW includes number of gabarit	2	Error or cause detected by the operator at the station by a visual/tactile	Visual / operator	7	84
			6	Mistake an operator (incorrect placing the part in the gabarit)	Workplace training. The construction of gabarit hindering the incorrect putting of detail visualization method of putting part in gabarit is include in SW	2	Error or cause detected by the operator after operation using attribute gauge	CSE Gauge / yhe operator of next operation	7	84
			6	Improper mounting tool (gabarit) in machine	Comparing axis marking with the index nominal on gabarit (gabarit must include an index of nominal)	3	Error or cause detected by the operator on station using attribute gauge	Assembly gabarit / operator	6	108
Positon of according to draw ng tolerances	Incorrect position of tampo (for the hose ends - the distance from the start of hose)	Leakage in the car or in the engine	8	Use wrong tool (gabarit) - operator mistake	S W includes number of gabarit	2	Error or causa detected by the operator on the position by measuring variables	Ruler / operator	5	80
			8	Incorrect setting of the trolley with the part	Adjusting the trolley with part to the correct position marking the beginning of production	2	Error or cause detected by the operator on the position by measuring variables	Rules / operator	5	80
			8	Mistake an operator (incorrect placing the part in the gabarit)	Workplace training. The construction of gabarit hindering the incorrect putting of detail. visualization method of putting part in gabarit is	2	Error or cause detected by the operator after operation using attribute gauge	CSE gauge ruler / the operator of next operation	7	112
Positon of according to *drawing* tolerances	Incorrect position of tampo (for marking on middle of hose)	Risk of contact with adjacent elements	6	Use wrong tool (gabarit)- operator mistake	SW includes number of gabarit	2	Error or cause detected by the operator at the station by a visual *i* tactile	Visual / operator	7	84
			6	Mistake an operator (incorrect placing the part in the gabarit)	Workplace training. The construction of gabarit hindering the incorrect putting of detail. visualization method of putting part in gabarit is	2	Error or cause detected by the operator after operation using attribute gauge	CSE gauge, ruler / the operator of next operation	7	84
Readability marking	Unreadable, incomplete marking (for standard marking)	Product incomplete	6	Badly prepared paint - the wrong proportions	Instructions for preparing the paint (the mixing operation is not less than 20 sec) The paint is prepared in a graduated burette.	3	Error or cause detected by the operator on the position by measuring variables.	Laboratory buret / warehouse operator	5	90
			5	Badly prepared paint - improperly mixed	Instructions for preparing the- paint (the mixing operation is not less than 20 sec). The paint is mixed by paint mixer.	3	Error or cause detected by the operator at the station by a visual/tactile	Visual / operator	7	105
			6	Expired paint delivered from the warehouse to the production line	Compliance with dates FIFO by a system during download component for the production, Cabinet for chemicals (access only by the leaders of change)	3	Cause detected by the automatic control of the position - nonconforming product will not be made	AS400 / automatic control	2	36
			6	The use of expired ink located in the area of the production line	All cans are labeled with an expiration date of paint	3	Error or cause detected by the operator at the station by a visual ' tactile	Visual / warehouse operator	7	126
			6	Wet/oily/dirty detail	Trolley between operation are cohered, visualization included in KWD	3	Error or cause detected by the operator at the station by a visual "tactile	Visual/ operator	7	121
			6	Improper lighting or position does not allow the operator to accurately assess the marking	Lack	5	Error or cause detected by the operator at the station by a visual/tactile	Visual/ operator	7	210

			6	The operator left paint uncovered for a long time and later used it (bad viscosity ink density)	Prohibition of the use beakers of paint	3	Error or cause detected by the operator at the station by a visual/tactile	Visual/ operator	7	126
Incorrect hose used in process	Incorrect hose used	mounting impossible	8	The operator did not check the number hose before starting work	SW includes number of detail and a photo, Additionally the operator based on a shaped part gage	2	Error or cause detected by the operator at the station by a visual/tactile	Visual/ operator	7	112
			8	Operator used wrong hose	SW includes number of detail and a photo, Additionally, the Operator based on a shaped part gage	2	Error or cause detected by the operator at the station by a visual/tactile	Visual/ operator	7	112
Marking dimension according with drawing	Wrong dimension of marking	Risk of contact with adjacent elements	6	Operator used incorrect cliche	Number cliche in specification SW /Photos or drawing printing in the specification SW / dimension marking specified in KZD	2	Error or cause detected by the operator on the position by measuring variables	Ruler/ operator	5	60

3 Conclusion

Marking process analysis determined production of the workplace. Their analysis was a basis of developing standard operating procedure of the process and guidelines of visual control conducted by the operator. The most important incompatibility type is illegibility of the printing pad, which results from few possible errors. Analysis of causes determined as many as 6 sources, half of which surpassed RPN value over 100 pts. A vital issue in this area is employee training on proper workplace preparation and following standard operating procedure.

The problem is illegible printing pad, which causes the product to be incomplete. There may be the following causes: incorrectly prepared ink, expired ink, damaged detail, incorrect lighting, using ink without cover, using incompatible cord, using incorrect film. In all of these cases apart from the incorrect film usage, RPN is higher than 100, which proves these are actual problems that can occur during production.

Described 6 production incompatibilities of marking process have been included in the Standard Operating Procedure (SW) and guidelines of workplace control.

References

1. R. Ulewicz, *International Conference on the Path Forward for Wood Products, A Global Perspective Location* (Baton Rouge, LA, 2016)
2. R. Wolniak, B. Skotnicka, *Metody i narzędzia zarządzania jakością: teoria i praktyka.* (Wydawnictwo Politechniki Śląskiej, 2010).
3. A. Tubis, Advances in Intelligent Systems and Computing, **637** (2018)
4. K. Knop, Production Engineering Archives **16**, 16-21 (2017)
5. M. Zasadzien, J.Zarnovsky, Management systems in production engineering **26(1)**, (2018).
6. P. Bartosz, M. Dudek-Burlikowska, M. Roszak, Production Engineering Archives **18**, 35-41 (2018)
7. J. Pietraszek, A. Gądek-Moszczak, T. Torunski, Adv. Mat. Res.-Switz., **874**, 139 (2014)
8. J. Nowakowska-Grunt, M. Mazur, *Metal 2016: 25th Anniversary International Conference on Metallurgy and Materials, 1956-1961* (2016).

Use of IT technologies in the management of production process quality

Elżbieta Milewska[1,*], and *Bartłomiej* Skowron[2]

[1]Silesian University of Technology, Faculty of Organization and Management, Institute of Production Engineering, ul. Roosevelta 26, 41-800 Zabrze, Poland
[2]UIBS Teamwork Sp. z o.o., ul. Por. K. Ogrodowskiego 27, 44-203 Rybnik, Poland

Abstract. When presenting the functionality of IPOsystem™, the author of the article described the manner of controlling the quality of products and calculating the technical cost of production in a selected production plant. In the article actions taken to remove product quality nonconformities have been discussed and a calculation by the subtractive factor method has been presented. Limitations of the described IT tools have also been illustrated with an example of implementation.

1 Introduction

Continuous improvement of the quality of products, services and production processes results from the consumer market requirements and prevailing trends. An important issue is the way of implementing and observing quality management standards in a company. A Quality Management Standard compliant with ISO 9001:2015 standard [1] is applied in all branches and institutions of various kinds and sizes. The process approach used when developing and improving the quality management system not only streamlines the circulation of information and documents in an institution, but it also brings in a precise division of tasks, competences and responsibilities of the employees [2] (Fig. 1). The organization achieves the result of effective business activity, enabling it to increase customer satisfaction, adjust to the changing conditions on the market and maintain the market position.

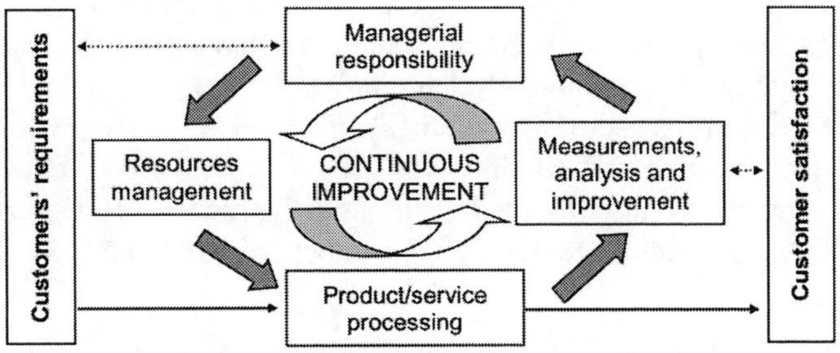

Fig. 1. Dynamic structure of a quality management system according to ISO 9001:2008 standard

* Corresponding author: elzbieta.milewska@polsl.pl

2 Product quality improvement

Manufacturing process control is an important aspect of enterprise activity [3-9]. Based on numerous studies regarding production costs analysis, one can conclude that failure to meet quality requirements is frequently a source of additional costs, causing unexpected financial losses for the enterprise [10, 11].

Actions that eliminate the identified nonconformities, but, at the same time, generate additional costs, include [12]:

- correction of errors, involving the necessity to use additional materials, raw materials or utilities, as well as the use of different tools, etc.;
- performing additional technological operations which restore the initial quality of a product;
- conducting another quality control;
- performing diagnostic tests to establish the source of nonconformities;
- utilization of damaged products which cannot be repaired or reused.

Popularization of quality management systems caused that contemporary enterprises base their activities on the functioning of IT systems [13, 14]. The reasons and effects of product nonconformities are defined and errors in the course of the production process are identified using production management support systems. Application of IT systems provides tangible benefits, which contribute to reduced costs of product manufacturing [15]. Following changes in the value of components of the technical production cost structure enables evaluating a defective product, taking into account both the time of the defect occurrence and the duration of its forming. This evaluation should include both the cost of direct production materials and direct wages. They are calculated on the basis of the number of man-hours devoted to the performance of technological operations involved in the order until a nonconformity is found. Particular attention in the calculation of a technical production cost should be paid to estimating the activities involved in the restoration of the initially planned quality of a manufactured element [16], utilization of defective products as well as diagnostics enabling identification of the cause or/and source of nonconformity. The above described analysis makes it possible to implement corrective or preventive actions, which allow reducing the identified problem [17].

In the context of the above considerations, an important issue is specifying the production process check points. They are an intermediate stage of the product manufacturing process, in which its technological parameters should be controlled or the conformity of half-product quality should be verified. For the final product check is an obvious procedure. Production process monitoring provides a basis for decisions controlling the flow of materials in a company. The most commonly applied measure of production process disturbances is the product quality level indicator. It represents a ratio of the number of defects to the total number of manufactured units, or to the number of checked products [1, 18-20]. Deviations between the expected and obtained value can be eliminated by changing the operation technological parameters, changing the material structure of the order or changing the workstation instructtions. Another variant of actions regulating the flow and aiming to offset the expected and obtained value is quantitative correction of the adopted standards. Correcting the conditions of the production process is the most radical action aimed at changing the algorithm of operational production plannning [13].

3 Example of implementation supporting quality management

The subject of investigations was a Polish producer of highly specialist tailor-made machines. The enterprise has implemented a system to support production management called IPOsystem™.

IPOsystem™ is a product of UiBS Teamwork Sp. z o.o., based on Microsoft SQL Server platform. The system functionality includes planning, controlling and recording of manufacturing activities. Planning takes into account the working time and the adjustment of resources to the processing of tasks, material availability as well as limitations in the performance of works in the form of retooling, brigade work and others. Tasks are allocated autonomously. Instructions for performing some work are given using a terminal, which is placed in the hall and connected to the server. The terminal, working in a Unique 125 kHz system, identifies the contractor by an RFID proximity card. It gives orders to do some work and receives reports from an employee in an alternate way, which is a solution providing systematic information on the course of manufacturing works. Apart from current recording of the time of commencing and finishing technological operations, IPOsystem™ allows for a possibility of recording the number of half-products leaving the workstation as well as the number of rejects, for which the previously adopted classification of nonconformity causes is applied.

Moreover, the functionality of IPOsystem™ supports users in activities related to auditing the quality of materials, half-products and final products, thus enabling them to check compliance of the obtained evaluation with the requirements contained in constructional or technological documentation. Control is determined by prior declaring an order for acontrol operation – either a workstation or walk-around one - in the production technology, and granting relevant system permission to the controller. Workstation control takes place at a controller's measuring station, whereas walk-around control is performed at a workstation of a direct production employee as a response to call.

The controller is a user of IPOsystem™, who, using the functions of Quality Control window, downloads tasks, records the result of control and, in justified cases, stops his activities. The tasks are sequenced in the window according to their urgency, with a division into workplace control and walk-round inspection.

IPOsystem™ allows for simultaneous performance of a few control works and continuing the operation by another controller. Noting a positive outcome of the control results in giving an instruction to perform another operation, whereas a negative outcome immediately stops the production order. The order can be resumed after modifying the technology, as the default setting of the system is attempting to repair the defects. For this purpose, the order can be supplemented with additional repair operations, restoring the required quality of products as well as with an operation of conducting another quality check. The recording of unrepairable defects in IPOsystem™ results in an autonomic reduction of the number of products in the order. In the event there are a few major products in the order, the number of recorded defects indicates the number of defective sets.

Evaluation of actions involved in the elimination of identified nonconformities complies with the adopted manner of production costs calculating. It has application in conjugated production (also known as combined or correlated), as a result of which three types of products are produced simultaneously from one raw material:

- a main product whose production is the primary purpose of the company's operation;
- by-product, created accidentally, having a certain utility value, that can be sold or used for own needs;
- production waste — a worthless product requiring disposal or repair.

Calculation of production costs in IPOsystem™ is conducted on an object, that is the production order. Acceptance of products from production is carried out in accordance with the proportion reflecting the mutual quantitative correlation between main products in the order. The employee's report may additionally include reporting an identified nonconformity. This causes an automatic evaluation of the product, which takes into account both the cost of materials used and direct wages, calculated on the basis of recorded hours worked by contractors. In the event corrective actions are taken, the technical cost of production is increased by the cost of operations which restore the initially planned quality of the product.

Improvement of the production process requires prior identification of defects and the causes of nonconformities. As IPOsystem™ forces continuous monitoring of events, which shows the occurrence of disturbances in the production process and corrective actions undertaken to remove a quality nonconformity, the technology diagrams are corrected on an ongoing basis. The updated technology is used in new orders. Improvement of production processses is also supported by advanced analytical tools of IPOsystem™. The scope of data collected in the data system enables the deviations from the plans and production costs to be analysed. It is also possible to conduct a comparative analysis of processes of various production orders which have an identical final effect but are started at different times. The obtained results are used to specify the most favourable conditions for running a business.

Conclusion

A measure of effective functioning of a production enterprise is an efficient flow of information integrated with a qualitative and quantitative stream of material flow. A solution increasing the flexibility of actions is implementation of IT systems to support production activity. Striving to achieve a product quality compliance with the customers' requirements is determined not only by the management support, but also by the awareness of employees with regard to eliminating the causes of production process disturbances. It is a basis for improvement of the company processes.

This article was created as part of statutory work 13/030/BK_18/0039 conducted at the Institute of Production Engineering at the Faculty of Organisation and Management of the Silesian University of Technology.

References

1. EN ISO 9001:2015 *Quality management systems – Requirements*
2. B. Berret-Koehler, J. Kenneth, A. Kiser, M. Sashkin, *Putting Total Quality Management to Work* (Wiley & Sons, 1993)
3. M. Assid, A. Gharbi, A. Hajji, Int. J. Adv. Manuf. Technol., **72** (2014)
4. S.M. Sajadi, M.M. Seyed Esfahani, K. Sörensen, Int J Adv Manuf Technol **53**, (2011)
5. M. Caramanis, A. Sharifnia, Int. J. Flex. Manuf .Syst., **3** (1991)
6. R. Akella, P.R. Kumar, *Optimal control of production rate in a failure prone manufacturing system*. IEEE Trans. Autom. Control., **31**, 116–126 (1986)
7. K. Dhouib, A. Gharbi, M.N. Ben Aziza, Int. J. Prod. Econ., 137, 126–136 (2012)
8. A. Gharbi, A. Hajji, K. Dhouib, Int. J. Prod. Res., **49(21)**, 6539–6557 (2011)
9. P. Lavoie, J.P. Kenne, A. Gharbi, IIE Trans., **4(3)**, 209–222 (2009)
10. R. Hlioui, A. Gharbi, A. Hajji, , Int J Adv Manuf Technol **81**, 335 (2015)
11. B. Skotnicka-Zasadzien, *Application of Quality Engineering Elements for the Improvement of Production Processes-Case Study, Proceedings of ICIEMS*, (DEStech Publicat Inc, 2013)
12. PN- EN ISO 9000:2006 System Zarządzania Jakością – Podstawy i terminologia; (PKN, Warszawa 2006)
13. J. Kimemia, S. Gershwin, IEEE, **191** (1983)

14. E. Milewska, *Modelling a cost structure for polish manufacturer: A case study of the implementation, Conference Proceedings SGEM 2016, Book 2, Political sciences, law, finance, economics and tourism. Vol. 5, Economics and tourism. Sofia: STEF92 Technology* (Sofia, 2016)
15. B. Skotnicka-Zasadzień, R. Wolniak, M. Zasadzień, *Use of quality engineering tools and methods for the analysis of production processes - case study, Proceedings of the Second International Conference on Economic and Business Management. FEBM 2017October 21-23, 2017. Eds. Piman Limpaphayom, Gordon Huang, (*Atlantis Press, Shanghai, 2017)
16. H. Rivera-Gomez, A. Gharbi, J.P. Kenné, Int J Adv Manuf Technol., **69**, 2111–2130 (2013)
17. R. Kolman, *Inżynieria jakości* (PWN, Warszawa 1992)
18. A. Hamrol, W. Matura, *Zarządzanie jakością. Teoria i praktyka* (PWN, Warszawa 2006)
19. A. Hajji, A. Gharbi, J.P. Kenné, Int. J. Prod. Res., **42(6)**, 1107–1130 (2004)

Quality improvement in Cleaner Production aspect

Joanna Rosak-Szyrocka[1]*

[1]Department of Production Engineering and Safety, Czestochowa University of Technology, Poland, joanna.rosak-szyrocka@wz.pcz.pl

Abstract. The aim of the paper is presented the Cleaner Production (CP) issue in the aspect of the benefits that it gives the company in the competitiveness productivity and efficiency aspect. The research object is a company that is a leader in the Cleaner Production area - cement plant CEMEX. The research was based on a comparative analysis the company's sustainable development in 2008-2012. It was shown that substitution of alternative fuels (an increase of 43% in 2014 compared to 2010) and the construction the first world's alternative fuels dryer enabled not only for carbon dioxide reduction, but also hard coal consumption reduction and as a consequence allowed to relieve landfills waste. Thanks to elimination a part of the water contained in alternative fuels up to 8%, the unit heat consumption in the clinker production process is reduced. This is due to avoiding evaporation of water at high temperatures. Analysis the presented data showed that the use of CP not only effectively protects the environment by reducing / minimizing waste streams, but also leads to increased profitability, productivity, efficiency and competitiveness of industry, which in turn leads to economic growth. The paper emphasizes that the CEMEX Rudniki cement plant has saved 132,000 Mg of carbon thanks to the use of alternative fuels over the last 5 years.

1 Cleaner Production - literature review

Cleaner Production (CP) is an environmental protection strategy based on continuous, integrated, preventive action in relation to processes, products and services, aimed at increasing production and service effectiveness and reducing risk for people and the natural environment [1]. The primary goal of Cleaner Production is to restructure the industry in an economically beneficial way, while improving the industry's environmental performance by reducing water and energy consumption, pollutant emissions and waste, while improving product quality and safety at the workplace. The foundation of Cleaner Production is the idea of sustainable development, i.e. the emphasis on preventing pollution at the source of their formation. CP emphasizes that it is important to apply **good pro-ecological practices,** the use of which may bring measurable benefits, including economic ones [2]:

Less water, energy = lower bills
Less emissions to air = less or no fees.

* Corresponding author: joanna.rosak-szyrocka@wz.pcz.pl

Authors [3] emphasize that Cleaner Production has a significant contribution to the reduction of environmental impact. Authors [4] emphasize that cleaner production, applied in a systemic manner, by means of an appropriate procedure, becomes a voluntary, informal environmental management system that allows any organization, regardless of its size and profile, to achieve in a short time measurable environmental and economic benefits and diversify your position on the market.

2 Cleaner Production in the CEMEX company

CEMEX is a leading manufacturer of cement, ready-mix concrete and aggregates in Poland. Chełm Cement Plant has been an active member of the Polish Cleaner Production Movement since 1997. CEMEX is a CP leader which means that not only fulfills all obligations resulting from environmental law and administrative decisions, but applies an appropriate system of responsible entrepreneurship principles and is successful in achieving particularly high production parameters while limiting emissions. CEMEX as a component of products uses process waste from other industries such as energy, metallurgy. These are wastes such as: slag, fly ash from coal combustion processes, fluid sands, which, in order to be introduced into the production process in a cement plant, are removed from landfills of industrial waste in the environment. Waste as alternative fuels (fraction of municipal and industrial waste properly processed and sorted) and sewage sludge can be safely burned in the production of clinker (instead of conventional fuels). Alternative fuels contain flammable fractions from renewable sources such as: paper, wood, rubber, natural textiles and are partly produced from municipal waste. The use of these wastes significantly reduces the impact of the company on the environment as it reduces the consumption of coal, and at the same time CO_2 emissions from combustion processes and relieves the local municipal waste site. The Rudniki Cement Plant, up to 3 long rotary kilns currently in use (dry method) uses grinded solid waste (including municipal waste) in the amount of 8Mg / hour as a fuel, as well as dry sewage sludge. The use of rotary kilns by the plant enables the recovery of energy and the total utilization of waste, as these furnaces are considered to be the safest and state-of-the-art equipment used for co-firing. At the same time, high combustion temperatures ensure a significant reduction of harmful emissions in the process itself, which is described as waste-free, because the ash produced during the combustion of alternative fuels is permanently embedded in the clinker structure and becomes part of the product [5-7]. The CEMEX Rudniki company eliminated the unorganized emission due to the construction of the clinker silo. The reduction of the clinker index by reducing its share in produced cements and increasing additives (ie: fly ash, blast-furnace slag, limestone) is associated with a decrease in the unitary CO_2 emission / tonne of the product. The plant aims to reduce the clinker index by increasing the production of cements with additives and by applying new solutions. The article analyzes the results of sustainable development in 2008-2010 and 2010-2014. The article also analyzes the situation of the company before the implementation of cleaner production and the implementation of the strategy of cleaner production. Table 1 presents the example of sustainable development in 2008 - 2010.

Table 1. Sustainable development results in 2008-2010.

GREENHOUSE GAS EMISSIONS					
Issue of NOX compounds (in relation to limit values from the integrated permit)					
Chełm (mg/Nm^3) NO$_X$ (800Limit NO$_X$)			Rudniki (mg/Nm^3) NO$_X$ (800Limit NO$_X$)		
2010	2011	2012	2010	2011	2012
468	506	480	510	349	297

Emission of SO_2 compounds (in relation to limit values from the integrated permit)					
Chełm (mg/Nm^3) **SO_2 (50Limit SO_2)**			**Rudniki (mg/Nm^3)** Derogation from the limit in accordance with the Regulation on emission standards for installations		
2010	**2011**	**2012**	**2010**	**2011**	**2012**
20	13	19	130	93	98
Dust emission (in relation to limit values from the integrated permit)					
Chełm (mg/Nm^3) **Dust limit 30**			**Rudniki (mg/Nm^3)** **Dust limit 30**		
2010	**2011**	**2012**	**2010**	**2011**	**2012**
6	6	5	13	5	10
TOTAL ISSUE					
Total net CO_2 emissions [thousand] tons] (in relation to the allocation of allowances in accordance with the National Allocation Plan - 1796 CO_2 allocation)					
Chełm			**Rudniki**		
2008	**2009**	**2010**	**2011**	**2012**	**2008**
1093	1068	1239	1297	1035	445

WATER							
Unit water consumption (liters / ton of cement)							
Cement (liters / ton of cement)			**Concrete (liter / m^3 concrete)**			**Aggregates (liter / ton of aggregate)**	
2010	**2011**	**2012**	**2010**	**2011**	**2012**	**2010**	**2011**
216	181	182	185	174	175	137	143

Use of recycled water for concrete production		
2010	**2011**	**2012**
15.6%	12.8%	17.4%

Substitution of alternative fuels in CEMEX Polska cement plants				
2008	**2009**	**2010**	**2011**	**2012**
34%	64%	69%	70%	72%

ALTERNATIVE FUEL	
Heat consumption from alternative fuels (GJ)	
2011	**2012**
$6.6 \cdot 10^9$	$5.4 \cdot 10^6$

Unitary fuel consumption by the CEMEX own fleet (liters / 100km)								
Cement			**Concrete**			**Aggregates**		
2010	**2011**	**2012**	**2010**	**2011**	**2012**	**2010**	**2011**	**2012**
32	32	33	62	63	62	34	33	32
Individual energy consumption								
Cement (kWh / ton of cement)			**Concrete (kWh/m^3 concrete)**			**Aggregates (kWh/tonę aggregates)**		
2010	**2011**	**2012**	**2010**	**2011**	**2012**	**2010**	**2011**	**2012**
124.5	124	125	3.4	2.9	3.2	5.8	6.2	5.6

Table 2 presents sustainability results at CEMEX in 2010-2014. Table 3 shows Rudniki Cement Plant before and after CP implementation.

Table 2. Sustainability results at CEMEX in 2010-2014.

Individual energy consumption					
Cement (kWh/ton)		**Concrete (kWh/m³)**		**Aggregate (kWh/tonę)**	
2010	**2014**	**2010**	**2014**	**2010**	**2014**
124.5	113.9	3.4	3.1	5.8	1.9
⇩ **10.6**		⇩ **0.3**		⇩ **3.9**	

Substitution of alternative fuels in CEMEX Polska cement plants	
2008	**2014**
34%	77%
43% ⇧	

Heat consumption from alternative fuels **(GJ)**	
2008	**2014**
$6.6*10^9$ = 719.4	$5.4*10^6$ = 572.4
⇩ **147**	⇩

Total emissions CO_2 **(thousands of ton)**			
Chełm		**Rudniki**	
2008	**2014**	**2008**	**2014**
1093	997.9	445	247
⇩ **95,1**		⇩ **198**	

Issue of compounds **NO_x (mg/Nm³)**			
Chełm		**Rudniki**	
2010	**2014**	**2010**	**2014**
468	530	510	589
62 ⇧		**79** ⇧	

Issue of compounds **SO_x (mg/Nm³)**			
Chełm		**Rudniki**	
2010	**2014**	**2010**	**2014**
20	19	130	194
⇩ **1**		**64** ⇧	

Dust emission **(mg/Nm³)**			
Chełm		**Rudniki**	
2010	**2014**	**2010**	**2014**
6	10	13	16
⇧ **4**		**3** ⇧	

Individual water consumption					
Cement (kWh/ton)		**Concrete (kWh/m³)**		**Aggregate (kWh/ton)**	
2010	**2014**	**2010**	**2014**	**2010**	**2014**
216	165	185	165	137	101
⇩ **51**		⇩ **20**		⇩ **36**	

Table 3. Rudniki Cement Plant before and after CP implementation.

Before CP introduce 2010	After CP introduce 2011	Difference
Issue of NO_X compounds (in relation to limit values from the integrated permit) **(mg/Nm^3) NO_X (800Limit NO_X)**		
510	**349**	**161**
Emission of SO_2 compounds (in relation to limit values from the integrated permit)		
130	**93**	**37**
Dust emission (in relation to limit values from the integrated permit)		
13	**5**	**6**

3 Conclusion

Analysis of issues related to Cleaner Production has shown that the use of CP not only effectively protects the environment by reducing or minimizing waste streams, but is also profitable because it leads to increased profitability, productivity (mainly by minimizing manufacturing costs) increasing the efficiency and competitiveness of industries that lead as a result of economic growth. Thanks to the use of alternative fuels, CEMEX Rudniki cement plant has saved 132,000 Mg of carbon in the last 5 years. The ideal of a Cleaner Production strategy is waste-free production. In the current economic and technological situation enterprises can only look for different ways of waste management and use closed cycles of supply streams, eg water. This often leads to the integration of production lines, where waste from one line is used to supply material or energy to another production line. There may also be cooperation between entities under the so-called the supply chain. The reduction of pollutant emissions can be reduced by considering the entire life cycle of the product, paying attention to the materials used, water and energy flow, and the economic impact of switching to CP. Enterprises that decide to implement CP on a voluntary basis constantly strive to reduce pollution, reduce energy consumption and costs associated with environmental charges by developing and implementing so-called CP projects. The benefit of implementing CP projects is the limitation of the amount of fees for economic use of the environment, as well as the cessation of paying penalties for non-compliance with environmental conditions, if such penalties were imposed [7-13]. The use of CP as a tool for environmental management ensures that the objective in the field of environmental management becomes the most important environmental effect. ISO 14000 does not take into account the tools used, while CP provides them. The analysis of the data showed that in the plant in 2010-2014 the following environmental actions were carried out:

- Individual energy consumption: cement 10,6 kWh / ton, concrete 0,3kWh / m3, aggregate 3.9 kWh /ton.
- Total CO_2 emissions: Chełm, 95,1 thousand tons, Rudniki 198 thousand tons. Emissions of SOx compounds: reduction by 1mg / Nm^3.
- Unit water consumption: cement 51 kwh / ton, concrete 20 kWh / m^3, aggregate 36 kWh / ton.

References

1. S. Borkowski, R. Ulewicz, *Zarządzanie produkcją, Systemy produkcyjne* (Oficyna Wydawnicza Humanitas, Sosnowiec, 2008)
2. B. Gajdzik, Problemy Ekologii, **11**, 43, (2007)

3. J. Lopes, M.D. Lima, A. Kiperstok, Journal of Cleaner Production, **170(1)** (2018)
4. J. Boba, A. Saratowicz. *Katalog produktów czystszej produkcji* (Katowice, 2010)
5. A. Chmielarz, Czystsza Produkcja i Eko-zarządzanie, Cleaner Production &Eco-management, **03** (2017)
6. E. Kardas, *The Effect of Quality of Ferrous Burden Materials on the Quality of Pig Iron. METAL 2012, 21st International Conference on Metallurgy and Materials (*Brno, 2012)
7. K. Jagieła, M. Gała, J. Rak, M. Kępiński, K. Szewczyk, *Analysis of Electric Energy Quality in Arc Furnace System with Follow-Up SVC Compensation, EPQU 2009.,10th International Conference on Electrical Power Quality and Utilisation (*Lodz, 2009)
8. J. Fresner, Journal of Cleaner Production, **6** (1998)
9. Ch. Hicks, R. Dietmar, Journal of Cleaner Production, **15** (2007)
10. Z. Nowak, M. J. Cichy, *Phenomenological model of Cleaner Production* (Springer, Netherlands, 2008)
11. L. Stone, Journal of Cleaner Production, **8** (2000)
12. A. Wiśniewska, *Strategie proekologiczne w zarządzaniu przedsiębiorstwem, M. Kistowski (red.), Studia ekologiczno-krajobrazowe w programowaniu rozwoju zrównoważonego. Przegląd polskich doświadczeń u progu integracji z Unią Europejską*, (2004)
13. R. Ulewicz R, Production Engineering Archives, **1/1** (2013)

Effect of the number of non-conforming samples on the Kappa indicator values

Pavel Klaput[1] and *David* Vykydal[2]

[1]VSB-TU Ostrava, Department of quality management, 17. listopadu 15/2172, Ostrava, Czech Republic

Abstract. Evaluate the quality of this measurement system is possible by using multiple methods, which are described in the methodology guides for evaluation of the measurement system quality. Within these methods, values of different indicators are evaluated. This paper deals with the cross tabulation method and foremost Kappa indicator. More specifically is examined the effect of the number of used nonconforming samples on explanatory power of this indicator. The effect on the kappa values is examined from several aspects that must be taken into account during evaluation of the quality of measurement system for attributive quality characteristics.

1 Introduction

From practical experience with the implementation of measurement system analyses (MSA) and from more and more frequent practice requirements, several problem areas can be defined in assessing the quality of measurement systems. Where a measurement system is used to control attribute quality characteristic, there are problem areas in performing and evaluating analyses of measurement systems used to products control. Obstacles to greater use of cross tabulation method in practice can be seen in particular in the lack of information about right sampling and in the analysis evaluation itself. For these reasons, a detailed examination of this issue can lead to a better understanding and wider application of MSAs in practice. For this reason, in Excel prepared and conducted simulations aimed at the calculation of kappa indicator. The influence of different numbers of nonconforming samples on the difficulty of the measurement system analysis evaluated by the cross-table method was evaluated. Consideration was also given to the explanatory power of the calculated values of the kappa indicator and also the fact, whether the conditions of the analysis correspond to the real conditions of the process [1].

2 Cross tabulation method

Evaluation of the measurement system for attribute quality characteristic is most often done by the cross-table method, which assesses the agreement between the different operators and the agreement between the operators and the reference value. An example of a cross table for conformity assessment between operator A and reference is shown in Table 1 [2].

Table 1. Setting Word's margins.

<table>
<tr><td colspan="3" rowspan="2"></td><td colspan="2">Reference</td><td rowspan="2">Sum</td></tr>
<tr><td>0</td><td>1</td></tr>
<tr><td rowspan="4">Operator A</td><td rowspan="2">0</td><td>Result</td><td>X_{00}</td><td>X_{01}</td><td>$X_{0.}$</td></tr>
<tr><td>Expected result</td><td>x_{00}</td><td>x_{01}</td><td>$x_{0.}$</td></tr>
<tr><td rowspan="2">1</td><td>Result</td><td>X_{10}</td><td>X_{11}</td><td>$X_{1.}$</td></tr>
<tr><td>Expected result</td><td>x_{10}</td><td>x_{11}</td><td>$x_{1.}$</td></tr>
<tr><td colspan="2" rowspan="2">Sum</td><td>Result</td><td>$X_{.0}$</td><td>$X_{.1}$</td><td>N</td></tr>
<tr><td>Expected result</td><td>$x_{.0}$</td><td>$x_{.1}$</td><td>n</td></tr>
</table>

The values in the result cells of the table labelled of capital letter X (X_{00}, X_{01}, X_{10}, X_{11}) indicate how many times the possible combinations of operator A results with the reference rank occurred. Values in cells in total ($X_{0.}$, $X_{1.}$, $X_{.0}$, $X_{.1}$) summarize the overall a decision on acceptance and rejection [1]. Calculation of expected results is performed in the following manner:

$$x_{00} = \frac{X_{0.}}{N} \cdot X_{.0} \tag{1}$$

$$x_{01} = \frac{X_{0.}}{N} \cdot X_{.1} \tag{2}$$

$$x_{10} = \frac{X_{1.}}{N} \cdot X_{.0} \tag{3}$$

$$x_{11} = \frac{X_{1.}}{N} \cdot X_{.1} \tag{4}$$

For the calculation of kappa indicators is needed to calculate values p_o and p_e value is p_0 share results when operator agree with the reference on a decision on acceptance or rejection of samples. Value p_o should ideally approximate to one and is calculated as follows:

$$p_o = \frac{X_{11} + X_{00}}{N} \tag{5}$$

The p_e value represents the share of expected results when operator agree with the reference on the decision to accept or reject the samples. It is calculated as follows:

$$p_e = \frac{x_{11} + x_{00}}{n} \tag{6}$$

The k (kappa) indicator is used to assess the agreement between the operator and the reference and is calculated according to the formula:

$$\kappa = \frac{p_o - p_e}{1 - p_e} \tag{7}$$

The values of the kappa indicator can be in the range from -1 to 1. Kappa is a measure rather than a test. A general rule of thumb is that values of kappa greater than 0,9 indicate excellent agreement (with a maximum kappa = 1). Values less than 0,4 indicate poor agreement. Based on this indicator, decisions and interventions are made to improve the quality of the attributive measurement system. All these decisions are made on the basis of faith in the flawless information contained in the value of kappa indicator. However, on the basis of earlier analyzes, it can be concluded that the used values of the kappa indicators may be affected by setting the input parameters of the evaluation used methods or other restrictive conditions [3, 4].

2 Influence of the number of nonconformities

The most important area using the value of the kappa indicator as a measure of the quality of the measurement system is the evaluation of the compliance between the operator and the reference assessment. The basic parameter of the cross-table method is the number of samples (parts) used for application of this method. As mentioned above, a sufficient number of samples (50) for which their reference values are known have to be used for the analysis of the cross-tabulation method. Among the selected samples should be a certain proportion of nonconforming samples to achieve the most relevant results of the analysis. None of the methodological manuals does not say the recommended number of non-conforming samples. Due to the method of calculating the value of the kappa (formula 5, 6 and 7), it can be assumed that the number of used non-conforming samples will affect the possible range of kappa value. At the half representation of non-conforming samples used for the analysis of attributive measurement systems, the value of the kappa indicator ranges from -1 (operator took no right decision) to a value of 1 (the operator made the right decision in all cases). The range of possible kappa values, depending on the number of correct decisions made by the operator for half number of the non-conforming samples (25 from 50), is clearly shown in Figure 1.

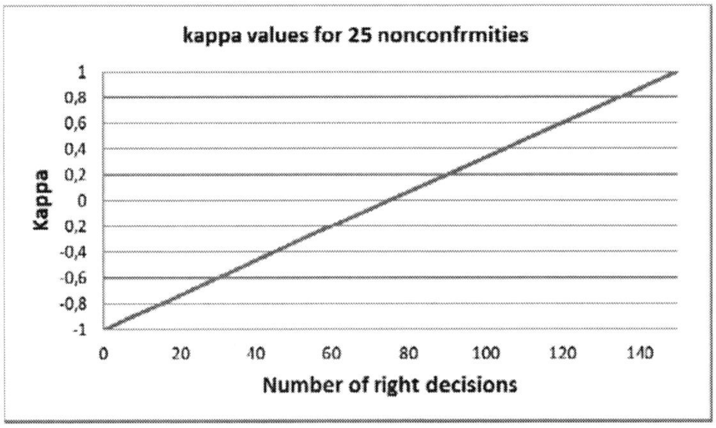

Fig. 1. Kappa values for 25 nonconformities.

Constructed graph shows that in the case of uniform proportion of conforming and nonconforming samples values of the kappa indicator changes uniformly. In practice, however, it is often possible to encounter a situation where for a variety of reasons it is not possible to ensure a uniform proportion of nonconforming samples. The changes of kappa values in the case of less than half number of nonconforming samples are clearly presented in Figure 2. Analysing the displayed data, it can be noticed that with low numbers of nonconforming samples used to analyse the attributive measurement systems, it is not possible to reach the full range of kappa values. For example, if ten nonconforming samples is used, the value of the kappa in the case of the zero match with reference is equal to -0.47. From the course of the constructed curves it is clear that the increase in the values of the kappa indicator with the increasing number of correct decisions taken by the operator is not linear. A special case is a situation where no one non-conforming sample was used to evaluate the quality of the attributive measurement system. In this case, the value of the kappa cannot be different from zero, so in this case the kappa value cannot be evaluated and the analysis of the attributive measurement system is in this case meaningless (see Figure 2).

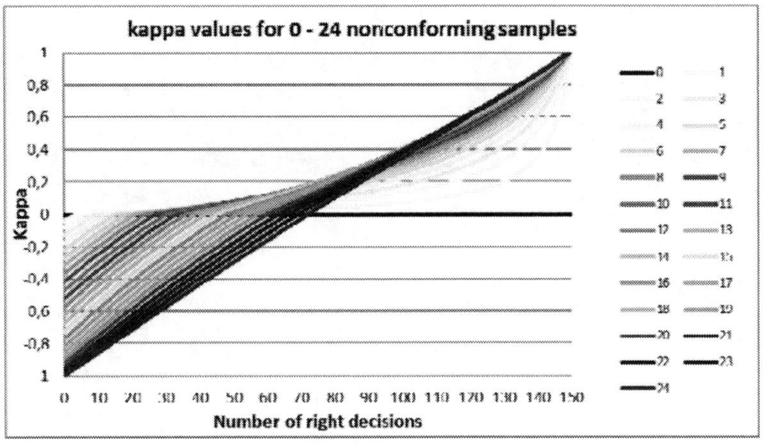

Fig. 2. Kappa values for 0 - 24 nonconformities.

In addition to lower fraction of nonconforming samples used for the analysis may occur converse situation, where (especially in the case of processes that produce a high percentage of non-conforming products) will be analysed majority proportion of nonconforming samples (26 to 50). This situation is illustrated in Figure 3.

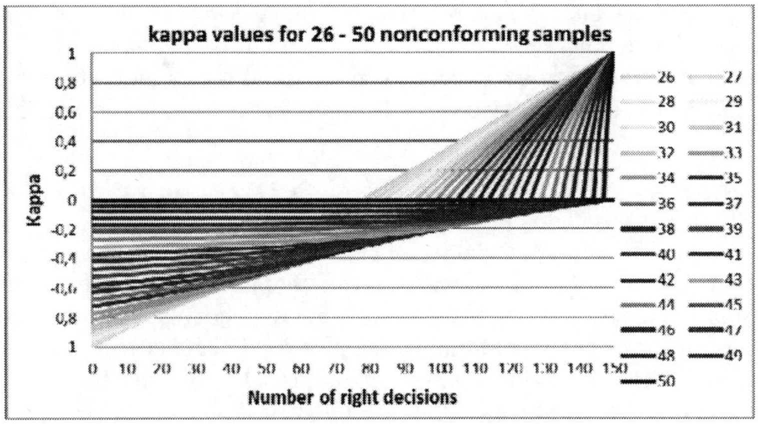

Fig. 3. Kappa values for 26 - 50 nonconformities.

Also, in this case, it can be noticed that with higher numbers of nonconforming samples used to analyse the attributive measurement systems, it is not possible to reach the full range of kappa values. For example, if forty nonconforming samples is used, the value of the kappa in the case of the zero match with reference is equal to -0.47. This is the same value as in the case of the ten nonconforming samples. From the above mentioned findings, it is clear, that to achieve the limit values of the kappa indicator for the conditionally acceptable (0.7) resp. acceptable measurement system (0.9), it will be necessary to achieve at a different number of correct decisions according to the reference for a different number of nonconforming samples used for analysis. Changes in the number of correct decisions in dependence on the number of nonconforming samples that must be taken to achieve the required values of the kappa indicator is shown in Figure 4. From the constructed curves, it can be observed that with the decreasing proportion of nonconforming samples used to analyse of attributive measurement systems, increases the number of correct decisions that the operator has to make to assess acceptable compliance with the reference. The same finding is also valid for the increasing proportion of non-conforming samples. Using too small or too large numbers of non-conforming samples, thus placing greater requirements on the accuracy of operator evaluation.

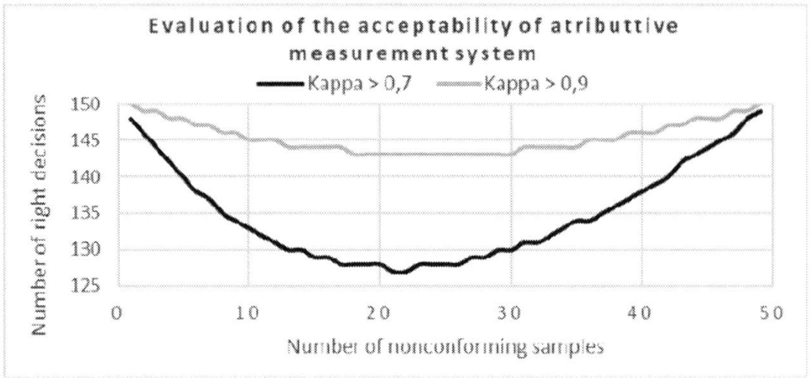

Fig. 4. Dependence of the number of correct decisions on the number of non-conforming samples.

For example, if five nonconforming samples are used, the operator will need to make 140 correct decisions so that the value of the kappa assess the compliance between the operator and the reference reaches the minimum required value of 0,7. Conversely, in the case of half-shares of non-conforming samples (25), the operator has to make only 128 correct decisions to achieve the required value of 0,7. Its error rate can be more than twice as large as 25 nonconforming samples than if 5 nonconforming samples were used. By appropriately choosing the number of non-conforming samples, the " difficulty" of the kappa indicator can be reduced. The smallest difficulty of the kappa indicator, and thus the entire cross-table method, can be achieved whit 15 to 30 non-conforming samples. Within this range, the number of necessary correct decisions varies from 127 to 130 [5, 6].

3 Conclusions

On the basis of the findings presented in the previous chapters, it is possible to define some recommendations that could reduce, respectively, eliminate the effect of different number non-conforming samples. Since methodological guides do not make clear recommendations regarding the number of non-conforming samples used for analysis, there is a need to devote increased attention to sample selection. In this selection must be taken into account a number of conflicting requirements, respectively. aspects of the composition of the samples used to measurement system analysis for attributive quality characteristics. These aspects are shown in Figure 5.

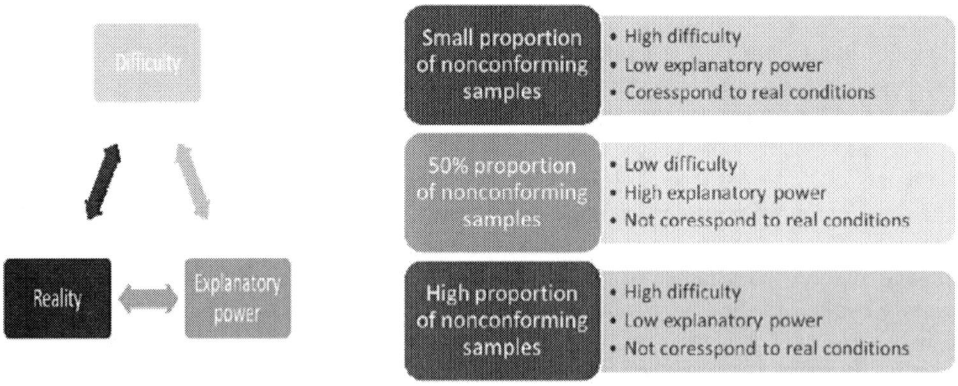

Fig. 5. Influence of aspects at different proportions of non-conforming samples

The first aspect to be taken into consideration is the difficulty or, the accuracy of operator evaluation. With a lower or higher number of non-conforming samples, the operator must make more correct decisions to achieve the required level of the kappa indicator than when the proportion of nonconforming and conforming samples is equal. Another aspect that needs to be consider is the relevancy, or the explanatory power of the whole analysis. If a small number of non-conforming samples is used, the full range of possible kinds of the defect (non-conformities) will not be covered and the obtained results may not be entirely relevant for this reason. The last considered aspect is the realisticality of performed analysis. Because the manufacturing process typically does not reach such a low quality level, which produces half of nonconforming products (half representation of non-conforming samples), such analysis conditions does not correspond with the reality of the production process. For certain types of attribute quality characteristic and certain types of control systems, it was observed that operators achieved less error in the case of low (realistic) proportion of non-conforming samples. [7]

This paper was elaborated in the frame of the specific research projects No. SP 2018/97 and SP 2018/109, which has been solved at the Faculty of Metallurgy and Materials Engineering, VŠB-TU Ostrava with the support of Ministry of Education, Youth and Sports, Czech Republic.

References

1. Chrysler Group LLC, Ford Company, General Motors corporation *Measurement Systems Analysis. Reference Manual. 4th ed.*, (2010)
2. D.C. Montgomery, *Statistical Quality Control: A modern introduction*, *7th ed.* 768, (2013)
3. H. Pačaiová, J. Sinay, R. Turisová, Z. Hajduová, Š. Markulík, Measurement, **109**, 359-365 (2017)
4. J. Petrík, Archives of Metallurgy and Materials, **61**, 4 (2016)
5. P. Klaput, R. Macek, A. Lubojacká, *Kvalita – Quality* (2014)
6. E. Vágó, A. Kemény. *Quality and Reliability Engineering International,* **27**, 6 (2011)
7. J. Petrík, P. Palfy, M. Havlík, International Journal of Engineering, **11,** 1 (2013)

Comparison of Approaches to Gauge Repeatability and Reproducibility Analysis

Pavlína Mikulová[1,*], and *Jiří* Plura[2]

[1]VSB - Technical University of Ostrava, FMME, Department of Quality Management, Ostrava-Poruba, 17.listopadu 15, 70833 Czech Republic
[2]VSB - Technical University of Ostrava, FMME, Department of Quality Management, Ostrava-Poruba, 17.listopadu 15, 70833 Czech Republic

Abstract. The paper presents approaches to Gauge Repeatability and Reproducibility (GRR) analysis. The aim of the paper is to compare three methods of GRR analysis – A&R method, ANOVA and EMP III. The first part gives insight to use of Measurement System Analysis (MSA) since properly planned and executed measurement systems can provide a good basis for decision making in companies. One of the new approaches will be presented that is different from the others described by AIAG or VDA guide. Compared to known A&R method and ANOVA, EMP III interpret the results in more detail. Beyond the information about variability of measurement process, the approach provides other useful parameters as attenuation of process signals, change of detecting error shift and an ability to track process improvements. Wider use of the results is illustrated by an example of measuring the dimensions of fasteners.

1 Introduction

Decision-making based on facts is one of the fundamental principles of contemporary quality management. These facts are in most cases obtained from measurements. The proposed or already used measurement system requires attention, since an inadequate measurement system can provide distorted information that can lead to wrong decisions. The suitability of a measurement system is an important part of quality planning and quality improvement [1] and, for example, in automotive industry, it is strictly required [2].

An ineffective measurement system can allow bad parts to be accepted and good parts to be rejected. MSA could have prevented the problem. Thus, it is a collection of experiments and analyzes performed to evaluate a measurement system capability, performance and amount of uncertainty regarding the values measured [3, 4]. The knowledge of measurement system properties is important in assessing the conformity of products. The conformity/nonconformity of a product can be clearly confirmed only if the measurement uncertainty interval lies within or outside the tolerance [5]. Due to the tolerance limits given by the specifications, this means that the possibility of making

* Corresponding author: pavlina.mikulova.st@vsb.cz

a wrong decision about the part exists only if the measurement system error showed out of the limits. This leads to three distinct areas (see Fig. 1).

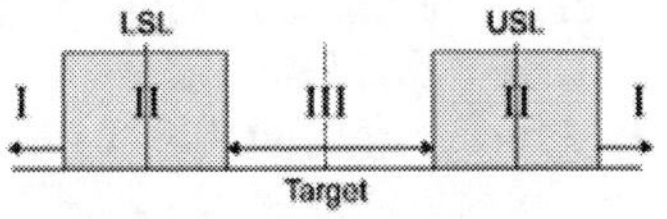

Fig. 1. Measurement uncertainty areas [3].

Area I shows bad parts will always be marked bad, area II shows a potential wrong decision can be made and area III represents good parts will always called good [3, 5]. There are many different factors that affect the measurement systems and influence the measured values including the causes related to the properties of measuring device itself, the measurement procedure, the operator performing measurements with a measured object, and a number of other factors [6]. The actions of these causes can be detected by analyzing the properties. The most used analysis of measurement system, which makes possible to evaluate repeatability and reproducibility of measurement system is GRR analysis.

2 Analysis of repeatability and reproducibility

Generally, it is not possible to ensure changeless measurement conditions in realization of the measurement. The real measurement conditions usually change, while it is mostly caused by the change of operator performing the measurement. Two major sources of measurement system variability are repeatability and reproducibility [7].

2.1 Methods of GRR analysis

Comprehensive GRR study can be performed by using one of the following techniques:

1. A&R method (Average and Range method)
2. ANOVA (Analysis of Variance)
3. EMP (Evaluating the Measurement Process).

A&R method uses subgroups of data to determine the variance due to the various sources of variation. It does not take into account the operator-part interaction. ANOVA is a technique that examines what sources of variation have a significant impact on the results. This approach actually adds another source of variation to the mix and it is able to identify the operator-part interactions [3]. Recently, the attention is focused on the third method known as EMP III approach characterized below.

2.2 Introduction of EMP III approach

D. J. Wheeler, an expert in statistical process control, introduces a new approach and points out the statistical errors (the use of less appropriate criteria) of AIAG approach, the most significant of which is the use of ratios based on the standard deviation. Wheeler introduces a number of notable differences considering: metrics based on the variance instead of standard deviation; Intraclass Correlation Coefficient (ICC); probable error providing an absolute characterization of measurement error; process signal attenuation; monitor classes classifying the measurement system [8]. A proper understanding of probable error and intraclass correlation coefficient will allow to quantify the uncertainty in the measurement. D. J. Wheeler classifies the measurement system as a first, second, third or fourth class monitor based on the ICC. It is the ratio of the part variance to the total variance that also can be simply calculated from the results of A&R or ANOVA method:

$$ICC=(PV)^2/(TV)^2 \quad (1)$$

In case the samples represent the production range and part variance can be determined, it is worth to calculate ICC. This ratio describes that proportion of the variation in the product measurements that is directly attributable to the variation in the product stream. The EMP results are interpreted by four categories which give insight into the measurement system (see Table 1): Firstly, how it can reduce the strength of a signal (out of control points) on a control chart. Secondly, give the chance of the measurement system detecting a large shift. Thirdly, are able to track process improvements. D. J. Wheeler claims those criteria give a good understanding of relative usefulness of the measurement system telling that the first, second and third class monitors include acceptable measurement systems [9].

Table 1. Four class monitors according to D. J. Wheeler [9].

ICC	Type of class monitor	Attenuation of process signal	Chance of detecting ±3σ std. error shift	Ability to track process improvements
0.8 to 1.0	First class	Less than 10%	More than 99% with rule 1	Up to Cp80
0.5 to 0.8	Second class	From 10% to 30%	More than 88% with rule 1	Up to Cp50
0.2 to 0.5	Third class	From 30% to 55%	More than 91% with rules 1, 2, 3, 4 (see [9])	Up toCp20
0.0 to 0.2	Fourth class	More than 55%	Rapidly vanishing	Unable to track

In Table 2, the acceptance criteria of measurement system are stated. Those are criteria prescribed by AIAG and compared with new evaluative criterion introduced by D.J. Wheeler, ICC. As seen, related to the acceptance of measurement system, the value of ICC corresponding to %GRR is allowed to be in the small range (0.91-1) if considering the AIAG criteria. However, observing the Table 1 Wheeler's criteria are significantly less strict (ICC∈(0,5-1)).

Table 2. Comparison of evaluative crtiteria.

%GRR	Corresponding ICC	Classification
0 to 10	0.99 to 1	Acceptable
10 to 30	0.91 to 0.99	Conditionally acceptable
> 30	<0.91	Unacceptable

2.3 Application of EMP III approach

Measurement system which is used to measure the height of the nuts is subjected to analysis. Measured values of ten nuts (each was measured twice by an operator), representing the entire production range obtained by three operators using the digital calliper (data from [1]). The procedure of EMP III approach is identical with the one of A&R method with the difference of calculating the values considering the variance instead of the standard deviation. The procedure of the approach and equations for the all calculations are summarized in [9].

2.3.1 Characterizing the relative usefulness of measurement system

The individual proportions of variance were calculated: proportion of repeatability, proportion of reproducibility, proportion of repeatability and reproducibility, proportion of part variation to total variance which is, at the same, the value of intraclass correlation coefficient (see Equation 1). After multiplying the ICC by hundreds, it simply represents percentage of total variance caused by part variance. The important data for calculating ICC were obtained: firstly, the height of ten nuts were measured twice by each operator (PV is calculated), secondly, single standard sample of nut is measured twice by each operator (TV caused by the measurement system can be obtained).

$$ICC=(PV)^2/(TV)^2 =[(TV)^2-(GRR)^2]/(TV)^2=1-(GRR)^2/(TV)^2 \quad (2)$$

The value of ICC tells the following: if only the random causes of variability affect the process, %ICC is the variability between measured parts and, on the contrary, the %(1-ICC) is the variability of measurement system. According to Wheeler's criteria (see Table 1) and from the value of ICC=0.968 (see Table 3), our measurement system belongs into first class monitor. As seen in Table 3, the the results obtained by A&R method and EMP III are identical and very similar to ANOVA results. While according to AIAG criteria measurement system is classified as conditionaly acceptable, in accordance to EMP III approach the measurement system belongs into first class monitor.

Table 3. Comparison of results obtained by three methods of GRR analysis.

Source	EV	AV	GRR	PV	TV	
A&R method						**ndc**
Standard deviation	0.02511	0.02151	0.03307	0.18244	0.18542	
Variance	0.00063	0.00046	0.00109	0.03329	0.03438	7
% of total std.	13.54	11.60	17.83	98.40	100	
ANOVA						**ndc**
Standard deviation	0.02443	0.02102	0.03223	0.18589	0.18866	
Variance	0.00060	0.00044	0.00104	0.03455	0.03559	8
% of total std.	12.95	11.14	17.08	98.53	100	
EMP III						**ICC**
Standard deviation	0.02511	0.02151	0.03307	0.18244	0.18542	
Variance	0.00063	0.00046	0.00109	0.03329	0.03438	0.968
% of total variance	1.83	1.35	3.18	96.82	100	

2.3.2 Interpretation of results

Characterization of measurement system use by classification in the first, second, third or fourth class monitors: after classifying the measurement system into one of the four categories, the evaluation of measurement system is assessed from several aspects:
a) attenuation of process signals: any signal of change in production process will be reduced (not fully recorded in control chart). This attenuation is characterized by:

$$\textit{Attenuation of process changes signals} = 1 - \sqrt{ICC} \quad (3)$$

Hence, this attenuation means how many percentages of process change signals will be reduced in control chart. In our case, these signals are reduced by 1.6%.
b) Chance of detecting ±3σ std. error shift: the chance of detecting a significant changes during process in control chart for the elected position (control chart for individual values or averages) by ±3σ is more than 99% (considering the first class monitor) using rule 1 (Western Electric rules apply if 10 samples are measured).
c) Ability of measurement system to track process improvements that means to reduce production variability up to the corresponding C_p level. Recognizing significant process changes is very important because if this does not happen, there is a problem concerning the control chart detection of incorrect measurements. Either the control chart should have

detected incorrect measurement, but did not do so, or the measurement was correct, but control chart identified it as an incorrect one. After calculating value of C_p80 (see the equation in [9]), we have got value 4.75. The value tells reduction in variability of the process, whose C_p index would prove to be more than 4.75, also indicates as for the assessment of measurement system, that the value of intraclass correlation coefficient will be less than 0.80. It is proven by moving the measurement system from one class monitor downwards. Capability of measurement system to capture an improvement of the process reflects how great an improving or worsening the process have to be.

2.4 Comparison of approaches to GRR analysis for simulated data

For a better comparison and understanding the differences between EMP III and AIAG approaches, we did a simulation of the data where following situation occurred – a measurement system is unacceptable as the results of A&R method and ANOVA, but acceptable by Wheeler's method.

Table 4. Comparison of three approaches to GRR analysis for simulated data

Source	EV	AV	GRR	PV	TV	
A&R						ndc
Standard deviation	0.08242	0.05919	0.10147	0.18244	0.20876	2
Variance	0.00679	0.00350	0.01030	0.03329	0.04358	
% of total std.	39.48	28.35	48.61	87.39	100	
ANOVA						ndc
Standard deviation	0.073137	0.057764	0.093197	0.180378	0.203032	2
Variance	0.005349	0.0033366	0.0086857	0.0325363	0.041222	
% of total std.	36.02	28.45	45.9	88.84	100	
EMP III						ICC
Standard deviation	0.08242	0.05919	0.10147	0.18244	0.20876	0.76
Variance	0.00679	0.00350	0.01030	0.03329	0.04358	
% of total variance	15.59	8.04	23.62	76.38	100	

After the data simulation, A&R method and ANOVA evaluated the measurement system (see Table 3) as unacceptable for next using. In compliance with EMP III study, the measurement system belongs to the second class monitor (due to ICC value equal to 0.76). For the illustration of excessively high variability of measurement system (see Fig. 2), both situations are compared (with original and simulated data), where the difference of variability can be observed in Fig. 2.

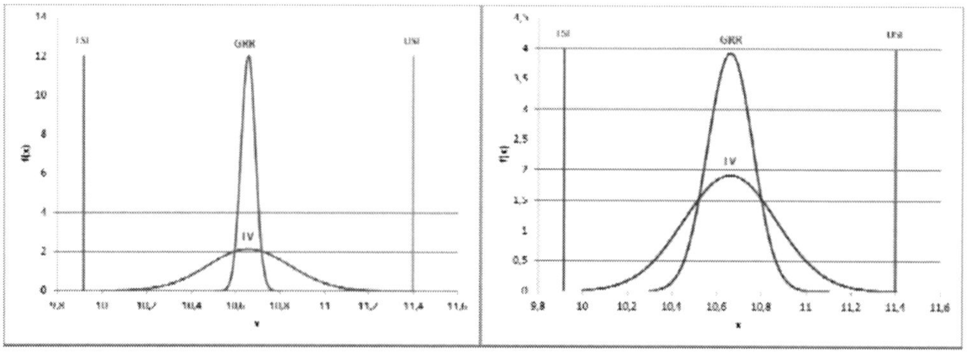

Fig. 2. Comparison of variability of measurement system (GRR) and total variability (TV) for a) original data, b) simulated data

These figures include the calculated tolerance limits that correspond to the centred process with $C_{pk} = 1.33$. In particular, Figure 3 shows that the measurement uncertainty interval for simulated data is very wide and creates almost half (48.61%) of the production range width. Related to the tolerance width, it means 36.5%. Due to the fact we have to calculate the uncertainty interval at each of the tolerance limits, the given measurement system would only be able to evaluate the parts as conforming if the measured value would lie in the area constituting 27% of tolerance width around the tolerance centre. It means for product conformity assessment this measurement system cannot be considered as acceptable.

3 Conclusion

Due to the above results of acceptance or rejection the measurement systems obtained by the three methods of GRR analysis, we have come to the following conclusion. Even though, methodology recommended by AIAG strictly refuses measurement system with %GRR = 48.61% and ndc = 2 (see Table 3), EMP III allows to classify the measurement system to the second class monitor, i.e. accepts it with any doubts. According to Wheeler this measurement system is still able to monitor the process changes (to detect the out of control points in control chart and to indicate process improvement), i.e. it is applicable only for cases when measurement system is used for process monitoring. For the measurement systems, which are used for decision making about products conformity, EMP III criteria cannot be applied.

This paper was elaborated in the frame of the specific research project No. SP 2018/109, which has been solved at the Faculty of Metallurgy and Materials Engineering, VŠB-TU Ostrava with the support of Ministry of Education, Youth and Sports, Czech Republic.

References

1. J. Plura, Quality Planning and Continuous Quality Improvement, **244** (2001)
2. AIAG, *Production Part Approval Processs (PPAP)*, (Chrysler Gp. LLC, Ford Motor Company, General Motors Corp., 2006)
3. MSA Work Gp., *Measurement Systems analysis, Reference Manual fourth edition*, (2010)
4. C. D. Montgomery, *Introduction to Statistical Quality Control*, 754 (2009)
5. ISO 10576-1, *Statistical methods - Guidelines for the evaluation of conformity with specified requirements - Part 1: General principles*, (2003)
6. J. Plura, P. Klaput, *The Possibilities of Confidence Improvement of Measurement Systems Analyses Results, 20th Anniversary International Conference on Metallurgy and Materials "METAL 2011"*, (Brno, Czech Republic, 2011)
7. J. Plura, *Quality Planning II*, 172 (2012)
8. Miner. *Intro to MSA of Continuous Data-Part 7* (Quality Forum Online, 2016)
9. D. J. Wheeler, *EMP III: Evaluating the Measurement Process & Using Imperfect Data*, 316 (2006)

Improvement of the production system functioning with using quality management tools in the metallurgical enterprise

Marta Daroń[1,*], *Monika* Górska[2] , and *Anna* Dunay[3]

[1]Czestochowa University of Technology, Faculty of Management
[2]Czestochowa University of Technology, Faculty of Production Engineering and Materials Technology
[3]Szent István University, Faculty of Economics and Social Science

Abstract. Quality management tools undoubtedly influence the production processes in the company, by the creation of favorable conditions for its improvement. The proper selection and proper use of these tools is essential to improve the functioning of the company. Therefore, the main goal undertaken at work was to assess the situation related to the selection and effectiveness of methods used by a selected company operating in the metal industry. Moreover, it was necessary to determine to what extent these methods are applied and which of them have the greatest impact on the effectiveness of the process of improving the functioning of the enterprise.
Key words : improvement, processes, production

1 Introduction

The metallurgical industry includes the production of pig iron, ferroalloys, iron, steel, steel products and the production of metal finished products (see Regulation of the Council of December 24, 2007 regarding the Polish Classification of Activities – PKD [1]). Thus, a specific product is created which constitutes the effect of the activities of entities in this industry. Its quality largely results from the functioning of the production system being at the same time the showcase of the company on the market.

Metal industry companies must pay attention to the efficiency and effectiveness of production processes. The most important goals of improving production efficiency include [2]: increase in productivity, improvement in the quality of manufactured products, reduction in the level of work in progress, shortening the time or number of retoolings, shortening production cycles, reducing the need for production space. In order to achieve these goals, the entities of the metal industry have at their disposal many methods or tools of quality management [3], which are described in the next part of the study. The application of them allows introducing improvement actions dedicated to a given production system. However, not all methods may contribute to achieve the same effect. Therefore, the research goal of the work was to analyze the applied methods of production

* Corresponding author: marta.daron@wz.pcz.pl

system improving in a selected company in the metallurgical industry. There were also presented results of implementing of these methods. Additionally, there was indicated the validity of the applied methods and there was assessed their suitability.

2 Improving production processes

The basis for improving the company's production system is the optimal set and use of production factors, which result from the need to meet all customer requirements and adaptation to changing environmental conditions, as well as increase of competitiveness and cost pressure, or increase of the role of internal and external flexibility. The goal of the undertaken activities is to achieve measurable economic and production effects such as the efficiency of the production process and timely performance of tasks, as well as identification of factors that affect not only the quality of products but also shortening of inter-operative times, minimization of inventory. An important role in the improvement of existing production systems or their redesign is mainly played by modern methods of organization and management of production processes that result from the desire to improve efficiency and reduce costs [4].

Process management is embedded into the activities of today's enterprises not only in the field of manufacturing. However, in order to be able to talk about process management, one should realize the necessity of measuring it – to evaluate its efficiency, flexibility, etc. Numerous process measures are indicated in the literature [5-7]. Some of them can be used in general, but there are also those dedicated to specific processes (production, logistics, customer service and others). Nevertheless, each process should be assessed by measuring the main attributes of the process [8]: process costs, length of process completion time, flexibility, quality, organization relevance, customer relevance and the ethical dimension of the process. What's more, there are a number of methods and tools described in the literature that are useful by improving each process, e.g. TQM, 5WHY, Pareto chart, the control chart, the Ishikawa diagram, the relationship diagram, process mapping, ABCD (Audience, Behavior, Condition and Degree Method - which is an expert tool that allows to identify the most important causes of the problem, according to the evaluators) and others. However, there are also those that are used to analyze individual categories of processes, e.g. production: FMEA (Failure Mode and Effects Analysis*)*, SMED (Single Minute Exchange of Die), TPM (Total Productive Maintenance), Poka Yoke, 5W2H method or 5S [9-13]. Assessment of a given process is often the basis for undertaking actions aimed at introducing improvement changes.

3 Main results

The presented research was carried out in a metal industry enterprise belonging to medium-sized business entities (according to the employment size criterion in accordance with the European Commission Regulation No. 364 of February 25, 2004 on SME [14]). The main production profile are rolling logs and other metal elements sold on the domestic and foreign market. The selection of the company for the study was partly random - related to the selection of the unit from the manufacturing industry, and partly deliberate - related to the knowledge of the industry and location-related, which involved the possibility of conducting direct research. The surveyed company plans to expand its activity, therefore the problem of process improvement is inscribed in its development strategy.

To achieve research goal, a direct survey was conducted in a selected company, covering the last four years (due to the low percentage of people working in the surveyed enterprise over 5 years). Respondents were employees of the production department and their superiors, engaged in key processes (production). The research tool was a questionnaire consisting of open, semi-open and closed questions. On the basis of the analysis of the answers from open and semi-open questions, some groups of answers were created (this concerned mainly the indication of the benefits from the implementation of specific improvement instruments). Some of the questions also made it possible to assess research feature in the form of a rank (from the least important to the most important / from useless to the most useful, etc.). Parallel, a direct interview was conducted with persons on managerial positions, in which the improvement activities undertaken within the field of production were analyzed over four years. On the basis of company's data from the selected research period there was also analyzed the level of selected indicators (production readiness of machines, their efficiency and production involvement). This made it possible to identify specific measurable benefits in the area of production.

First of all, the knowledge of the respondents regarding the methods and techniques of improvement of processes used in the surveyed enterprise was assessed. Then, the respondents were asked to indicate which of these instruments are used by them in practice (Fig. 1).

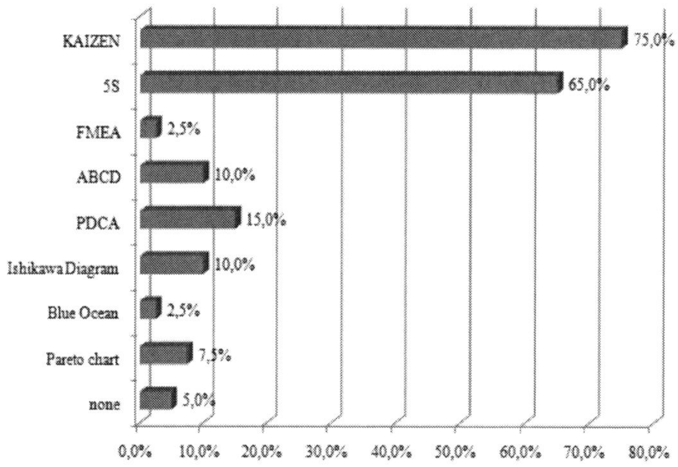

Fig. 1. Improvement instruments used in practice by respondents in a selected company.

The results of the study confirm that, from among many available methods in the process of improving the processing system, in practice, two are mainly used, i.e.: Kaizen and 5S. Only 5% of respondents (2 production employees) stated that they did not use any improvement instruments.

According to Grajewski's research on the use of tools for improving processes [7] there is shown that in many cases the methods are misused or have a low level of suitability. Therefore, it was reasonable to test the level of suitability of the methods used in the research entity in the respondents' opinion. Due to the experience and practical knowledge of the respondents participating in the study, their assessment of the level of usefulness of methods and improvement tools is correct in substance. The results of the study on determining the level of suitability of selected improvement instruments in processing processes are presented in Table 1.

Analyzing the results from Table 1, the surveyed employees consider the 5S method as the most useful improvement technique for the company. As many as 45% of respondents awarded "7" in this case. In turn, 47% of respondents consider the Kaizen technique (score 6) to be important in this respect. On the other hand, 35% of respondents considered the FMEA method (assessment 5) to be helpful in improving their activities. In the case of the analyzed method, the obtained result proved that despite the using of this tool in practice by only one person, other employees had knowledge about it. Although they did not use this method, they participated in the development of relevant data, which was a key source of information for FMEA process analysis. The decisions that were made on the basis of FMEA were disseminated among the employees of the processing area, which was reflected in the research results.

Table 1. Assessment of the usefulness of improvement tools [% of respondents].

Improvement tools	Assessment (Grade 1 - unhelpful tool / method of improvement, assessment 7 - improvement tool useful in the highest degree)						
	1	**2**	**3**	**4**	**5**	**6**	**7**
Kaizen	0	0	5	5	20	**47**	23
5S	7	4	4	7	13	20	**45**
FMEA	0	7	15	25	**35**	13	5
PDCA	0	7	18	**33**	22	15	5
ABCD	18	20	**33**	12	5	5	7
Ishikawa Diagram	30	**40**	15	10	0	0	5
Pareto Diagram	**45**	22	10	8	5	0	10

The analysis of the remaining results of the study shows that the respondents positively assess the effectiveness of tools and techniques used in the enterprise. As many as 80% of employees believe that information received using selected methods or techniques were helpful by identifying and improving those areas that should be subjected to improvement processes. In the examined case, the area that needed changes the most over the period (i.e. four years), was the area of production including such elements as: the technology used, machines, equipment and technical equipment. Therefore, the company invested financial resources in the modernization and extension of the machine park. As a result of the actions taken, there were eliminated some of the reasons that were responsible for the quality deficiencies of the finished product resulting from the use of outdated production technology. The largest amount of financial capital was invested in machinery and technical equipment, which resulted in not only improving the quality of the production process itself, but also the finished product. Currently, the company has one of the latest technological lines for the production of bearings. Confirmation of the legitimacy of introducing improvement actions was reflected in the growth of indicators, i.e.: production readiness of machines (machines installed / machines owned * 100%), machines' efficiency (active machines on the largest change / machines owned * 100%) and production commitment (active machines on the largest change / machines installed * 100%). The indicators' increase is the preferred situation. The analysis of collected data allows to state that the indicator of production readiness of machines and devices remains constant, i.e. 100%. Index of machine and equipment efficiency from 68.84%, in 2014 increased to the level of 92.68% in 2017. The index of production engagement of machines and devices oscillated in similar limits as the index of machines and devices efficiency. The machine and equipment efficiency index improved as a result of the modernization of the machine park. It is also important to increase the skills of employees in the field of management and control of machines and devices. The above means that improvement actions have had a positive effect

Undoubtedly, every enterprise striving to improve its operations and to match more completly to the market needs should undergo a detailed analysis of the benefits resulting from the introduction of improvement actions using appropriate instruments. Therefore, the issue was analyzed in the next stage of the study (Table 2).

Table 2. Benefits from the implementation of improvement tools [% of respondents].

Advantages	Assessment (Grade 1 - no benefits, grade 7 - the occurrence of a given benefit to a very appreciable extent)						
	1	**2**	**3**	**4**	**5**	**6**	**7**
Improving efficiency and modernization of the production process	10	17	**33**	20	13	7	0
Increased demand for offered products	**40**	23	25	7	5	0	0
Improvement of quality, including work safety	25	**46**	13	5	8	0	3
Improving the control of the production process	5	0	5	8	17	25	**40**
Increased competence and responsibility of employees	5	3	7	10	10	**38**	27
Increased competitiveness of products	10	11	17	**32**	15	10	5
Organization of documentation	5	0	0	18	**32**	20	25

The analysis of the results contained in Table 2, allows to conclude that the benefits that result from the implementation of improvement tools and techniques, the respondents mainly consider improving the control of the production process. As many as 40% of respondents chose the indicator "7" in this case. Undoubtedly, streamlining the control of the production process is essential in a properly operating company. Wrongly, the respondents assess very high the benefits of increasing the demand for offered products. The least important factor was indicated by 46% of respondents. Increasing demand means the necessity to increase production and increase profits for the company. 33% of the respondents rated increasing the efficiency and modernization of the production process on 3 on a seven-point scale. This is partially explained by the recently introduced technological changes in the surveyed enterprise. However, improving the efficiency of processes should be continuous, therefore emphasis should be placed on the greater involvement of employees in this area. Respondents paid little attention to improving work safety, by this indicator "2" was marked by 46% of respondents.

Summary

The selection of appropriate methods and tools is of fundamental importance in the improvement process. They are used to collect and process information, supervise processes, detect errors and defects, finally allowing to check the effectiveness of decisions made. They are therefore necessary for the proper functioning of the enterprise.

The research (case study) concerned the production system of a selected company in the metallurgical industry. It is obvious that each entity should have its own set of improvement tools. However, the results obtained allowed to draw specific conclusions and shed light on further research directions. The presented methodology can be successfully applied in other production units. It would be reasonable to extend this type of research to a larger group of entities in a given industry. Perhaps it would be possible to create a model for the selection of specific methods / tools, also in the context of specific benefits from application of them, which would be dedicated to enterprises in a given industry. The obtained results of the research allowed to determine which of the tools are the most popular in the studied enterprise and at the same time which and to what extent they allow to achieve specific benefits. The analysis of the respondents' answers also showed that the improvement tools used do not always allow achieving the assumed benefits (they are not always useful or do not correspond to the expected benefits).

The results are presented as a whole for the research sample. For a more in-depth analysis, it should be presented in a breakdown into groups of employees (production and their superiors). For example, production workers evaluated different methods, whereas they actually used only 2 (5S and Kaizen), while their superiors used much more improvement methods at work which was combined with the specificity of their work - analysis and evaluation of processes, etc. Unfortunately, the volumetric framework of this study does not allow to present such a detailed analysis.

References:

1. Rozporządzenie Rady Ministrów z dnia 24.12.2007 r. w sprawie Polskiej Klasyfikacji Działalności - PKD, Dz. U. z 2007 r. nr 251, poz. 1885 z późn. zm.
2. A. Koliński, Log., **5**, 1083-1091 (2011)
3. R. Ulewicz, Manuf. Tech., **14**, 104-111 (2014)
4. M. Brzeziński, *Organizacja produkcji w przedsiębiorstwie* (Difin, Warszawa, 2013)
5. M. Flieger, Przeg. Org., **9**, 38-42 (2009)
6. K. Bors, B. Nogalski, *Strategie wzrostu produktywności firmy* (Akad. Ek. w Krak., Kraków, 2000)
7. S. Suss, V. Thomson, J. Eng. Des., **23**, 829–851 (2012)
8. P. Grajewski, *Organizacja procesowa* (Pol. Wyd. Ek. S.A., Warszawa, 2016)
9. B. Cs. Illés, Cs. Szuda, (A. Dunay (ed.) *Proceedings of the 5th International Conference on Management 2015: Management, Leadership and Strategy for SMEs' Competitiveness,* (Szent István Publ. H., Gödöllő, Hungary, 2015)
10. B. Cs. Ilés, Cs. Szuda, A.Dunay, (B. Cs. Illés, M. Nowicka-Skowron, E. Horská, A. Dunay, (eds.) *Management and Organization: Concepts, Tools and Applications,* (Pear. Ed. Ltd Harlow UK, 2017),
11. S. Kot, K. Grondys, *Applied Mechanics and Materials,* **309**, 324-331 (2013)
12. N. Baskiewicz, M. Kadłubek, *Prace Naukowe UE we Wrocławiu,* **463**, 121-131 (2017)
13. K. Nowakowska, M. Ingaldi, Archiwum Wiedzy inżynierskiej, **1(1),** 39-41 (2016)
14. Rozporządzenie Komisji (WE) nr 364/2004 z dnia 25 lutego 2004r. o MSP, Dz. U. UE 28.2.2004

The 3x3 matrix as a tool for evaluation of technological position of the enterprise

Manuela Ingaldi[1,*], and *Marina* Zhuravskaya[2]

[1]Czestochowa University of Technology, Faculty of Management, Poland
[2]The Ural State University of Railway Transport, Russia

Abstract. Technology is the way we do things. It is the application of knowledge to create products that serve human needs and aspirations. It decides about the success of the enterprises. The technological position of every enterprise can be treated as an element of its strategy. This position depends on existing technologies, equipment and on element deciding about its market position. For this evaluation a 3x3 matrix can be used. This matrix consists of 9 fields that correspond to the respective technological positions of the enterprise and help to identify its strategy in terms of technology. The research was shown on example the foundry. To evaluate its technological position it was necessary to indicate factors - those that determine the technological possibilities of the company and those that determine its position on the market.

1 Introduction

The technology can be defined as a practical application of the knowledge in a specific method of producing goods or achieving a particular effect; that is the collection of techniques, skills, methods and processes used in the production of goods or services or in the accomplishment of objectives, such as scientific investigation. This definition suggests that, in order to produce products of the right quality level, which will find buyers on the market, the enterprise must have adequate knowledge of their production.

The word "technology" comes from the Greek words "techne" and "logos" (collection, reason). It means "a method of processing material goods into another useful good; also knowledge of this process" [1]. It is processing material goods in a useful goods in a targeted and cost-effective way," but also "technology" means "knowledge of the process.

Many products require the use or may be produced with different technologies. Their choice depends on many technical, economic, social or cultural factors, such as costs, time, legal restrictions, safety, comfort, etc. Thus, the technology and product quality, and at the same time the final result of production will be affected by production machines owned by the enterprise. All resources of enterprises are increasingly considered as potential strategic elements and the conditions of their development.

Very large impact on the effect of the technology therefore have a variety of resources. Many of these resources do not depend directly on the enterprise, but on its suppliers,

* Corresponding author: manuela@gazeta.pl

and/or its environment. Lack of those resources often contributes to the loss of positions on the market. Therefore, actions for their development or their acquisition are such an important part of the strategic thinking of enterprise's management [2-4].

Competitiveness is the process by which one entity strives to outperform another. For an enterprise, being competitive implies that it has the ability to produce a product, in a timely and cost effective manner, which meets the test of the marketplace and the needs of customers [5]. The activities of enterprises are to gain and increase market share, which in turn determines their profits. And the profit is the main goal of every enterprise.

Technological strategy does not include only the selection of procedures and products, but the skills which are necessary to maintain the competitive advantage. It is not only about the ability how to produce or about knowledge of technology, but also about developing the skills of the sales, supply, financing, understanding the environment and involved processes and their consequences [6].

The research was shown on an example of a chosen enterprise. The purpose of this article is to evaluate the technological position of the enterprise from the foundry industry. For this purpose a little known and little-used 3x3 matrix was chosen. At the beginning those factors that determine the technological possibilities of the company and those that determine its position on the market were indicated. Then, based on their evaluations and averages of each group, the technological position in the 3x3 matrix was indicated.

2 Literature review

Many managers are aware of the strategic importance of technology in delivering value and competitive advantage of their enterprises and the industrial networks in which they operate [7, 8]. This technology management framework includes some important elements such as technology management processes (identification, selection, acquisition, exploitation and protection), business processes (strategy, innovation and operations), highlighting the dialogue that is needed between the commercial and technological functions in the business to support effective technology management. It points to the link between the technological possibilities of the enterprise and its competitiveness, which in turn is used in a 3x3 matrix presented in the paper.

Scientists of the world have used the idea of a matrix to describe the various enterprise's scenarios, e.g. the technological one. Waters proposed to define the relationships between the customer and supplier based on a 2x2 matrix [9].

Zhuravskaya et al. have structured the problems solved in the logistics systems of different levels using the 3x3 matrix [10]. Khalil and Ezzat also [5] used the 3x3 matrix. This matrix is presented in Fig. 1. They used the matrix reflecting economic and technology momentum and competitive posture. Each of the entries in the matrix (dots) can be viewed as an option, deliberately chosen, to capitalize on. Furthermore, the clustering of the dots can be an indication of (and should reflect) the vision and general strategy of the organization. To exemplify, two clusters are shown. In this case the matrix can be used to compare position of one enterprise in comparison to position of other enterprises.

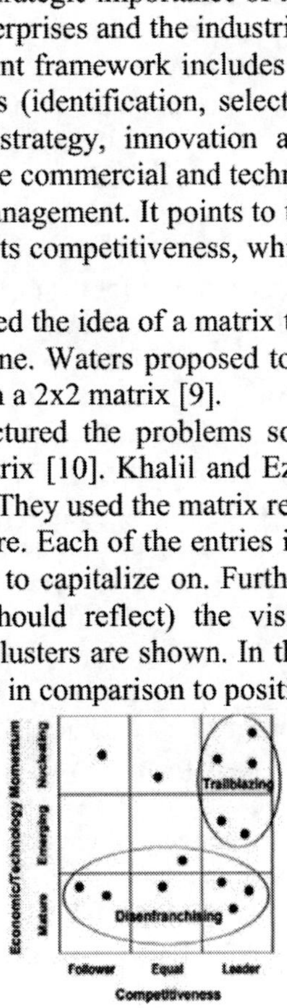

Fig 1. Profiling opportunities [5].

A bit different variant of the 3x3 matrix was proposed by Lowe [11]. The 3x3 matrix should be used to evaluate the strengths and weaknesses of the technology. Lowe called it as "the matrix of linkages between enterprise's market position and its technological possibilities". He also proposed the use of the matrix that would define the relationship of technological uncertainty and threats of the enterprise. Technology evaluation is often subjective, but with use of the 3x3 matrix this evaluation becomes more accurate. In this case, the matrix was used for a single enterprise. Thus, different elements deciding about position of the enterprise can be included. The basic version of the 3x3 matrix is shown in Fig. 2.

POSITION ON THE MARKET				
	7-9	Buy the ready technology	Develop your technological potential	Focus on the revealed chance
	4-6	Keep in the background	Search for occasions	Improve the marketing
	1-3	Keep in the background	Discover the incidental market	Search for partners
		1-3	**4-6**	**7-9**
		TECHNOLOGICAL POSSIBILITIES		

Fig 2. A scheme of the 3x3 matrix [11-13].

The 3x3 matrix consists of 9 fields that correspond to the respective technological positions of the enterprise and help to identify the way to proceed in terms of technology. It can therefore be used as a part of defining the strategy of the enterprise. The objective of any enterprise is the field marked as "Focus on the revealed chance".

The first changes to the basic version of the 3x3 matrix, presented in paper [12], are related to the axis. Its new version included the relation between the technological possibilities and competition of the product. This research was a part of the questionnaire which was filled by production workers of the enterprise, who often do not fully know the reality of enterprise. Respondents were asked for the evaluation both factors on a scale from 1 to 9 (1 – low, 9 – high evaluation). The matrix showed that the adopted scale of evaluation must be divided by 3. It was necessary to change locations of the borderlines to avoid problem with interpretation of the results. After changes they were placed between the possible answers. In the 3x3 matrix the individual pairs of answers were marked. The field with the most pairs of responses was treated as the position of the enterprise.

Other suggestions, related to the matrix, were concerned with the research method [13-14]. The first stage of the research was to determine the factors that would evaluate the technological position of the enterprise: these that determine the technological possibilities of the enterprise and these that determine its position on the market. All these factors were evaluated on a scale from 1 to 9. However, due to the presence of negative factors that change the technological positions of the enterprise, it was decided to change the scale, where 1 indicates a negative impact, 5 is a neutral or very small impact, but there is a factor, while 9 indicates a positive impact. It was also assumed that the borderlines of individual fields have the half-values. Then, average values of both groups of factors were calculated. Those averages indicate the technological position of the enterprise. In case of this research the executives and managers took part. The procedure for the use of experts is more justified, since these are generally people associated directly with the research technology, they carry out the evaluation in many ways.

3 Metodology

The research enterprise was founded in the 90. of the XX century, is located in the Lodzkie Voivodship. The main direction of its activities is the production of castings from Al-Si alloys, by the pressure method. The volume of production per year is estimated at approx. 2400 tons of castings. The product assortment reaches almost 400 different types of castings. The main customers of products are enterprises operating in following industries: automotive, electrical, household appliances, mechanical equipment, construction, for the gas industry, measurement and control equipment.

At the beginning, all the factors, which can have any influence on the research enterprise were listed and divided into two groups: determining the technological possibilities of the enterprise and determining its position on the market. The factors that did not fit into any of the groups have been omitted. All factors were evaluated on a scale presented in papers [13-14]. Finally, average of both groups were calculated and then, based on them, the position in the 3x3 matrix was indicated.

4 Results

Individual factors, determining position of the research enterprise on the market and its technological possibilities and their evaluations are presented in Table 1. The relations of both averages are presented in Fig. 4.

From Fig. 3 it results that the research enterprise is located in the field of the 3x3 matrix, which is marked as "improve the marketing". Factors determining the position of the enterprise on the market reached the average evaluation of 4.8, while its technological possibilities 7.3. Such position means that the research enterprise should put more emphasis on improving the factors related to its position on the market. Some factors had positive influence on results.

Table 1. List of factors for the 3x3 matrix [own study].

Factors deciding about technological possibilities		Evaluation	Factors deciding about position on the market		Evaluation
1.	Implemented management systems	8	1.	Localization	7
2.	Technical equipment	6	2.	Position on the market	5
3.	Machine park	5	3.	Regular customers	7
4.	Production technology	7	4.	Competition	2
5.	Products quality	8	5.	New markets	5
6.	Operation of quality department	8	6.	Export	6
7.	Transport base	6	7.	Marketing department	2
8.	Assortments and its possible extension	8	8.	Advertising	3
9.	Production flexibility	9	9.	Demand and its possible growth	6
10.	Experienced staff	8	Average		**4.8**
Average		**7.3**			

POSITION ON THE MARKET			
7-9	Buy the ready technology	Develop your technological potential	Focus on the revealed chance
4-6	Keep in the background	Search for occasions	Improve the marketing
1-3	Keep in the background	Discover the incidental market	Search for partners
	1-3	**4-6**	**7-9**

TECHNOLOGICAL POSSIBILITIES

Fig 3. The 3x3 matrix for the research enterprise [own study].

Among the factors that determine the technological possibilities, the special attention should be given to production flexibility. Castings are often made on request, according to customer needs, therefore so high evaluation. 5 factors out of 8 received 8, which means positive influence on the technological position: implemented management systems, products quality and operation of quality department, assortments and its possible extension, experienced staff. The assortments and its possible extension are connected with the production flexibility. If it comes to the operation of quality department, it has a partially positive effect on the products quality. Properly selected measuring machines, prepared control procedures and corrective actions allow for a well-conducted quality control. This can also indicate the main production problems and counteract them. The research enterprise has implemented an integrated management system, which has an impact on the process and the products quality. Employees of the research enterprise are people over 40, most well educated, but not necessarily in foundry, not always after studies.

It should be noted that among the factors that determine the technological possibilities, the machine park was evaluated as the lowest. The enterprise uses machines that have been bought in the late 90. Those machines work without bigger breakdowns, but their efficiency is not as very high as it should be.

Among the factors that determine the position on the market, the special attention should be given to regular customers and localization. The research enterprise is located in the Lodzkie Voivodship, i.e. in the central Poland. This allows for easier access to potential customers from the country. The research enterprise has also signed long-term contracts with several customers, which allows the production also in times of lower demand.

In case of the factors determining the position of the enterprise on the market, several factors that negatively affected the position of the enterprise in the 3x3 matrix, were noted. Those are: the competition, marketing department and advertising. It should be kept in mind that in Poland there are many enterprises dealing with casting, hence such an evaluation. If it comes to marketing department and advertising, they are linked together. The research enterprise does not have the marketing department, its managers are dealing with it. It does not carry any advertising campaign, only a few posters with information and sometimes it advertised its products and services in professional journals.

It should be noted that many of the factors, that influenced the enterprise's position on the market, are external factors. So the enterprise cannot do anything to change them. Therefore, it should deal only with those factors, which were evaluated as quite high.

5 Summary and conclusion

In the paper, the 3x3 matrix, which includes elements associated with the technological possibilities of an enterprise and its position on the market, is presented. The technology used by individual enterprise determines the quality of the finished products, and thus the customer satisfaction. Its position on the market depends on the competitiveness of products. Therefore, it is important to determine an appropriate strategy, which will decide about further development of the enterprise.

In the paper, the results for the designation of technological position of the enterprise from foundry industry with regard to its strategy, were presented. For this purpose, the 3x3 matrix was used. It turned out that the research enterprise is located in the field of the matrix labelled as "improve the marketing". This means that the enterprise should put more emphasis on improving the factors related to its position on the market.

Among the factors that determine the technological possibilities of the enterprise the special attention should be given to production flexibility, implemented management systems, products' quality and operation of quality department, assortments and its possible extension, experienced staff, while among the factors that determine the position of the enterprise to the regular customers and localization.

In the article it was also shown that simple tool like the matrix of relations between position of the enterprise on the market and its technological possibilities, also known as 3x3 matrix, can easily be used by the enterprises to evaluate their operation and to determine the elements of their strategies. In addition, it was shown that easy tools, such as the 3x3 matrix, that are available in the literature and used by other scientists, can be subjected to small modifications, which allows for their better use.

Reference

1. K. Bialecki, A. Dorosz, W. Januszkiewicz, *Słownik handlu zagranicznego* (PWE, Warszawa, 1986)
2. E. Gwarda-Gruszczynska, Acta Univer. Lodz., Folia Oecon. **234**, 265-275 (2010)
3. J.B. Barney, *Gaining and Sustaining Competitive Advantage* (Addison Wesley, New York-Paris, 1997)
4. P. Sygut, M. Krynke, Production Engineering Archives, **17**, 32-35 (2017)
5. T.M. Khalil, H.A. Ezzat, Selected Discussion Papers presented at the Vienna Global Forum (Vienna, 2002)
6. J. Wisniewska, Studia I Prace W. N. E. i Z., **39 (1),** 101-111 (2015)
7. R. Phaal, C.J.P. Farrukh, D.R. Probert, Techn. For. & Soc. Ch., **71**, 5-26 (2004)
8. D.R. Probert, R. Phaal, C.J.P. International Association for Management of Technology (IAMOT) Conference (Lausanne, 2000)
9. D. Waters, *Logistics: An Introduction to Supply Chain Management* (Palgrave Macmillan, New York, 2003)
10. M. Zhuravskaya, A. Lempert, E. Smorodintseva, Innotrans J. **5**, 6, 20-23 (2012).
11. P. Lowe, *The Management of Technology: Perception and Opportunities* (Chapman and Hall, London, 1995)
12. S. Borkowski, R. Ulewicz, J. Selejdak, M. Konstanciak, D. Klimecka-Tatar, Proceedings of 21st International Conference on Metallurgy and Materials 1722-1728 (Brno, Czech Rep., 2012)
13. M. Ingaldi, Logistyka, **6**, 12133-12136 (2014)
14. M. Ingaldi, Hutnik-Wiadomości Hutnicze, **81**, 10, 714-717 (2012)

Management of the subsystem of production logistics in an enterprise

Marta Kadłubek[1,*], and *Agata* Mesjasz-Lech[1]

[1]Czestochowa University of Technology, Faculty of Management, Al. Armii Krajowej 19 B, 42-200 Częstochowa, Poland

Abstract. The logistic system emerged through the implementation of a systemic concept in logistics. In a production company the system can be very complex and it often consists of individual subsystems such as procurement and production subsystems. The article discusses the management of a logistic subsystem of production in terms of cost effectiveness on an example of a selected enterprise.

1 Introduction

Looking at the development of logistics so far one thing stands out. It is the shift in the focus of the concept – from sheer transformation of activities in the flow of goods towards an important and integrated concept of management which integrates and optimizes logistic systems. Systemic approaches determine different effects of thinking with logistic categories. Major emphasis is placed on the dimension of flow phases in a logistic system, where logistics is understood as a separate category of tasks within an enterprise [1]. Systemic references are key in distinguishing the logistic subsystem of production in an enterprise.

The logistic system of a company, including all logistic subsystems, should be effective. The concept of the effectiveness of a logistic system and its subsystems can be seen as a tendency to form a desired relationship between outlays and effects in the area of the system. The basic premise for the analysis of the effectiveness of a logistic system or its subsystems can be willingness of a business to improve its effectiveness.

The goal of the article is to present the aspects of the management of a logistic subsystem of production in terms of its effectiveness on an example of a selected business entity.

2 The concept of production logistics in a logistic system

The role of logistics in managing production is determined by its systemic interpretation. The range of application of the theories of systems reveals solid links with many fields of science, including logistics, which normalizes the basis for the determination of the notion of a logistic system. A systemic approach towards logistics enables enterprises to realize

* Corresponding author: martakadlubek@wp.pl

many strategic tasks such as improving customer service in order to counterbalance the competitive edge of others or to maintain their own strong position on the market.

The logistic system of an enterprise can be very complex and consist of a number of subsystems such as the subsystem of procurement, production, distribution.

Scholars like: P. Blaik [2], G. Ghiani et al.[3], M. Nowicka-Skowron [4], H. Ch. Pfohl [5], indicate that the basic task of production logistics is twofold. On one hand it is the synchronization of the flow of goods (resources, semi-finished products, finished products) in the logistic system of an enterprise the moment they enter the enterprise, and on the other the creation of best possible environment for an effective planning and controlling production. Therefore, the subsystem of production logistics deals with all activities connected with the process of procurement of materials necessary for production and then moving the semi-finished products and finished products to the warehouse. The function of such a logistic activity in a company is increasing effectiveness of production processes, speeding up material flows and minimizing their cost. The subsystem of production logistics, which interfaces production technology and material procurement, is to reduce stock and improve customer service the through production of products in the right quantities, time and structure [6]. Production and logistic processes and activities are interrelated and often integrated.

In a systemic approach one must keep these relationships in mind when considering the cost effectiveness of the logistic system of an enterprise [2,4]. The evaluation of effectiveness is based on a global theory of systems which determines the search for a holistic concept of an effective organization of a given process or activity. The functional dimension of a system is examined in close consideration of the problems in the effectiveness of individual subsystems, including the subsystem of production. The starting point is distinguishing the tasks of the subsystem as a separate segment in the activity of the enterprise. The next sub-chapter presents the tasks of the production subsystem of the selected enterprise which later on are confronted with the notion of enterprise effectiveness.

3 The subsystem of production logistics in the selected enterprise

The chosen enterprise is a limited company - a production factory in the foundry industry [7]. The examined cast iron foundry has a machining division and an assembly line of cast iron products. The enterprise does both serial and custom production. It offers grey iron castings, cast iron radiators, household appliances, low voltage box-type switchboards, cast iron sewage splitters, and other products such as: cast iron stoves, elements of garden furniture, dumbbells, plant shelves, etc.

The activities in the subsystem of the production of manual meat mincers encompass the following:

– analyzing cast iron constructions in the light of technological requirements and the implementation of changes,

– determining the basic production parameters,

– dividing the system processes into operations and activities, organizing work and workstations,

– creating technological processes for the production of individual parts,

– determining time norms for the production of individual parts,

– developing temporary technical conditions for product acceptance,

– producing a sample batch for evaluation,

– performing an economic analysis of the production process,

– introducing changes to the construction and technology of individual parts,

– making comparisons with the benchmark,

– developing the final technical conditions of acceptance.

The main task of the subsystem of production is to launch the production of new or modernized products technically similar to the ones produced in the world in a quick, proficient and economical way.

The initial stage of the process is control. It is realized through:

– checking the quality of procured input, moulding and auxiliary materials (e.g. chilled iron, hydrofluoric acid, coke, fluxes) as well as steel and tin,

– comparing the production quality with the current technical documentation,

– preventing shortages through identifying their causes and implementing solutions,

– initiating, coordinating and supervising the realization of all organizational and technical measures to improve the quality,

– controlling the quality and precision of technical support,

– controlling the precision of used machines and appliances.

The next stage is collecting raw materials for production and supplying them in right quantities to individual workstations. Production processes in the foundry, the hot galvanizing plant, the treatment plant and the tool shop necessitate moving considerable masses of different materials.

The majority of machines and appliances in the factory used in the production of manual meat mincers is also used for the production of other product. A significant part of the machinery for the production of mincers is also used in the production of different household appliances, cast irons, radiators, box-type switchboards, splitters, etc. Therefore the anticipated workload on some machines and appliances in the production of meat mincers is rather low which, however, does not mean they are unused. Considering the total production of the examined enterprise, the production capacity of the machinery is exploited in 90-95%.

The number of machines and appliances necessary for the production of manual meat mincers was calculated on the basis of expected loads on machines and appliances with the volume of mincers production on the level of 279 400 in 2017. The production structure in a technological view according to the number of machines, appliances and workstations is depicted in Figure 1.

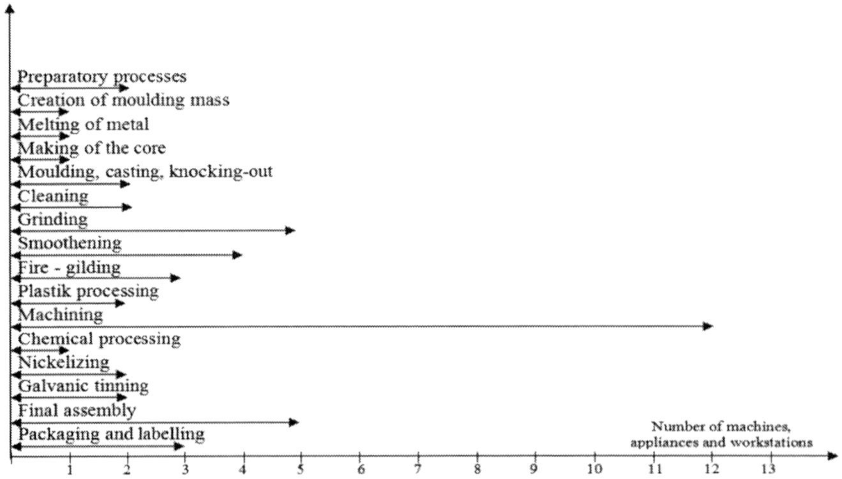

Fig. 1. The structure of the production of a manual meat mincer in a technological view according to the number of machines, appliances and workstations.

4 The model of production control for the analyzed enterprise

Efficient and effective production is possible only if all necessary resources, including stock resources, are supplied. It should be pointed out that the analyzed enterprise is characterized by limited resources at the way in to the production logistics. The article suggests a SILSP (Single-item capacitated lot-sizing problem) mathematical model of production management. It's an LSP (Lot Sizing Problems) model for planning the size of a production batch which helps determine an optimal size of production realized for one product in a situation of demand intermittence and limited production resources [8]. The model was built for the enterprise described in the previous chapter. It will inform about the stock which should be procured for the production process and the number of products which should be made for further sales. This solution is to minimize the cost of both resource stock procurement and production processes with the use of the company's production factors. It will also reduce the cost of untapped output capacity. Therefore the function of the objective minimizes the total cost of the procurement of resources necessary for processing and production. Moreover, it was assumed that at the beginning and at the end of the discussed period the level of raw materials and finished products ready for warehousing is zero and the capacity for warehousing raw materials and distributional stock as well as processing capacity are the same in every period. The limiting conditions were:

- volume of output that can be achieved by the enterprise through the production process with the use of its own production factors in the analyzed period,
- capability to realize warehouse processes and utilize the space for storing products and materials,
- capability to realize production processes.

The model assumes the following decision-making variables:

- size of a given type of stock put through a production process,
- stock level of a product,
- size of a production batch,
- level of a given stock type,
- binary variables concerning the start of a production process,
- amount of a given stock type procured from suppliers.

All decision-making variables were applied to the discussed period which yields a dynamic representation of the problem. The mathematical formula of the model is as follows:

The function of objective:

$$\sum_{t=1}^{T}\sum_{i=1}^{I}(BC_i \cdot BP_{it} + USC_i \cdot I_{it} + UBC_i \cdot IV_{it} + MC_i \cdot MP_{it} + OC_i \cdot IM_{it}) +$$
$$+ \sum_{t=1}^{T} USCP \cdot IP_t + \sum_{t=1}^{T}\sum_{i=1}^{I}(1 - MP_{it}) \cdot NMC \rightarrow min \qquad (1)$$

Conditions:

- connected with quantity of product obtained in the process of production:

$$PP_t = \sum_{i=1}^{I} VP_i \cdot MP_{it} \qquad \forall i, t \qquad (2)$$

-

onnected with level of given stock type:

$$I_{it} = I_{i,t-1} + IV_{it} - MP_{it} \qquad \forall i, t \qquad (3)$$

–

onnected with level of product stock:

$$IP_t = IP_{t-1} + PP_t - DP_t \qquad \forall t \tag{4}$$

–

onnected with capabilities in terms of storage and production:

$$\sum_{t=1}^{t} AP \cdot IP_t \leq SP_t \qquad \forall t \tag{5}$$

$$IP_t \leq SP_t \qquad \forall i, t \tag{6}$$

$$IM_{it} \leq MP_i \qquad \forall i, t \tag{7}$$

–

onnected with concordance of the result with regard to variables:

$$IM_{it} \leq M \cdot MP_{it} \qquad \forall i, t \tag{8}$$

$$IV_{it} \leq M \cdot BP_{it} \qquad \forall i, t \tag{9}$$

–

onnected with the type and sign of variables

$$MP_{it}, BP_{it} \in \{0,1\} \qquad \forall i, t \tag{10}$$

$$IV_{it}, IM_{it}, I_{it} \geq 0 \qquad \forall i, t \tag{11}$$

$$PP_t, IP_t \geq 0 \qquad \forall t \tag{12}$$

–

onnected with sets of objects:

$$i = \{1, 2, \dots, I\} \quad t = \{1, 2, \dots, T\} \tag{13}$$

Parameters:

DP_t – required quantity of product in time t
VP_i – quantity of product obtained from the processing of a stock unit i
SP_i – capability to warehouse stock i
MP_i – capabilities in terms of processing stock i
AP – area taken by a unit of a product
SP_t – capability to warehouse a product in period t
USC_i – unit cost of stock i storage
MC_i – cost of starting the process of stock i processing
OC_i – unit operational cost of processing stock i
BC_i – cost of buying stock i
UBC_i – unit cost of buying stock i
$USCP$ – unit cost of product storage
NMC – cost of untapped production capacity
Variables:
IM_{it} – quantity of stock i to be processed in period t
I_{it} – level of stock i in period t
PP_t – quantity of product obtained in the process of production in period t
IP_t – level of product stock in period t
MP_{it} – binary variable (1 – processing of stock i in period t is started, 0 – is not started)
BP_{it} – binary variable (1 – process of buying stock i in period t is started, 0 – is not started)
IV_{it} – quantity of stock i bought from suppliers in period t.

The function of objective (1) includes the cost of stock supply, cost of storage, cost of procurement of raw material stock, cost of starting production, operational cost of processing stock, cost of storage of a finished product, cost of untapped production capacity. Condition (2) determines the quantity of a product which can be obtained through processing of different stock types. Conditions (3) and (4) refer to the stock levels of raw material and finished product. Condition (5) provides assurance that the warehouse space occupied by the store does not exceed the available warehouse space. Condition (6) guarantees that the amount of products planned to be stored will not exceed the possibilities in this area. Condition (7) takes into account the limits for the processing of individual types of raw materials stock. Conditions (8), (9) guarantee the concordance of the result with regard to binary and decision-making variables. Conditions (10) - (12) are connected with the type of variables, and condition (13) determines the sets of objects considered in the model.

The proposed model considers the coordination of the processes of procurement logistics and production logistics. The level of raw material stock is determined by the demand for the product made of that material. Therefore, the model couples the processes of procurement and production. This means it is a solution which considers the needs of the enterprise in terms of stock according to the demand for the product and the output capacity of the enterprise which gives flexibility in the control of stock levels of raw materials.

5 Conclusions

To conclude on the topic of logistic systems it should be underlined that their scope extends from the issues of procurement all the way to the consumption of a product. The multiplicity of systems is determined by the perspective of logistics and the issues it deals with. A systemic approach towards the activity of businesses is characterized by a holistic look at the problems and an analysis of the importance of individual elements of systems (subsystems) and their relationships.

Activities realized in the framework of production logistics should be focused not only on ensuring the efficiency of the production system, but on its effectiveness in the first place. The proposed model considers all aspects of the functioning of production logistics: efficiency, proficiency and cost-effectiveness.

References

1. P.M. Price, N.J. Harrison, *Looking at logistics: a practical introduction to logistics, customer service, and supply chain management* (Access Education, United Kingdom, 2013)
2. P. Blaik, *Efektywność logistyki. Aspekt systemowy i zarządczy* (PWE, Warszawa, 2015)
3. G. Ghiani, G. Laporte, M. Musmanno, *Introduction to logistics systems management* (Wiley-Blackwell Pub., New Jersey, 2013)
4. M. Nowicka-Skowron, *Efektywność systemów logistycznych* (PWE, Warszawa, 2001)
5. H. Ch. Pfohl, *Logistiksysteme* (Springer Berlin Heidelberg, Berlin, 2010)
6. D. Bowersox, D. Closs, M.B. Cooper, *Supply Chain Logistics Management* (Mc Graw–Hill, United States, 2012)
7. M. Nowicka-Skowron, R. Ulewicz, METAL 2015: 24th International Conference On Metallurgy And Materials, 1707-1712, (2015)
8. N. Brahimi, N. Absi, S. Dauzere-Peres, A. Nordli, European Journal of Operational Research, **263(3)**, 838-863 (2017)

Selected aspects of technological innovations management in district heating companies

Aleksandra Rak[1,*]

[1]Department of Production and Safety Engineering, Czestochowa University of Technology, Armii Krajowej 19 B, 42-200 Czestochowa, Poland, EU

Abstract. The accession of Poland to the EU results in the realization of EU directives, including the field of energy efficiency and CO_2 emissions. The regulations oblige enterprises that produce and distribute heat to technical activities involving the use of existing heat sources and modernization of thermal networks, as well as economic and environmental optimization of the heat supply process. The implementation of these projects requires a change in the way the energy sector is managed and thoroughly restructured by implementing and applying technological and product innovations. The article discusses the legal and environmental conditions of the heat market in Poland, the key challenges for the heating industry and selected innovations in heating companies regarding two areas: technical infrastructure and management of the heat production and transmission.

1 Introduction – the innovative activity

Changes in the conditions of the company's operation result from the influence of different factors, e.g. economic, social, political, demographic and technological. Among them the most important are: diversification of clients' needs, increase of competitiveness, increasing influence of knowledge and information and development of IT technology [1].

The concept of innovation was introduced for the first time by an Austrian economist J. Schumpeter in the literature of economics at the beginning of the 20th century. However, the paradigm of economy that based on knowledge was created in the 1990s and, as a result of which, the theory of innovation management was created. Five main areas of innovation can be distinguished - product, service, process, business process and business model innovation [2]. In general, companies can make changes that apply to all the areas and levels of their operations [2-4]. However, companies need to identify which areas are important for them.

2 The technological innovation management process

Nowadays, innovation management can be defined as seeking, based on possessed resources, such types of innovations that make the innovation process more effective in confronting the challenges placed before the organization by the market, competition and the client. It includes the selection of new techniques, technology, work organization, issues of acquiring new solutions and the possibilities of their use, considering legal, economic and financial, administrative, structural and process, environmental and strategic matters. Multidimensionality and complexity of innovative processes requires a systematic approach

* Corresponding author: aleksandra.rak@wz.pcz.pl

to managing this area of activity, including all levels of business management, i.e. the level of strategic and operational management, personnel, organizational structures and processes [1, 2, 5, 6].

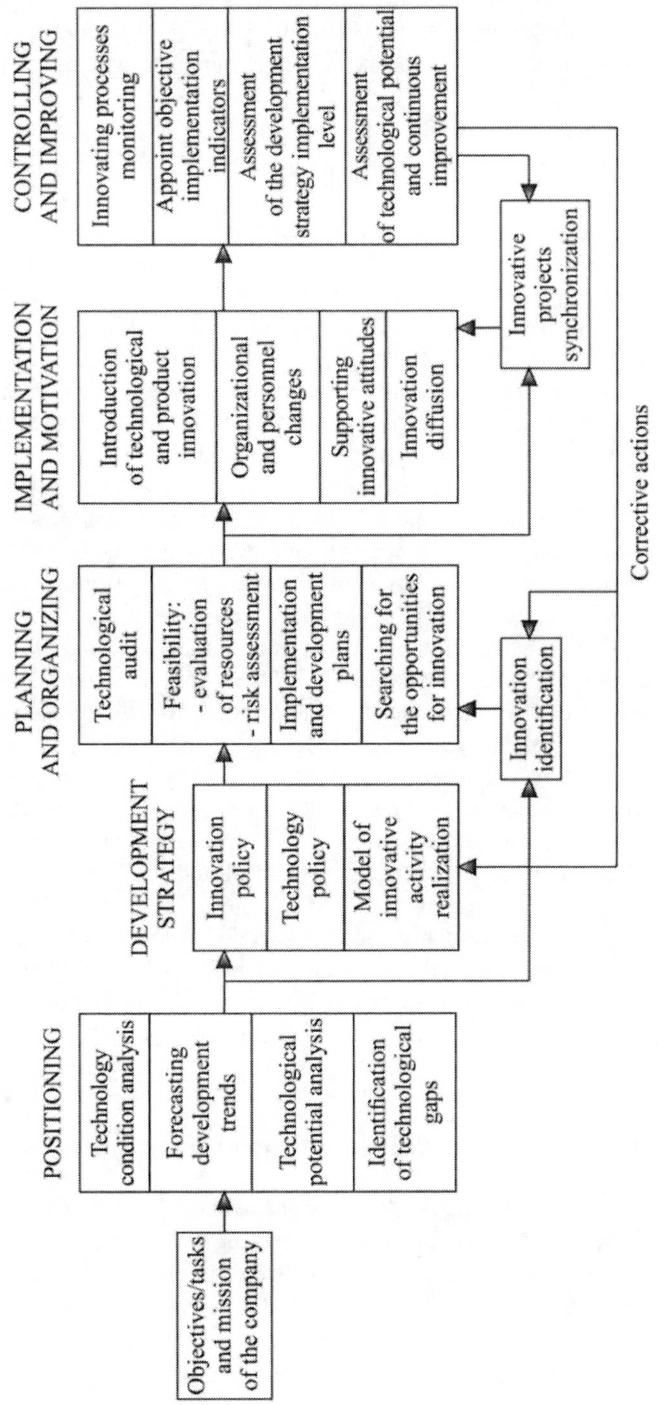

Fig. 1. Model of technological innovation management process in the companies.

Innovation management in companies includes the following activities aimed at effective achievement of the assumed objectives [3]:

- change of the company's strategy,
- making decisions in the field of innovative activity of the enterprise,
- organization, including searching for opportunities for innovation, financing innovative activities, implementation of research and development projects, ensuring the security of intellectual property,
- controlling the innovative activity of the company,
- supporting innovative attitudes by building competences for innovation and developing organizational culture.

There are many external and internal factors that have a significant impact on the innovation management. These determinants are divided into internal or organizational (related to the company) and external ones resulting from the enterprise environment. External determinants are primarily: the market and demand shaping, competition, globalization and regionalization, technological progress, the functioning and development of industry. However, the determinants of innovation management at the organizational level are the result of changes taking place in the global economy, which significantly change the way in which innovative activity is implemented and formed the development of innovative potential. The main factor determining the effective management at the level of the organization is the skillful adjustment of material and immaterial resources (knowledge, competencies of employees) of the company to achieve the assumed objectives [1].

In the literature on the subject, management of technological innovations has not been precisely defined. This concept can be derived from the combination of technological innovations definitions (technical, process) and models of innovation management described in the literature. The management of technological innovations contains all management activities that decide about innovation and technology policy, objectives and responsibilities and their realization by operation, such as: planning, providing resources, organizing and controlling the development of technology, improving innovation in the processes of their implementation and exploitation [7]. In the model approach, the technological innovation management process can be presented in the way shown in Figure 1. Important elements of this process are building the innovation strategy and implementation and development plans based on the analysis technology condition, the technological potential of the company and forecasted development trends. The decision of innovations implementation should be preceded by a technology audit and resource and risk assessment. In the phase of innovation implementation it is important to monitor the innovation processes, set indicators of realization objective and evaluate the implementation of the company's development strategy.

3 Determinants and trends in heating systems development

The development of heating systems is influenced by actions resulting from the energy policy of the country, implemented under the European Union regulations. The main directions of this policy, relating to heating sector in Poland are [8]:

- improving energy efficiency and increasing the efficiency of systems,
- balanced development and increased use of renewable energy sources,
- energy security and development of competitive fuel and energy markets,
- limiting the impact on the environment

The increase in the energy efficiency is a priority binding for the implementation of other objectives of the Polish energy policy, which was reflected in the adoption of the law on energy efficiency in 2011 [9]. The Directive 2012/27 / EU on energy efficiency is very important for the district heating development [10]. It defines an "efficient heating and cooling system" as a system in which at least 50% of energy from renewable sources or waste heat is used for heat or cold production, or at least 75% of heat from cogeneration, or in at least 50% of the combination of such energy and heat. This implies supporting the use and expansion of existing infrastructure, heat sources and heating networks, in terms of their adaptation to the use and development of high-efficiency cogeneration, including waste heat and energy from renewable sources [11].

District heating systems must guarantee the continuity of heat supply to consumers, despite of the influence of external factors, usually independent of the producer, and internal ones comprising its management as well as the processes of producing and distributing heat [12]. Production and supply system heat is a complex process, which should take into account a number of problems in the fields of technology, ecology and economy, as well as the existence of a number of threats. The basic factors affecting the directions and opportunities for the development of heating systems are given in Table 1.

Table 1. Factors influencing the directions and possibilities of development of heating systems.

Internal factors	External factors
– economic and financial situation of the company – development strategy of the company and implementation of technological innovations – heat generation technology – the level of reserves in the heat source – technical condition of the infrastructure for generating and transmitting heat – implementation of IT tools for monitoring and management of the heating system – involvement of the management cadre – skills, experience and competences of employees operating the heating network	– macroeconomic conditions – technical and environmental requirements – programs and funds promoting the improvement of energy efficiency – financial support of the state for innovative activity – development of energy-saving construction technologies – development of prosumer energy and micro-generation – climatic conditions – reliability of fuel supply and electricity – fuel price level (impact on costs and profitability of operations)

4 Technological innovations in district heating companies

Fulfilment of priorities, regarding the improvement of energy safety and efficiency as well as ecological indicators in the area of heat supply, requires comprehensive modernization and implementation of innovative technological solutions. These activities are necessary in all segments creating the local heat market [11-14]:

- heat production - development of cogeneration and trigeneration systems (combined heat and power production), change of energy production technology, implementation of fuels with lower environmental load, heat storage,
- in the area of heating networks - construction of intelligent heating networks (smart grid), exchange on pre-insulated pipes networks, automation systems, monitoring and management of network operation, hydraulic regulation, new regulatory tables,
- construction and installation systems at the recipients - implementation of the low-energy construction standard, thermo-modernization of buildings, modernization of nodes (weather sensors and shut-off valves), individual heat consumption accounting systems, increased use of renewable energy sources.

Realization of technology improvement processes as well as the introduction of technological innovations in district heating enterprises are dictated by market needs (reduction of ordered power, lower costs for recipient) and competition pressure (prosumer energy and RES development). Improvement of heat generation technology in cogeneration and trigeneration systems requires large investment expenditures, which is why innovative changes are most often inducted. Technological innovation in district heating companies concerns two areas: technical infrastructure and management of the heat production and transmission system [12]. Typical technological innovations implemented in heating enterprises are given in Table 2.

Table 2. Typical technological innovations in heating companies.

Technical infrastructure	Managing the production system and heat transfer
– modification of the heat and electricity production method, – implementation of fuels with a lower environmental burden, – applying new technical solutions in the form of pre-insulated pipes with an alarm system.	– market models for the regulation of heat supply in heating systems, – using IT tools to manage the network operation, taking into account weather conditions, – use of teletransmission systems for monitoring and controlling the operation of the network.

Innovative solutions in the production infrastructure concern, i.a. changes in the method of heat and electricity production from classical coal combustion to combustion and fluid gasification or oxygen combustion technologies (TST) and the introduction of fuels with lower environmental load (dust, CO2, SO2 and NOx emission levels). In the transmission infrastructure are applied new technical solutions in the form of multilayer pipe systems insulated with PUR foam and placed as a pipe assembly together with an alarm system in a common casing [12, 13].

In the area of management of heat production and transmission systems, technological innovations are based on the use of IT tools and teletransmission systems. The implementation of modern IT systems requires the installation of heat network metering and data transmission systems for further analysis and processing. The use of distributed control systems and SCADA type packages enables visualization of the heating system operation and remote control of the operation of individual devices. Visualization systems facilitate current operation and improve the economic efficiency of network management [12, 15].

The presented range of innovations possible to introduce is so extensive, that in the conditions of limited material and financial resources, the process of their implementation must be spread over time.The priorities result from the adopted development strategy, which depends on the market position and potential of a given heating enterprise. Due to the achieved effects, mainly economic and environmental, large heating companies in Poland initially has introduced technological innovations in the area of generation, while small and medium-sized companies were focused on improving heat transfer (11).

The introduction of market models for the supply of heat, based on technical parameters and a cost account, enables improving the operational efficiency of the heating system and optimization of investment processes, both on the scale of a single company and in the whole sector. Such a comprehensive approach to the innovation management process in heating must include not only technological aspects, but also organizational solutions and decision-making processes [18].

The implementation of software packages dedicated to the heating sector not only provides current control and control of the operation of heating systems, but also enables modelling of static and dynamic state occurring in real conditions. Advanced programs for modeling and optimizing the operation of heating networks taking into account weather conditions are available on the market, such as Bentley SisHYD and TERMIS [15].

5 Summary

The changes taking place in the national economy, the requirements of the climate and energy policy, trends in heat demand and the increase in energy security force heating companies to induct modern and innovative technologies as well as a new way of business management. Management of technological innovations in heating is an important element of the development strategy of the companies in this sector, that determines the competitive position as well as the operational and economic efficiency. It should be emphasized that the introduction of new technical solutions and technological innovations also results in organizational and personnel changes, consisting in the formation of interdisciplinary teams of employees with high qualifications and competences.

References

1. P. Mielcarek, *Procesy zarządzania innowacjami w przedsiębiorstwach przemysłowych* (2016)
2. K. Goffin, R. Mitchell, *Innovation Management: Effective Strategy and Implementation* (2017)
3. M. Karlik, *Zarządzanie innowacjami w przedsiębiorstwie. Poszukiwanie i realizacja nowatorskich projektów* (2014)
4. J. Tidd, J. Bessant, *Management Innovation. Integrating Technological, Market and Organizational Change* (2009)
5. K. Szatkowski, *Zarządzanie innowacjami i transferem technologii* (2016)
6. M.A. White, G.D. Bruton, *The Management of Technology&Innovation. A Strategic Approach* (2011)
7. J. Łunarski, *Innowacje technologiczne w przedsiębiorstwach. Projektowanie, wdrażanie i użytkowanie* (2016)
8. Polityka energetyczna Polski do 2030 roku. Załącznik do obwieszczenia Ministra Gospodarki z dnia 21 grudnia 2009 roku (MP z 2010, nr 2, poz. 11)
9. Ustawa z dnia 15 kwietnia 2011 r. o efektywności energetycznej (Dz.U. z 2011 r., nr 94, poz. 551)
10. Dyrektywa Parlamentu Europejskiego i Rady 2012/27/UE z 25 października 2012 roku w sprawie efektywności energetycznej, ... (Dz.U. UE L 315 z 14.11.2012 r.)
11. M. Turski, R. Sekret, Rynek Energii, **4(119)**, 27-34 (2015)
12. A. Rak, R. Kucęba, R. Sekret, *Techniczne i materialne aspekty bezpieczeństwa* 102-113 (2017)
13. A. Jachura, R. Sekret, Instal, **7-8**, 7-11 (2013)
14. T.M. Mróz, Ciepłownictwo, Ogrzewnictwo, Wentylacja, **10**, 403-407 (2012)
15. A. Rak, Zeszyty Naukowe UE w Katowicach, **308**, 115-127 (2016)
16. U. Sarma, G. Bazbauers, Energy Procedia, **84**, 401-407 (2016)
17. K. Sipilä, J. Pietiläinen, A. Nuorkivi, *The building level substation – the innovation of district heating system* (2016)
18. J. Birkinshaw, G. Hamel, M.J. Mol, *Management innovation* (2010)

Applications of linear programming to optimize the cost-benefit criterion in production processes

Marek Krynke[1,*], and *Krzysztof* Mielczarek[1]

[1]Czestochowa University of Technology, Department of Production Engineering and Safety, ul. Armii Krajowej 19B, 42-200 Czestochowa, Poland

Abstract. In this paper linear programming was shortly characterized. The implementation of some basic functions for Octave have been presented. It is the GNU General Public License program for the optimization the cost-benefit criterion in production processes. A problem of dividing production tasks between workstations was presented in way that the cost of the realization was the smallest. The final score of the course was obtained the computational environment Octave using the GLPK (GNU Linear Programming Kit) function.

1 Introduction

Management is the administration of an organization, whether it is a business, a not-for-profit organization, or a government body. Management includes the activities of setting the strategy of an organization and coordinating the efforts of its employees to accomplish its objectives through the application of available resources, such as financial, natural, technological, and human resources. The necessity to cope with the problem of the economic practice, limitations, the risk and the uncertainty are connected with it [1].

In production companies the process of directing refers not only to the attitude to an employee, but also to the attitude to sources [2]. It leads to the fact that correct application to the production system demands the professionalism and practical abilities of the interdisciplinary attempt at analysis of the manufacturing process. Decision-making is an integral part of modern management. Decisions are made to sustain the activities of all business activities and organizational functioning with the help of owned sources [3].

Taking important decisions many times causes a lot of problems. Decision making process can be regarded as a check-and-balance system. When one problem is solved another arises and so on, so decision making process is continuous and dynamic [4]. In making beneficial choices, mathematical modeling can be helpful. To present many of them it is possible to use the mathematical description called the task of linear programming [5]. Linear programming is a mathematical technique used in a variety of practical fields to maximize the useful output of a process for a given input. In general form the extreme of the function (the function of the objective) has to be found. Linear programming is

* Corresponding author: marek.krynke@wz.pcz.pl

a method to achieve the best outcome in a mathematical model whose requirements are represented by linear relationships. Linear programming is a special case of mathematical programming (mathematical optimization) [6, 11-13].

2 Problem of the linear programming. Computational environment Octave

The linear programming is the most often applied model of the optimization on account of the existence of finding efficient algorithms of optimal solution. The intuitiveness of linear relations appearing in the mathematical model [7].

A few methods of solving tasks of the linear programming exist. A graphic method is simplest, however, its practical application is limited to tasks with two decision-making variables. Mathematically it is possible to prove, that searching the task of the linear programming by extremes of function (if it exists) will be acting in one of the tops of the figure, defined through imposed limitations (called the polyhedron of solutions). Rarely is the problem so simple, and, while increasing the number of searched parameters a number of dimensions of the answers, polyhedron is increasing. Now, there are very effective algorithms of the solution to this problems. A popular algorithm for linear programming is the simplex method. George Dantzig, while describing the „Origins of the Simplex Method, „noted that it was the availability of early digital computers that supported and invited the development of LP models to solve real-world problems. When Dantzig introduced his method in 1947, it was somewhat easier to sort out the details of the simplex method than to deal with the "where are we in the domain space" questions that are the core of interior-point approaches. The solution of a linear program is accomplished in two steps. In the first step, known as Phase I, a starting extreme point is found. Depending on the nature of the program, this may be trivial, but, in general, it can be solved by applying the simplex algorithm to a modified version of the original program. The possible results of Phase I are either that a basic feasible solution is found or that the feasible region is empty. The method allows to find solutions in short time and is applied in many popular solver [5, 8].

It is possible to solve problems where there is a large number of variables with the help of programs for numerical calculations e.g. MATLAB or Octave [9]. Octave is a free programme for numerical calculations (mathematical and engineering calculations). This language is intuitive and friendly (for a mathematician). With basic functionality in Octave there are operations on matrices and number of numerical methods solving linear and non-linear problems. Octave is software featuring a high-level programming language, primarily intended for numerical computations. Octave helps in solving linear and nonlinear problems numerically, and it is used for performing other numerical experiments with the use of a language that is mostly compatible with MATLAB. In addition to being used on desktops for personal scientific computing, Octave is used in academia and industry. Octave has been built with MATLAB compatibility in mind, and shares many features with MATLAB. [10]. For many applications, where speed or generating 3D animation is not essential, Octave is more appropriate. Both programs are compatible with Linux, Windows, MAC OS and Android system.

3 Analysis of the production tasks segmentation

Production tasks segmentation is an important part of a production process. It is connected with the size of the profit and costs. Production tasks should be allocated in order for the profit made by the production system to be maximized while, at the same time, the costs are minimized.

In this article a problem of dividing 5 production tasks between 4 workstations was presented in such a way that the cost of the realization was smallest. The following research problem was considered.

There are 4 machines whose available standard hours of work were given in Table 1. With the help of these machines the production program of five products should be realized. Manufacturing cost (1 hour) of the machine j for the one product i are shown in Table 2. The production of every product can be divided in randomly between the machines. The hourly productivity of machines depends on which product is produced on a given machine. Individual value was given in Table 1. The production tasks should be allocated to machines in such a way so that the total cost of the realization of the production program is minimal.

Table 1. Productivity of individual machines depending on products [piece/hour].

Products	Machines				Planning for production [piece/month]
	M1	M2	M3	M4	
P1	200	200	500	300	20000
P2	700	250	300	200	25000
P3	450	300	400	400	18000
P4	400	300	200	300	15000
P5	100	400	600	200	20000
Available standard hours of work [hour/month]	30	90	60	90	

Table 2. Cost per piece of the machine j for the product i [PLN/hour].

Products	Machines			
	M1	M2	M3	M4
P1	10	25	10	40
P2	10	30	35	25
P3	10	30	10	10
P4	30	15	15	25
P5	25	30	50	10

In order to describe this situation with mathematical model a way to characterise the allotted tasks with the help of numbers should be found. The first parameter is cost connected with the realization of individual products, and the second parameter is available standard hours of work. It is possible to describe cost per piece of the machine j and the working time. An assessment of the effectiveness of work is a more difficult problem. In this example, the effectiveness of work is expressed through the productivity of individual machines. In the case of products which were already made on a given machine, determining their productivity is straight and quite accurate. However, in the case of a new product their preliminary estimation, and later correction will be necessary.

Designation:

k_{ji} – cost for one working hours of the machine i for product j,

p_i – number of work units required at planning for production,

h_i – available standard hours,

w_{ji} – productivity of machines - number of work units (1 hour) for machine i and product j,

The monthly total cost of the realization of the production program are described as:

$$cost = h_{11} \cdot k_{11} + h_{12} \cdot k_{13} + h_{13} \cdot k_{13} + h_{14} \cdot k_{14} + h_{21} \cdot k_{21} + h_{22} \cdot k_{22} + h_{23} \cdot k_{23} + h_{24} \cdot k_{24}$$

$$+ h_{31} \cdot k_{31} + h_{32} k_{32} + h_{33} \cdot k_{33} + h_{34} \cdot k_{34}$$
$$+ h_{41} \cdot k_{41} + h_{42} \cdot k_{42} + h_{43} \cdot k_{43} + h_{44} \cdot k_{44}$$
$$+ h_{51} \cdot k_{51} + h_{52} \cdot k_{52} + h_{53} \cdot k_{53} + h_{54} \cdot k_{54} \quad (1)$$

It is a function that should be minimized. At the same time, the available standard hours of work:

$$h_{11} + h_{21} + h_{31} + h_{41} + h_{51} \leq 30 \quad (2)$$
$$h_{12} + h_{22} + h_{32} + h_{42} + h_{52} \leq 90 \quad (3)$$
$$h_{13} + h_{23} + h_{33} + h_{43} + h_{53} \leq 60 \quad (4)$$
$$h_{14} + h_{24} + h_{34} + h_{44} + h_{54} \leq 90 \quad (5)$$

and requirement concerning production order – limitations of work units (size of the batch):

$$h_{11} \cdot w_{11} + h_{12} \cdot w_{13} + h_{13} \cdot w_{13} + h_{14} \cdot w_{14} = 20000 \quad (6)$$
$$h_{21} \cdot w_{21} + h_{22} \cdot w_{22} + h_{23} \cdot w_{23} + h_{24} \cdot w_{24} = 25000 \quad (7)$$
$$h_{31} \cdot w_{31} + h_{32} w_{32} + h_{33} \cdot w_{33} + h_{34} \cdot w_{34} = 18000 \quad (8)$$
$$h_{41} \cdot w_{41} + h_{42} \cdot w_{42} + h_{43} \cdot w_{43} + h_{44} \cdot w_{44} = 15000 \quad (9)$$
$$h_{51} \cdot w_{51} + h_{52} \cdot w_{52} + h_{53} \cdot w_{53} + h_{54} \cdot w_{54} = 20000 \quad (10)$$

Defined task is possible to be solved in the environment Octave. Fig. 1 shows a numeric code of the task to the solution into the console of Octave program. A structure of the code and individual elements was explained.

```
clear all
w=[200 200 500 300;
   700 250 300 200;
   450 300 400 400;
   400 300 200 300;
   100 400 600 200]

k=[10 25 10 40;
   10 30 35 25;
   10 30 10 10;
   30 15 15 25;
   25 30 50 10]
```

A column array containing the objective function coefficients

```
for i = 1:4
  for j = 1:5
    c(j+5*(i-1))=k(j,i)
  endfor
endfor
```

A matrix containing the constraints coefficients

```
a = [1 1 1 1 1 0 0 0 0 0 0 0 0 0 0 0 0 0 0 0;
     0 0 0 0 0 1 1 1 1 1 0 0 0 0 0 0 0 0 0 0;
     0 0 0 0 0 0 0 0 0 0 1 1 1 1 1 0 0 0 0 0;
     0 0 0 0 0 0 0 0 0 0 0 0 0 0 0 1 1 1 1 1;
     w(1,1) 0 0 0 0 w(1,2) 0 0 0 0 w(1,3) 0 0 0 0 w(1,4) 0 0 0 0;
     0 w(2,1) 0 0 0 0 w(2,2) 0 0 0 0 w(2,3) 0 0 0 0 w(2,4) 0 0 0;
     0 0 w(3,1) 0 0 0 0 w(3,2) 0 0 0 0 w(3,3) 0 0 0 0 w(3,4) 0 0;
     0 0 0 w(4,1) 0 0 0 0 w(4,2) 0 0 0 0 w(4,3) 0 0 0 0 w(4,4) 0;
     0 0 0 0 w(5,1) 0 0 0 0 w(5,2) 0 0 0 0 w(5,3) 0 0 0 0 w(5,4)]

ogr=[30;90;60;90;20000;25000;18000;15000;20000]
```

limitations resulting from the condition of the available machine time fund and the condition regarding the size of the production batch

type of limitation (U - An inequality constraint with an upper bound S - An equality constraint, L - An inequality with a lower bound)

```
ogr_typ="UUUUSSSSS"
```

start of the glpk solver (in the ***h_opt*** variable the optimal ***h*** values will be placed)

```
[h_opt,  cost, status, extra]=glpk (c,  a, ogr, [], [], ogr_typ, [], 1, struct)

h=reshape(h_opt,5,4) % zmieniony rozmiar macierzy
sum (h) % sumowanie każdej kolumny
```

type of optimization
1 - min, -1 - max

Fig. 1. Source code of the analysed problem prepared for solving in the environment Octave.

After doing calculations in vector *h_opt* optimal value of working hours for individual operation was written. A cost that was minimized is determined in the variable *cost*. More information about parameters can be obtained when *help glpk* is inscribed into the Octave console.

In Fig. 2 optimal assigning production tasks to machines was described. For a planned production process, the production of the product P1 will be the most effective only on the machine M3, the production of the product P2 on the machine M1 and M2, the production of the product P3 on the machine M3 and M4, the production of the product P4 on the machine M2 and the production of product P5 on the machine M2 and M4.

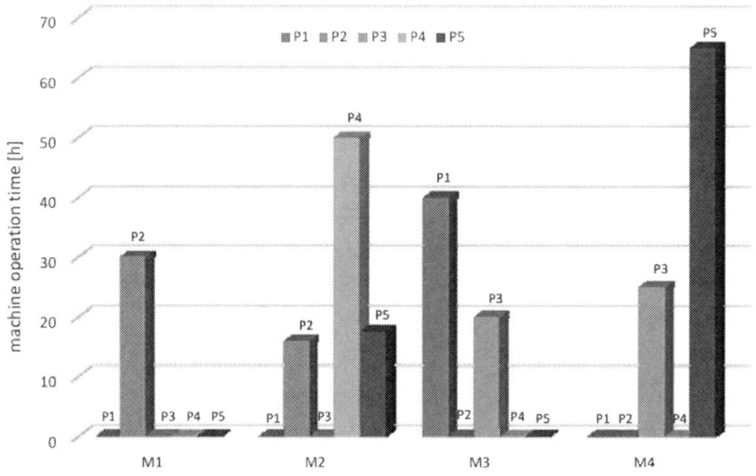

Fig. 2. Optimal assigning tasks to individual machines.

In Fig. 3a machine time fund against the background of available standard hours of work was presented. The most diversified production is on the machine M2. On this machine available standard hours of work were not fully exploited. It is beneficial because on this machine a large number of rearming can appear.

In Fig. 3bO a graph that compares the total cost of the realization of the production program for two extreme cases was presented. The first situation concerns the optimal solution, when costs are subjected to the minimization. For comparison, the second one shows a situation when costs were maximized.

a) b)

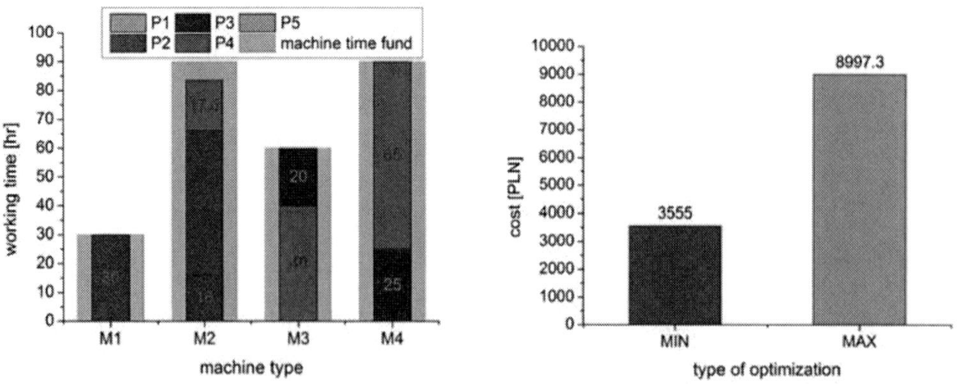

Fig. 3. Available standard hours of work (a), total cost of the realization of the production program in the case of the criterion the minimization and the maximization (b).

4 Summary

At present, the approach towards matters of production process improvement must answer to contemporary developmental tendencies in a scope of company management in conditions of the market economy. Linear programming is one of the simplest ways to perform optimization. It helps solve some very complex optimization problems by making a few simplifying assumptions. It is possible to use the linear programming for the minimization of production costs as well as for the maximization of profits.

The presented example can, by analogy, be widened to a large number of machines and tasks as well as months. This mathematical model find application in many engineering problems e.g. the problem of optimal using stores, realization the cheapest order, etc.

References

1. R. Ulewicz, M. Nowicka-Skowron, *26th International Conference on Metallurgy and Materials, Total Quality Management in practice of polish metallurgical companies* (Tanger Ltd., Ostrava, 2017)
2. D. Klimecka-Tatar, Production Engeenering Archives, **17** (2017)
3. K. Knop, J. Rosak-Szyrocka, *25th Anniversary International Conference on Metallurgy and Materials, Evaluating and improving the effectiveness of the rolling mill in the production of medium steel sections in the selected company from the metallurgical industry* (Tanger Ltd., Ostrava, 2016)
4. M. Zasadzień, Production Engeenering Archives, **11** (2016)
5. M. Afsharian, V.V. Podinovski, EJOR, **268** (2018)
6. P. Saengudomlert, *Optimization for communications and networks* (Boca Raton, CRC Press, 2012)
7. A. Piechna, PiKI, **55** (2012)
8. S. Kreter, A. Schutt, P.J. Stuckey, EJOR, **266** (2018)
9. I. Azzini, R. Muresano, M. Ratto, CLS&S, **52** (2018)
10. S. Nagar, *Introduction to Octave: For Engineers and Scientists* (Apress, NewYork, 2018)
11. N. Radek, J. Pietraszek, B. Antoszewski, *Adv. Mat. Res.-Switz.* **874**, 29 (2014)
12. P. Szataniak, F. Novy, R. Ulewicz, 23rd International Conference on Metallurgy And Materials, 778-783 (2014)
13. T. Lipinski, A. Wach, 23rd International Conference on Metallurgy And Materials, 23rd International Conference on Metallurgy And Materials, 778-783 (2014)

Sustainable Development in Accordance With the Concept of Industry 4.0 on the Example of the Furniture Industry

Anna Wiśniewska-Sałek[1,*]

[1]Czestochowa University of Technology, Faculty of Management, Poland

Abstract. Sustainable development, as an integral part of building an economically rational market, is a challenge which is faced by many branches of the economy. The paper is an attempt to assess the opportunities provided by Industry 4.0 in the process of sustainable development of the furniture industry in Poland, making use of the experiences resulting from the participation of the Author in the project co-funded by the Erasmus+ Program of the European Union "Curriculum Development of Master's Degree Program in Industrial Engineering for Thailand Sustainable Smart Industry - MSIE4.0".

1 Introduction

Global economic development brought about that the industry should respond relatively quickly to the market requirements while complying with high quality standards. Economic criteria clearly depict the pattern which technology, employees and resources ought to adjust to. Production processes adapted to the dynamics of changes imposed by the economic leader (regardless of who they are) should be possibly the most innovative and automated. The human being, wishing to be a part of this process of changes must "be a lifelong learner" simultaneously having an interdisciplinary experience". The environment ought to "produce" infinite amounts of resources and help (taking into account, e.g. clean air) the human existence. The idea of sustainable development allows for maintaining a relative balance between the listed criteria.

The economy and more specifically its development largely depends on production capacities. The processes of automation and digitization lead to the transformation of the currently existing factories into the so called Smart Factories, which allow for creating intelligent value chains. The industry operating in this way is described with the term of "Industry 4.0", which refers to "the fourth industrial revolution" taking place at the moment. The revolution is the reference to the previous stages in the socio-economic development of humankind which may include [1]: *revolution 1.0:* mechanization of production using water and steam driven machines; *revolution 2.0:* mass production based on the division of labor and electrification of machinery and equipment; *revolution 3.0:* use of electronics and information technology to automate production.

* Corresponding author: anna.wisniewska-salek@wz.pcz.pl

The origin of the term "Industry 4.0" comes from the government of the Federal Republic of Germany when in 2011 it was used by the initiative group "Industrie 4.0" (representatives of the world of business, politics and science) who identified the name with the concept of increasing the competitiveness of the German economy. Meanwhile, the government, while providing the support, acknowledged that the name of "Industry 4.0" would be an integral part of the program of the development of the country. The key objective of the program is leadership in the field of innovative technologies on a global scale [2].

The essence of "Industry 4.0" is brought to creating intelligent value chains, based on sociotechnical systems which are dynamic, self-organizing and optimizing, perceived as smart factories [3]. They are created by spontaneously emerging virtual networks: Internet of Services [4], Internet of Things [5], cloud computing [6], which include both employees, machinery and equipment: augmented reality [7], artificial intelligence [8] Human Machine Interface [9], and supporting IT systems: robotics [10], intelligent production [11], digital transformation [12], cyber-physical systems [13]. These value chains are to a certain extent a dynamic network which is centered around a unified object of cooperation. The network is subjected to constant reconfiguration, which depends on the current objectives and conditions, ensuring high flexibility and production efficiency.[1] The solutions changing the modern industry into the one operating in accordance with "Industry 4.0" make the Internet a driving force in the organization of operational work [14].

2 Sustainable development

In accordance with the Polish Environmental Protection Act, sustainable development is understood as "[…] socio-economic development in which there takes place the process of integrating political, economic and social activities maintaining the natural balance and durability of basic natural processes in order to guarantee the possibility of satisfying basic needs of individual communities or citizens of both modern generation and future ones [15]." The definition of sustainable development is associated with the UN since the report of this organization recorded it for the first time in 1987 in Our Common Future as a result of the work of the World Commission on Environment and Development [16]. Satisfying the needs of each of the key factors of sustainable development, being in the mutual correlation, is a very difficult process. "The fourth industrial revolution", being an integral part of economic development and processes taking place in a globalized world, is also a challenge in the field of its significance in balancing the needs of all participants of economic processes. From the point of view of the organization, it is very important to maintain a balance between innovation which is the measure of the efficiency of the organization and practices of sustainable development which is aimed at achieving efficiency in the field of sustainable development due to a gradual improvement in its individual processes. [17]

The economic situation of each country is generated by the emphasis on continuous economic growth, which is very often associated with an increasingly high consumption of already limited natural resources. This is related to the lack of synthesis between the exhaustibility and renewability of raw materials, which recently has been the source of many environmental and social problems. Sustainable development, as an idea practiced for many years, encounters numerous economic, political and cultural barriers, which i.e. lead to crossing the limits of tolerance of the nature. [18] The industry depending on renewable natural resources is the furniture production. In accordance with the report by PDF (Polish Development Fund), the furniture industry (taking into account manufacturing industries), in terms of net export value, occupies the second position and at the same time constitutes a significant part of the Polish economy. In 2016, exports of goods in the CN 94 group (the descriptive version of the Combined Nomenclature 2016 - Chapter 94: furniture; bedding, mattresses, mattress supports, cushions and similar stuffed furnishings; lamps and lighting fittings, not elsewhere specified or included; illuminated advertising, illuminated signs and the like; prefabricated buildings) exceeded PLN 48 billion. The share of sales of this industry (according to PKD – Polish Classification of Activities 2007, Chapter 31) in GDP for furniture producers (over 49 employees) remains at the level of about 2% and more than 130 thousand people are employed in this industry. [19]

2.1 The industry - automation

The development of the furniture industry brought about that it began to face some constraints resulting at least from the globalization of distribution or problems with human capital resources. Striving for production automation and robotics will allow for gaining a competitive advantage using economies of scale. This is associated with an increase in investments in the use of the latest generation solutions in the field of the automated interoperational transport; additionally, the implementation of the methodology of the operation of Lean principles (e.g. Lean Production), the use of autonomous and collaborative robots as well as or above all the use of all available IT resources in the correlation of subsequent stages of production. [19]
"Industry 4.0", in its concept, includes the areas closely related to IT. Its main elements include: *1. Industrial Internet of Things – IIoT*, originating from Internet of Things – IoT: formulated by Kevin Ashton in 1999. It consists in the use of data transmission via the Internet network using the radio identification of RFID to manage the value chain [20]. *2. Smart Factory*: constitutes the key characteristic of Industrie 4.0. It is assumed that factories are managed comprehensively thus being less susceptible to disturbances. In this way, they are able to produce more efficiently and humans, machinery and resources communicate with each other following the principles of social network [21]. *3. Cyber-Physical Systems (CPS):* consists in combining the computational layer and physical processes, where the system of the review of the manufacturing process functions in the feedback loop. Physical processes are identified as the data source used for calculating the control signal of the selected executive objects [22].

2.2 The society - education

Rapid economic growth, among others in Poland, brought about an increase in the need for employees. Increased production caused by a high share of sales in GDP, also in the furniture industry, brought about the same effect. The statistics by Central Statistical Office show that, in Poland, in 2017 there were registered 1081746 unemployed people with the following educational background: higher education - 149491; post-secondary, upper secondary vocational education - 234599; general secondary education - 117266; vocational education - 281679; lower secondary education and below - 298711. The number of the unemployed steadily indicates a downward trend. [23] In turn, the analysis of the higher education market conducted by Wiśniewska-Sałek or Ulewicz [26], indicated that the number of engineering graduates, which are of particular significance in industrial sectors, varied [24]. The detailed listing is presented in Figure 1. A drop in the interest in engineering studies in the context of other types of studies showed the smallest declines and in the academic year of 14/15-13/14 – even large increases. Obviously, the situation of a declining population among young people is the interpretation of such a downturn. Managerial skills, which freely allow for keeping up intellectually with communication technologies, may also become significant. Such people, while excluding practical knowledge, have the ability to use "computer technologies" at the same time constituting the so called senior manager [25].

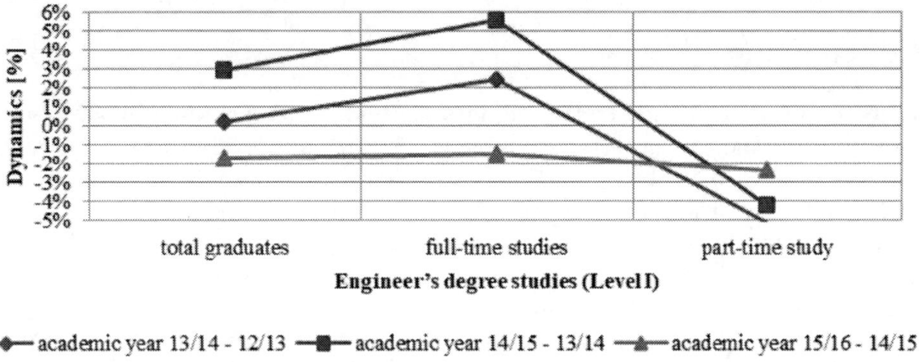

Fig. 1. The dynamics of changes in the number of graduates in the Poland from the academic year of 2012/2013 to 2015/2016.

Social Product Development (SPD) is one of the elements of "Industry 4.0" from the social point of view. This idea has no unambiguous definition. It is identified with the involvement of the team (which includes the qualified group of people from the inside or/and the outside of the company) in the development of the product using: technologies, social tools and media. Users have impact on the product life cycle (subsequent stages of design), due to which a new product, introduced into the market, has a shorter time of development and is burdened with lower investments. [22] From the point of view of the furniture industry, the fact that, with such a large share of this industry in the national economy, only 3 universities provide education in the field of tree farming and 4 universities – in the field of forestry, seems to be disturbing.

2.3 The environment - natural resources (wood)

The wood industry, as relating to the renewable natural raw material, is one of the key and most difficult to implement parts of sustainable development. The situation depends on the time which is needed by forest stands to "recover" its resources and its general usability in purifying the air and, on the other, the rapid development of the industry indicates an increased demand for this raw material. According to the statistical data published by Central Statistical Office, the area of forest land in Poland in recent 6 years (2010-2016) increased on average by 0.19% with an increase of 0.2% in woodiness, at the same time occupying less than 30% of the area of the country. In accordance with the Regulation by the Minister of Environment, the minimum felling age for individual species ranges from 30 years, e.g. for poplar and up to 120 years for oak and ash, it is connected with the problem of sustainability. On the basis of the data by Central Statistical Office, it can be observed that in years 2010-2016 the dynamics of changes (chain index) of restoration and a forestation in Poland was the following: 8.7%; 2.1%; -3.8%; 2.2%; 3.4%; -1.0%. At the same time, in the research period, there was received the following dynamics of changeability of the acquired large timber: 3.90%; 0.29%; 2.34%; 5,21%; 1.77%; 2.09% due to which there was acquired 33568291 m^3 of wood in 2010 and as much as 39129329m^3 in 2016. The restoration of forest land may take place naturally or artificially. With a significant increase in the acquisition of large timber in 2016 compared to 2015 there was artificially restored only 1.4% more of forest land with no natural restoration (-13.4%). The assumptions of "Industry 4.0" do not directly translate into the environment but indirectly, through more efficient use of raw materials for production, it is possible to reduce production waste or utilizing a more appropriate technology use it more significantly.

3 Conclusions

Industry 4.0 very clearly and legibly outlines the bridge between the economy and social expectations. On the one hand, it creates the grounds for the development of technological innovation but, on the other, shows how important the human being is in this whole process. On the one hand, it fosters economic development and, on the other, satisfies staff shortages with automation. On the one hand, it streamlines and optimizes production time and, on the other, allows the transfer of dangerous and burdensome tasks from the human being to the machinery. It allows for using the mental potential of the human being instead of crushing it. All these activities bring about that there may occur the co-integration correlation between sustainable development and "the fourth industrial revolution". An example of the implementation of the assumptions of "Industry 4.0" in accordance with the social dimension of sustainable development is the project MSIE4.0. Project assumes an increase in opportunities and capacity of universities in Thailand (based on the experience of European countries, including Polish) to provide a high-quality syllabus – based on competencies for Masters courses in the field of industrial engineering, which supports sustainable smart industry (Industry 4.0) and is compliant with European Qualifications Framework.

Acknowledgements
This publication is a partial outcome of project „Curriculum Development of Master's Degree Program in Industrial Engineering for Thailand Sustainable Smart Industry (MSIE4.0)" that has been funded with support from the European Commission (Project Number: 586137-EPP-1-2017-1-TH-EPPKA2-CBHE-JP).

Co-funded by the
Erasmus+ Programme
of the European Union

References

1. J. Bendkowski, Zeszyty Naukowe Politechniki Śląskiej, **112** (2017)
2. H. Kagermann, W. Lukas, W. Wahlster, VDI nachrichten, **13** (2011)
3. P.N. Koustoumpardis, J.S. Fourkiotis, N.A. Aspragathos, International Journal of ClothingScience and Technology, **19**, 2 (2007)
4. J. Cardoso, K. Voigt, M. Winkler, ICEIS 2008, LNBIP, **19** (2009)
5. T. Qu, S.P. Lei, Z.Z. Wang, D.X. Nie, X. Chen, G.Q. Huang, Int. J. Adv. Manuf. Technol., **84**, 1-4 (2015)
6. F. Scavo, B. Newton, M. Longwell, Computer Economics Report, **34**, 8 (2012)
7. V. Paelke, IEEE Emerging Technology and Factory Automation (ETFA), (2014)
8. S.P.L. Kumar, Engineering Applications of Artificial Intelligence, **65** (2017)
9. D. Gorecky, M. Schmitt, M. Loskyll, IEEE International Conference on Industrial Informatics (INDIN), (2014)
10. B. Kehoe, S. Patil, P. Abbeel, K. Goldberg K, IEEE Transactions on Automation Science and Engineering, **12**, 2 (2015)

11. R.Y. Zhong, X. Xu, E. Klotz, S.T. Newman S.T, Engineering, **3**, 5 (2017)
12. A. Ustundag, E. Cevikcan, *Industry 4.0: Managing The Digital Transformation* (Springer, Berlin, 2017)
13. N. Jazdi, IEEE Conference on Automation, Quality and Testing, Robotics (AQTR), (2014)
14. W. Torbacki, data odczytu 27.04.2018: http://www.ptzp.org.pl/files/konferencje/kzz/artyk_pdf_2018/T2/2018_t2_716.pdf
15. Prawo ochrony środowiska z dnia 27 kwietnia 2001 r., art.3, pkt.50.
16. A. Pawłowski, Środkowo-Pomorskie Towarzystwo Naukowe Ochrony Środowiska, **11** (2009)
17. F. Susanto, W. Arafah, Z. Husin, Polish Journal of Management Studies, **16,** 2 (2017)
18. A. Pabian, Zeszyty Naukowe Politechniki Częstochowskiej Zarządzanie, **28**, 1 (2017)
19. Polskie Meble. Raport specjalny. Inwestycje dla Polski, Data odczytu 27.04.2018: https://www.paih.gov.pl/files/?id_plik=30887
20. K. Ashton, Data odczytu 29.04.2018: http://www.rfidjournal.com/articles/view?4986
21. H. Kagermann, W. Wahlster, J. Helbig J., Data odczytu 27.04.2018 : http://www.acatech.de/fileadmin/user_upload/Baumstruktur_nach_Website/Acatech/root/de/Material_fuer_Sonderseiten/Industrie_4.0/Final_report__Industrie_4.0_accessible.pdf
22. P. Wittbrodt, I. Łapuńka, Data odczytu 29.04.2018: http://www.ptzp.org.pl/files/konferencje/kzz/artyk_pdf_2017/T2/t2_793.pdf
23. Bank Danych Lokalnych: https://bdl.stat.gov.pl/BDL/start
24. A. Wiśniewska-Sałek, Knowledge For Market Use 2017: People In Economics – Decisions, Behavior And Normative Models. International Scientific Conference Proceedings, Pavla Slavíčková (Ed.), Palacký University, Olomouc (2017)
25. L. Kiełtyka, Przegląd Organizacji, **8** (2016)
26. R. Ulewicz, Polish Journal of Management Studies, **8** (2013)

Quality of project financed by the European Union structural funds

Ewa Kozień[1,*]

[1]Cracow University of Economics, Faculty of Economics and International Relations, Department of Management Strategy and Organisation , ul. Rakowicka 27, 31-510, Cracow, Poland

Abstract. The European Union structural funds even out the chances of development of particular regions in the European Union. Entities applying for additional financing from structural funds for various actions implemented in a form of projects must meet specific criteria. For this reason, the quality of prepared and implemented projects is of key importance. This paper draws the attention to meeting the requirements of a project at a stage of its preparation and implementation, taking into account the formal, contents-oriented and technical criteria and application of methodology of management, for effective absorption of European Union structural funds.

1 Introduction

The development of modern organizations [1] is determined by dynamic changes in the surrounding environment. Therefore the organizations, in order to maintain their competitive edge, undertake the actions implemented in a form of projects, which are assumed to contribute to achievement of beneficial change. However, due to limited resources, organizations apply for financing from European Union structural funds in order to implement their projects. One of essential elements in competition among organizations for financial support for their projects from structural funds is the quality of their preparation and implementation. This paper formed the following thesis – the quality of the project related to the formal, contents-oriented and technical preparation and well-developed methodology of its implementation increases the chances for additional financing for the project from the European Union structural funds.

2 Structural funds as the aid from the European Union in co-financing projects

Currently Poland is the greatest beneficiary of the European Funds. In the years 2014-2020 the European Union earmarked 82,5 billion Euro for Poland. The funds in Poland are spent for projects related to transport infrastructure, innovativeness, environmental protection, power generation, education, culture, employment and counteracting social exclusions [2].

* Corresponding author: koziene@uek.krakow.pl

In the period of programming for the years 2014-2020 the instrument of implementation of the European Union structural policy are the structural and investment funds, including the following funds which can be used for co-financing the projects [3]:

- *European Regional Development Fund* finances the investments in infrastructure and development of potentials of the regions by implementation of projects regarding: building infrastructure of educational nature and in the scope of natural environment, construction and modernization of roads as well as supporting competitiveness of small and medium businesses.
- *European Social Fund* supports activities connected with development of employment market and human resources by preparation of projects related to the improvement of quality of educational system, training of specialists and business community as well as training unemployed persons.
- *European Agricultural Fund for Development of Rural Areas* supports agriculture as a result of implementation of projects concerning the improvement of conditions of life in rural communities and their development, differentiation of production and assistance for agricultural and food production sector.
- *European Marine and Fishing Fund* supports projects connected with renovation of fishing fleet, improvement in fish processing and marketing actions popularizing seafood.

Within the European Union structural policy the objectives and legal framework have been developed, which have been specified in the Regulation of the European Parliament and the Council (EU) No. 1303/2013 of 17 December 2013, establishing common legislation concerning the European Regional Development Fund, European Social Fund, Cohesiveness Fund, European Agricultural Fund for the Development of Rural Areas and European Marine and Fishing Fund as well as the legislation establishing general provisions concerning the European Regional Development Fund, European Social Fund, Cohesiveness Fund and European Marine and Fishing Fund and abrogating the regulation of the Council (EC) No. 1083/2006 [4].

The objectives of the EU structural policy strengthen the connection between structural and investment funds within the „Europe 2020" strategy. The connection of European structural and investment funds for the 2014-2020 programming period with the „Europe 2020" strategy, implemented in the years 2010-2020, involves coordination of actions in time as well as convergence of designated goals. Within the formed „Europe 2020" strategy three goals have been designated [3]:

Goal 1 – Intelligent development including the following thematic goals:
- support of scientific research, technological development and innovations,
- increase of availability, application and quality of IT and communication technologies,
- strengthening of competitiveness of small and medium business sector.

Goal 2 – Sustainable development focused on:
- preservation and protection of natural development and support of effective management of resources,
- promotion of adjustment to climate changes, prevention of hazards and risk management,
- supporting the conversion to low-emission economy,
- promotion of sustainable transport.

Goal 3 – Development promoting social inclusion, such as:
- promotion of lasting and high quality employment as well as supporting workforce mobility,
- promotion of social inclusion and fighting poverty and all forms of discrimination,
- investing in education, training (including vocational training) aimed at improvement of skills and continuous learning,

- strengthening the skills of institutional public authorities and interested parties as well as effectiveness of public administration.

Effectiveness of entities applying for support from structural funds requires an applicant to understand the nature of the project and possess knowledge from the scope of quality management at particular stages of preparation and implementation of a project.

3 Project quality

Total Quality Management philosophy finds practical application while preparing and implementing a project. A project, as an action with unique and temporary character, [5] is described by basic parameters i.e. meeting requirements, time and cost. Meeting requirements refers to quality requirements pertaining to a project. Interpretation of quality in specialist literature varies. Most generally, quality is perceived as a degree of adjustment of a given item to object's expectations. P.B. Crosby [6], E.W. Deming [7], defining quality, draw the readers' attention to such elements as: compliance with requirements, reliability, cost. More precisely, quality is interpreted by product, usability, value and normative approach. One should note that normative quality is of universal nature, it combines the remaining approaches, explaining quality as a collection of inherent meeting of requirements [8]. What is the quality of a project then? – one may directly define the project quality as the ability to meet requirements specified as each stage of its preparation and implementation. Many interconnected elements have impact on project quality, such as:

- surrounding milieu – political, economic, institutional, technical, social and cultural,
- preparation and implementation of a project, hence a definition of methodology of project management and actions in the field of management of scope, integration, time, costs, quality, human resources, communication, risk, orders, involvement, knowledge.

Moreover, the level of project quality may be also viewed in the aspect of effectiveness of implementation of the goal of the project, economy, benefits and social effects.

From the point of view of effectiveness in obtaining EU structural funds it is crucial to meet requirements connected with preparation and implementation of a project. Therefore it is necessary to consider the issue of project quality with a view to requirements concerning the following three criteria:

- formal,
- contents-oriented and technical,
- methodology of management.

Based on the author's own research regarding projects in the field of infrastructure, environmental protection, science and education, culture, sport as well as activization and social integration, implemented in selected communes in Poland and co-financed from the EU structural funds, the following requirements, essential from the formal point of view have been pointed out, which improved the quality of the projects being prepared:

- cohesiveness of project goals with the formulated strategy (of the state, commune or a company),
- compliance of the norms of EU and domestic law,
- innovativeness concerning methods of implementation and results of the project,
- completeness of an application (compliance of a hard copy of an application form with an electronic form,
- completeness of attachments,
- correct specification of qualified costs,
- structure of sources of project financing compliant with the amount and level specified in the documentation concerning a given action.

However, in the scope of contents-oriented and technical criteria, the qualitative requirements of the project concern:

- specification of goals and actions and descriptions of anticipated results,
- specification of added value and continuity of benefits following the completion of the project. Added value means actual benefits and usability of a project, subject to verification (economic and functional continuity), i.e. *ex post* evaluation after several years from the moment of project completion.
- system of monitoring and assessment which serves the verification of preparation and verification of the project. Qualitative and quantitative indicators are evaluated at each stage of project implementation. Therefore one must evaluate the project quality *ex ante, on – line* and *ex post.*
- resources. In the project the coordination in time of human resources, asset, finance, information – is crucial. Quality of the project in a special way refers to human resources, i.e. the possessed specialist knowledge connected with subject matter of the project as well the knowledge connected with project management as well as skills and experience.
- structure of project financing, which specifies the percentage share of the entity's own contribution (special purpose domestic subsidies, loans, credits) as well as money coming from structural funds. Moreover, the requirements of the developed cost estimate concern an important issue of division of costs into qualified and non-qualified costs, which cannot be covered from structural funds,
- project implementation schedule, i.e. the necessity of coordination of project tasks in time in order to eliminate delays (EVM method [9]).

The last criterion concerning the issue of quality and the constant element of structure of project management methodologies was considered based on analysis of selected methodological approaches. It is worth stressing the fact that the European Committee noticed a problem of effectiveness of participation of particular member states in implementation of aid policy. As a result of identification of committed errors during the preparation of projects for implementation, it proposes the application of methodological solutions improving this process, and which the beneficiaries and executive staff may take advantage of. The European Union does not impose on the beneficiaries any specific methodology assisting project management, but it encourages them to choose proven methodologies. In the Project Cycle Management [10] method which had been approved by the European Committee the aspect of project quality is stressed in the scope of assessment of validity and compliance of the problem with EU development policy, a method of its solution as well as continuity and quality of project results.

However, in the methodology: A Guide to the Project Management Body of Knowledge, PMBoK® [8] the field of quality management assures that as a result of the process of planning, quality assurance and monitoring, the project will meet quality expectations perceived in a category of client satisfaction, product usability as well as compliance with regulations, norms and legal provisions. Quality requirements of the project concern many standards, sometimes exceeding the subject matter of the project, and including the sphere of management, access to information, technical norms as well as ecological norms.

However, in a group of agile methodologies (i.e. Scrum), which are based on iterative and incremental building of results, the care for project quality is based each time in increases [11]. The significant advantage of agile methodologies means continuous presence of a client who evaluates the current quality of particular project results. It should be emphasized that agile methodologies treat exceeded deadlines and costs of project implementation as a certain „norm", stressing decisively the project scope and quality what is confirmed by research conducted by a company called VersionOne [12], where the increase of benefits arising from application of agile approach to project management in the scope of improvement of project results quality was defined at the level of 80%.

Analyzing selected project management methodologies one may find that they include a key parameter of meeting qualitative requirements concerning project results. Due to the implementation of hard (infrastructural) as well as soft (e.g. trainings) projects which are practically complicated as well as varied – with a view to their subject matter – connected with the investment in human development, qualitative requirements constitute a vast catalogue specified in relation to a given type of project. Currently the directive of effective project implementation is narrowed down to implementation of a project well (meeting qualitative requirements), quickly (time) and cheaply (cost). Taking into account the benefits arising from agile approach in the scope of project quality one must recommend this approach in implementation of not only innovative and IT projects. It should be emphasized that both in traditional and agile methodologies the mutual interdependencies of parameters are focused with a view to qualitative requirements of the project

4 Conclusions

Project quality is of key importance in the aspect of obtaining funds from the EU structural funds. The principles of launching structural funds (i.e. principle of concentration, partnership, compliance, programming and additionality) translate into requirements concerning the implementation of projects which were granted subsidies from the European Union structural funds. It should be emphasized that only correctly prepared and implemented projects, i.e. the projects which meet basic EU requirements in practice stand a chance for additional financing. In order to avoid the rejection of valuable projects on formal grounds one should approach their preparation and implementation holistically, taking into account the issue of formal requirements, contents-related justification and methodological approach to project quality.

Meeting qualitative expectations in relation to the final product is possible when the quality is not omitted but becomes an integral element of the project management process, what can be perceived in the proposed methodological approach. It should be stated that the project quality makes it possible to obtain money from EU funds, to achieve success from implementation of the project itself, it also impacts the potential for development of organizations and regions.

Quality management is one the most important factor impacting on the value of scientific investigation results, especially in these areas, where a laboratory scale research meets industrial requirements e.g. materials science [13-26], machining [27, 28], biotechnology [29] or huge enterprises [30-32]. This is clearly recognized in scientific publications in the field of management [33-35].

References

1. E. Kozien, Identification of stage phase growth in the checklist method using different statistical parameters, *Economic and Social Development. Book of Proceedings*, 538 (Prague, 27-28 April 2017)
2. https://www.funduszeeuropejskie.gov.pl/strony/o-funduszach/zasady-dzialania-funduszy/fundusze-europejskie-w-polsce/
3. *European structural and investment funds for years 2014-2020: Official texts and comments* (European Commission, 2016)
4. *Dz.U.*, **L 347** (20.12.2013)
5. P. B. Crosby, *Running Things: the Art of Making Things Happen* (Mc Graw-Hill, New York, 1986)

6. E. W. Deming, *Quality, Productivity and Competition Position* (MIT Press, Cambridge, Massachusetts, 1982)
7. *ISO 9000:2000*
8. *A Guide to the Project Management Body of Knowledge*, 5th Edition (Project Management Institute, Newtown Square, PA, 2013)
9. *Practice Standard for Earned Value Management* (Project Management Institute, Newtown Square, PA, 2011)
10. *Project Cycle Management Handbook* (European Commission, March 2004)
11. K. Schwaber, M. Bendle, *Agile Software Development with Scrum* (Prentice Hall, 2001)
12. *VersionOne. The 10th Annual State of Agile Report*, www.versionone.com (2015)
13. J. Korzekwa, P. Wawrzala, R. Skulski, *Eur. Phys. J.-Spec. Top.* **154**, 127 (2008)
14. A. Dudek, C. Kolana, *Solid State Phenom.*, **165**, 25 (2010)
15. T. Pieczonka, J. Kazior, A. Szewczyk-Nykiel, M. Hebda, M. Nykiel, *Powder Metall.*, **55**, 354 (2012)
16. N. Radek, J. Pietraszek, B. Antoszewski, *Adv. Mat. Res.-Switz.*, **874**, 29 (2014)
17. R. Ulewicz, F.R. Novy, *J. Balk. Tribol. Assoc.*, **22**, 1147 (2016)
18. A. Gadek-Moszczak, *Image Anal. Stereol.*, **36**, 151 (2017)
19. A. Gadek-Moszczak, *Image Anal. Stereol.*, **36**, 207 (2017)
20. M. Ulewicz, U. Lesinska, M. Bochenska, *Physicochem. Probl. Mi.* **44** 245 (2010)
21. T. Lipinski, *Solid State Phenom.*, **163** 183 (2010)
22. M. Ulewicz, E. Radzyminska-Lenarcik, *Physicochem. Probl. Mi.*, **46**, 119 (2011)
23. T. Lipinski, A. Wach, *Proc. 23rd International Conference on Metallurgy and Materials*, Brno, Czech Republic, May 21-23, 738 (2014)
24. D. Klimecka-Tatar, S. Borkowski, P. Sygut, *Arch. Metall. Mater.*, **60**, 735 (2015)
25. D. Klimecka-Tatar, G. Pawlowska, M. Sozanska, *Arch. Metall. Mater.*, **60** 153 (2015)
26. R. Dwornicka, N. Radek, M. Krawczyk, P. Osocha, J. Pobedza, *Proc. 26th Int. Conf. on Metall. and Mater. – METAL*, Brno, Czech Republic, May 24-26, 1252 (2017)
27. J. Pietraszek, *Proc. 6th Int. Conf. Neural Net. Soft Comput.*, Zakopane, Poland, Jun 11-15, 2002, 250 (2003)
28. I. Dominik, J. Kwasniewski, K. Lalik, R. Dwornicka, *Proc. 32nd Chinese Control Conf. (CCC)*, Xian, China, July 26-28, 7505 (2013)
29. J. Pietraszek, E. Skrzypczak-Pietraszek, *Adv. Mat. Res.-Switz.* **874**, 151 (2014)
30. P. Osocha, *Adv. Mat. Res.-Switz.* **874**, 89 (2014)
31. E. Baussan, *Phys. Rev. Spec. Top.-Accel. Beams* **17**, 031001 (2014)
32. R. Dwornicka, *Adv. Mat. Res.-Switz.* **874**, 63 (2014)
33. R. Ulewicz, J. Selejdak, S. Borkowski, M. Jagusiak-Kocik, *Proc. 22nd International Conference on Metallurgy and Materials*, Brno, Czech Republic, May 15-17, 1926 (2013)
34. R. Ulewicz, *Proc. Int. Conf. Path Forward for Wood Products: A Global Perspective*, Baton Rouge, LA, Oct 05-08, 51 (2016)
35. A. Pacana, R. Ulewicz, *Polish J. Environ. Studies*, **16**, 165 (2017)

Development of human capital in the context of improving processes in the enterprise

Rafal Prusak[1,*]

[1]Czestochowa University of Technology, Faculty of Production Engineering and Materials Technology, Department of Production Management and Logistic, Poland

Abstract. Human is the key resource of the organization, and proper human resources management, manifested primarily by matching knowledge, skills and attitudes of employees to the strategic goals of the company should lead to their transformation into human capital and the synergy effect of combining the qualities and strengths of employees intensifying the frequency of the innovative concepts. The main purpose of the article is to present the results of research conducted on a sample of 70 enterprises regarding the analysis of the extent of enterprises' use of the potential accumulated within human capital to improve the implemented processes, including quality management. The research covered issues related to the impact of the functioning of the human resources management system on the company efficiency, the main factors supporting and hindering the functioning of this system, as well as problems and difficulties occurring both at the stage of implementing the system and its subsequent operation. The basic research tool was a questionnaire consisting of 23 questions of a closed nature supported by interviews conducted with representatives of the management. The article also includes a literature review on selected aspects of shaping and managing human capital and increasing the professional potential of employees as a result of systemic management of their competences.

1 Introduction

According to modern management theories [1], the reality in which enterprises operate is characterized by such statements as: enterprises need people, intellectual advantage (including human capital) determines the competitive advantage, employees are mobile, and their relationship with the enterprise is not long-term, employees expect to meet diverse needs. Properly motivated people are able to use their knowledge and experience to actively participate in the process of achieving the goals of the company as an important source of building market advantage [2]. In this context, it is important to treat all employees as a source of potential opportunities and to make them aware of their strengths and weaknesses, indicating the areas in which their work will be the most valuable for the company [3]. The company must strive to make the employee voluntarily focus his efforts on the company's goals [4]. The role of employees' involvement was clearly emphasized in

* Corresponding author: prusak.rafal@wip.pcz.pl

the works by Beer and Spector, according to the model [5]: people are able to develop only when they operate in a favorable work environment, the key to building trust and commitment is to ensure open communication, greater employee engagement is achieved when they take part in identifying problems and finding solutions, it is important to have a personnel strategy combining employee goals with the company's goals.

Human resource management is understood as a strategic, coherent and integrated approach to the employment, development and quality of life of employees employed in the enterprise [6]. It is a kind of a system in which people's problems and goals are combined with the goals of the organization [7] and a human is treated as the key assets of the company and the source of its competitiveness occupying a central place in management processes [8]. Proper management of human resources therefore requires constant actions aimed at improving the knowledge about the skills and potential of employed people in order to be able to fully use their experience and competences for the benefit of the company and employees. Proper management of human resources is a key aspect of the process in which these resources are transformed into human capital [9].

In literature, human capital is defined in a rather broad way, taking into account different criteria, perspectives and scale. An overview of available theories allows to conclude that human capital:

- is a set of specific features and properties embodied in employees that have a certain value and are a source of future income for both the employee - the owner of human capital, and for the organization that uses certain capital on certain terms [9],
- is the main source of value and the factor determining the competitive position and is the only type of capital that can add to itself and by itself [10],
- includes combined knowledge and innovation of employees [11], competencies important in business [12], social and personal skills [13], physical, mental, intellectual and moral characteristics shaped by predispositions and internal motivation [14], innate human abilities, specific behavioral patterns and personal energy of the employee [15],
- is used as a result of synergic impact of ability, experience, creativity [6],
- relates strongly to performance, especially when the human capital in question is not readily tradable in labor markets and when researchers use operational performance measures that are not subject to profit appropriation [16], requires shaping appropriate attitudes of employees [17], takes into account intellectual attributes, commitment, behaviour ethics, leadership and other social predispositions to specific organizational behaviours [8],
- uses both the ability to analyze and synthesise [18] leading to the development and introduction of innovations in the enterprise [19],
- cannot become the property of an enterprise because it is closely related to specific persons, it only remains in its temporary disposal [11].

The value of human capital depends to a large extent on the level of employees' involvement, the degree to which they want to and can use their own potential to achieve the company's goals. Therefore, the key factor in this context is to ensure an appropriate level of employee motivation and professional satisfaction, which is a consequence of the synergistic impact of organizational culture, management style, terms and content of work, interpersonal relations and incentives within the incentive system.

Human capital can be perceived in three basic perspectives [20]:

- economic one, according to which human capital is specific, it is uncertain, and the actions of enterprises (connected with incurring costs) should focus on the selection of appropriate employees,
- human capital - placing emphasis on decisions regarding the level of outlays on investments in human capital,

- resource based, based on the study of competitive advantage resulting from the proper use of human resources.

Human capital management consists in making decisions (strategic, investment, operational, adding value) concerning the management of people based on acquired, analyzed and reported data [6]. At the same time, it is a set of activities aimed at the productive use of the company's abilities perceived as key to achieving current and future successes [21].

In the context of effective human capital management, literature often raises the issue of proper development of employees and the proper use of their knowledge, experience or - generally speaking - competence. Competences are defined as modifiable and adjusted to the requirements of the position, predispositions [9], and distinctive behavior of employees in various professional situations, determined by the knowledge and skills they use, and by attitudes and motivations [22] that allow him to effectively carry out professional tasks [23].

The main objectives of competence management can be included [24]:

- defining competence standards for workstations,
- examining the real competences of employees and job candidates in terms of job selection,
- preparing people in the organization for current and future tasks,
- increasing the value of the organization by increasing the value of human capital,
- increasing the level of employee satisfaction,
- maximum use of employees' strengths while minimizing their weaknesses,
- optimization of the relation: expenditures on human capital - effects.

2 Methodology and analysis of results

The main purpose of this study is to attempt to characterize selected relationships between the company's activities undertaken in the context of raising the level of employees' competences and selected strategic variables. The research was conducted using a questionnaire and a structured interview in 70 companies representing various sectors and characterized by different market situation. The survey consisted of 23 questions. The questionnaire was filled by representatives of the top management in each of the research facilities in the presence of the person conducting the research (some issues were discussed, answers were given to possible doubts of management representatives). The selection of the research sample was deliberate.

In this study, only selected elements of the study were characterized, the results of which are presented in Table 1 concern:

- impact of the introduction of the competence management system on the organization's efficiency: imperceptible (1.1), small (1.2), significant (1.3), large (1.4), very large (1.5),
- factors that have the greatest impact on the functioning of the competency management system: involvement of senior management (2.1), connection with the incentive system (2.2), awareness of employees about the role and importance of the system (2.3), properly developed training system (2.4), visible effects of the system (2.5), allocation of the appropriate amount of resources (2.6),
- elements that hinder the functioning of the system to the greatest extent: no management involvement (3.1), insufficient employee involvement (3.2), system mismatch (3.3), insufficient resources (3.4), poor selection of training units (3.5), no consequences in activities (3.6), no connection with other systems in the enterprise (3.7),

- elements contributing to problems during the development of the system: lack of staff motivation (4.1), inability to identify connections with other systems (4.2), difficulties in finding a consulting company (4.3), inability to select methods and techniques (4.4), incorrectly defined ranges liability (4.5), previous negligence and lack of action (4.6).

Table 1. Summary of research results.

No.	Intensity of competition			The bargaining power of suppliers			The bargaining power of recipients.		
	low	avg	high	low	avg	high	low	avg	high
1.1	2	6	1	0	1	8	1	5	3
1.2	2	4	1	0	1	6	0	4	3
1.3	8	12	2	1	12	9	2	19	1
1.4	0	11	7	1	15	2	2	10	6
1.5	1	2	11	1	11	2	0	4	10
2.1	0	5	15	1	17	2	0	7	13
2.2	2	6	0	2	3	3	1	6	1
2.3	2	1	0	0	0	3	0	3	0
2.4	3	2	0	0	1	4	0	5	0
2.5	2	9	4	0	10	5	1	10	4
2.6	5	11	3	0	8	11	3	11	5
3.1	7	10	2	0	8	11	2	15	2
3.2	3	10	5	1	12	5	1	9	8
3.3	3	4	4	0	7	4	0	8	3
3.4	1	5	8	1	6	7	1	6	7
3.5	0	1	1	0	2	0	0	1	1
3.6	0	2	1	0	2	1	1	1	1
3.7	0	2	1	1	2	0	0	2	1
4.1	1	9	5	1	9	5	1	9	5
4.2	4	12	1	1	9	7	0	13	4
4.3	3	1	0	0	1	3	0	4	0
4.4	2	5	4	0	6	5	2	6	3
4.5	1	2	4	0	5	2	1	3	3
4.6	3	5	8	1	9	6	1	7	8

The obtained results allowed to state the following basic trends and phenomena:

- Regardless of the strategic criterion applied, the introduction of a competency management system - in accordance with the declarations of the management personnel - had a significant impact on the efficiency of the organization's functioning. Among others, there were: fewer mistakes and errors of employees, reduced number of defective products (according to company data from several to several dozen percent), reduced number of customer complaints and general slight increase in their satisfaction level.
- The most frequently indicated factors affecting the functioning of the competence management system were: engagement of senior management (both in the supervision dimension and active inclusion in the consultation process and improvement), visible

effects of the system (significantly improving morale and the level of employee motivation), appropriate the amount of resources (both tangible and intangible).

- Among the most-difficult elements of the system's functioning, the following were pointed out, among others: insufficient employee involvement (especially in the context of consolidation of implemented practices and feedback), system mismatched to the company's needs (as a result of excessive reliance on standard, model or theoretical solutions).
- The most important elements contributing to the problems during the development of the system were: lack of staff motivation (mainly following an ineffective organizational culture, inappropriate management styles or too low involvement of employees in the development and implementation of the system), inability to identify connections with other systems (lack of experience in implementing such solutions, inadequate management methods), and previous negligence and lack of actions (a number of elements, including: fixed erroneous patterns of actions, disregard of intangible assets, high employee turnover, ineffective remuneration system).

3 Conclusions

According to contemporary theories in the area of organization and employment, people are a key resource of enterprises. It results from the accumulated potential in them which is a resultant of the effects of the education process, previous professional experience and personality traits (including openness to changes, willingness to participate in pro-development activities). Effective human capital management is able to bring measurable and satisfying benefits to the company. The basic ones include: more effective use of knowledge and competences of employees, increasing the number of pro-innovative activities (as a result of better knowledge management, effective team building, functioning of a stimulating organizational culture), improving the market situation, building unique market advantages, improving market image and customer relations.

The conducted research allowed to show that in almost half of the analyzed enterprises, the effect of implementing the human capital management process were significant changes affecting the efficiency of the company's operation, in some cases leading to a change in the strategic approach. Nearly one-third of the respondents pointed to the occurrence of significant effects in the area of quality in the production process. Research has shown that the efficiency of using the human factor in enterprises operating in difficult market conditions is greater and that there is a possibility of occurrence of large potential benefits in the area of interpersonal relations and the effectiveness of work of employee teams. At the same time, enterprises operating in more difficult market conditions declared greater interest in targeted development of individual and group employee competences.

Undoubtedly, the statement of the existence of clear statistical dependencies will require conducting additional, extended research using a more complex research apparatus. A more detailed recognition of individual elements and a more precise characterization of existing relationships and dependencies of both an internal and external character will be necessary.

Human capital and its management processes are becoming an everyday reality for modern enterprises. Obtaining unique market advantages due to them can ensure the strengthening of the market position and increase the profitability of functioning. It is important, however, that issues related to this area be permanently integrated with the basic strategy of the company. From the point of view of the interests of the company, human capital at the team level is particularly interesting. in this area, the competence of employees and the exchange and development of knowledge are combined. Synergic impacts of potentials of individual employees can lead at this level to obtaining above-average and unexpected results and performance. The task of the company is to create a working environment that allows for the most free and uninterrupted exchange of ideas and the integration of employees.

References

1. E. Michaels, H. Handfield – Jones H., B. Axelrod, *The War for Talent* (Harvard Business School Press, Boston, 2001)
2. E. Michaels, H. Handfield – Jones H., B. Axelrod, *Strategic HRM: The route to improved business performance* (IP, Londyn, 2002)
3. M. Adamiec, B. Kożusznik, *Zarządzanie zasobami ludzkimi* (Akade, Kraków, 2000)
4. B. A. Campbell, R. Coff, D. Kryscynski, Academy of Management Review **37**, 3 (2012).
5. O. Lundy, A. Bowling, *Strategiczne zarządzanie zasobami ludzkimi* (Dom Wydawniczy ABC, Oficyna Ekonomiczna, Kraków, 2001)
6. M. Armstrong, S. Taylor, *Zarządzanie zasobami ludzkimi* (Wolters Kluwer, Warszawa, 2016)
7. N. McKenna, E. Beech, *Zarządzanie zasobami ludzkimi* (Gebethner i Spółka, Katowice, 1998)
8. A. Pocztowski, *Zarządzanie zasobami ludzkimi, Strategie – procesy – metody* (PWE, Warszawa, 2006)
9. H. Król, A. Ludwiczyński, *Zarządzanie zasobami ludzkimi*, Wydawnictwo Naukowe PWN, Warszawa, 2014)
10. J. Fitz-enz, *The ROI of Human Capital* (AMACOMNowy Jork, 2000)
11. L. Edvinsson, M. Malone, *Intellectual Capital* (Harper Collins Publishers, New York, 1997)
12. *Human Capital Investment. An International Comparision*, Centre for Educational Research and Innovation, OECD, 1998)
13. J. K. Mawdsley, D. Somaya, Journal of Management 42, **1** (2016).
14. A. Pietruszka – Otryl *Wyzwania wartościowania kapitału intelektualnego organizacji*, in: E. Skrzypek (ed.), *Uwarunkowania sukcesu przedsiębiorstw w gospodarce opartej na wiedzy* (Wydawnictwo UMCS w Lublinie, Lublin, 2004)
15. T. O. Davenport, *Human Capital: What It Is and Why People Invest It* (Jossey-Bass, 2008).
16. T. Crook, T. Russell, Y. Samuel, J. G. Combs, D. J. Woehr, D. J. Ketchen Jr., Journal of Applied Psychology 96, **3** (2011).
17. B. A. Campbell, M. Ganco, A. M. Franco, Strategic Management Journal **33,** 1 (2012).
18. J. Roos, G. Roos, N.C. Dragonetti, *Intelectual Capital – Navigating in the Business Landscap* (Macmillan Press, London, 1997)
19. A. Baron, A. Armstrong, *Zarządzanie kapitałem ludzkim. Uzyskiwanie wartości dodanej dzięki ludziom* (Woltes Kluwer, Warszawa, 2012)
20. L. Yaping, J. Jingfang, Management Science and Engineering 1, **2** (2007)
21. D. Whitaker, Personnel Today, **2** (2007).
22. S. Borkowska, *Strategie wynagrodzeń* (Oficyna Ekonomiczna, Dom Wydawniczy ABC, Kraków, 2001)
23. C. Woodruffe, *Ośrodki oceny i rozwoju* (Oficyna Wydawnicza, Kraków, 2003)
24. T. Oleksyn, *Zarządzanie kompetencjami. Teoria i praktyka* (Wolters Kluwer Polska, Warszawa, 2010)

Identifying key criteria in development of Industrial Engineering education

Tomasz Nitkiewicz [1,*], and *Zuhal* Ayen[2]

[1]Częstochowa University of Technology, Faculty of Management, 42200 Częstochowa, Poland
[2]Sakraya University, Production Engineering Department, Sakraya, Turkey

Abstract. The objective of this study is to define criteria in current development of Industrial Engineering (IE) education on the basis of studies and concepts published in commonly accessible databases and knowledge exchange platforms. General assumption of the study is that the universal criteria used in developing IE study programs are already defined and some experiences on their use are already published. The expected outcome of the study is to identify the criteria that could support the development of curriculum for MSc in IE. The approach used here is content analysis with the publication related to the higher level education in the field of Industrial Engineering.

1 Introduction

The concept of Industry 4.0 is rapidly entering into many areas of economic processes and societal life. Certainly, it is an ongoing process already, especially, when industrial processes are considered, but also, it is a catchword to cover all the technological changes that enter to the society through new products, services and new ways they are offered and serviced. The problem that we want to address is the readiness of the higher education system to support Industry 4.0 challenge. For a start, we choose to take a close look at the development directions of Industrial Engineering Master Programs that should be, basically, preparing new adapts of Industry 4.0 practices. The scope of the study is defined within the project titled „Curriculum Development of Master's Degree Program in Industrial Engineering for Thailand Sustainable Smart Industry (MSIE4.0)" and its output is geographically limited to Thailand but opening phase addresses the global context of IE education that should face the challenge of Industry 4.0.

In short, Industry 4.0 concept could be defined as: "technical integration of cyber-physical systems into manufacturing and logistics and the use of the Internet of Things and Services in industrial processes" [1]. It will influence significantly working environments as a whole, addressing its purchase, production, manufacturing, sales or maintenance processes by including concepts as smart manufacturing and maintenance, with high degree of automation and integration in all enterprise processes [2]. Moreover, it will increase the role and the participation of customers in all the life cycle phases, starting from Design by customer approach [3].

* Corresponding author: tomasz.nitkiewicz@wz.pcz.pl

2 Methodology

In order to find appropriate resources documenting the approaches to development of Industrial Engineering education and indicating critical issues in the process we have used the simplest possible keywords: "industrial engineering" and "education". Basic search gave us numerous resources and most of them were not relating exactly to IE education and need further narrowing and filtering. Our primal search could be narrowed by some additional keywords in order to filter the results to the ones addressing the issue of our interest. The filtering phrases used are:

- "Management and Production Engineering"
- "Manufacturing Engineering"
- "Quality Management"
- "Product Management"
- "Master of Science (MSc)"
- "Studying Program"

In addition, we used the resources that were published in 2010 or later. The database we have used at first is Science Direct database, and it gave us most of the results used here. We have also added some additional resources from another databases (Scopus, Springer and Emerald). Automatic filtering has limited the resources to several thousands of articles, and we have manually searched through about 250 entries altogether in order to get the final set of resources. Manual filtering was based on removing the sources that were not referring to IE education on higher level, or to too detailed issues. Since the goal of the paper is to identify the major criteria for IE programs development we stopped the publication review process. But the process could be continued for the purposes of numerical analysis and referring the development criteria to the geographical, industrial, academic or cultural specificity of countries and regions addressed. Finally, we left 25 resources for our further analysis.

3 Defining the criteria for development of IE programs

Basing on the scope of the publications the following issues are identified as framework for basic criteria for IE education development:

- General program
- Educational methods and tools
- Cooperation and networks
- Industry based training
- Competences and skills
- Quality assurance and accreditation

The issues identified as possible framework for development criteria are overlapping and, usually, are coming together in order to get more holistic approach to IE programs development. It is worth to notice that most of the issues raised in the publications fall into "General program" category, but we limit it to the issues that address IE program as a whole. Distinction between the categories of "Cooperation and networks" and "Industry based training" could be also difficult since the trainings come from the cooperation and perhaps networking solutions between academia and industry. We have decided to single out the category of "Industry based training" since it is very strongly represented issue in publications. On the other hand, we decided to leave also category of "Cooperation and networks" since it could be just an organizational approach towards industrial training but also cooperation with other actors in the surroundings of IE university, like other universities, public authorities, R&D sector or just its graduates.

Table 1. Description of development criteria for IE programs.

Criteria	Description
General program	Sustainable value creation [4]; outcome-based education structure [5-9]; cooperative and participative education [10]
Methods and Tools	Experimental and problem based learning [11-13]; Interactive tools [14-17]; sustainability assessment tools [4], participative education [4,7]
Cooperation and networks	Cooperative and participative education [4,7,18], incorporation of international exchange [8]; experience sharing [18,19]
Industry based training	Cooperative training [7,10,19]; industrial talks and visits [7,18]; industry pulled MSc theses and projects [6,13,20]; evaluation of competences and skills [2,6,21,22]
Competences and skills	Industry-based evaluation [10,23]; soft skills (p.e. communication and language) [2]; ICT and software skills [14,15]; managing innovation skills [13,22,24]
Quality assurance and accreditation	Benchmarking of accreditation and quality assurance systems [8,25,26]; certifying and adapting QMS [26,27]

4 Conclusions

The set of identified criteria is a first step towards the development of IE curriculum that will provide graduates ready for Industry 4.0 challenges. The criteria should be used in development but also in the process of adopting IE curricula to the changing environment. It is important to notice that the criteria should be also adopted to the local conditions of implementing universities. This would include the references to the national higher education system, national and regional industry specificity, market development trends, candidates and students preferences and predispositions, possible external financing options or educational offer of a university. Finally, it is important to mention that development and adoption of any university curriculum, including IE programs, is an ongoing process and it constantly demands attention, intervention, work, sources and goodwill to make it a successful story of universities facing the challenge of Industry 4.0.

Acknowledgements

This publication is a partial outcome of project „Curriculum Development of Master's Degree Program in Industrial Engineering for Thailand Sustainable Smart Industry (MSIE4.0)" that has been funded with support from the European Commission (Project Number: 586137-EPP-1-2017-1-TH-EPPKA2-CBHE-JP).

Co-funded by the
Erasmus+ Programme
of the European Union

Disclaimer

This publication reflects the views only of the authors, and the Commission cannot be held responsible for any use which may be made of the information contained therein.

References

1. W. Wahlster, J. Helbig, A. Hellinger, M. A. V. Stumpf, J. Blasco, H. Galloway, and H. Gestaltung, *Securing the future of German manufacturing industry. Recommendations for implementing the strategic initiative INDUSTRIE 4.0. Final report of the Industrie 4.0 Working Group* (2013)

2. L. Prifti, M. Knigge, H. Kienegger, and H. Krcmar, A Competency Model for 'Industrie 4.0' Employees, in *Proceedings der 13. Internationalen Tagung Wirtschaftsinformatik,* 46-60 *(WI,* 2017)
3. Risdiyono and P. Koomsap, J Intell Manuf, **24**, 295-311 (2013)
4. P. Bilge, F. Badurdeen, and I. S. Jawahir, Procedia CIRP, **40**, 516-523 (2016)
5. O. Akir, T. H. Eng, and S. Malie, Procedia - Soc. Behav. Sci., **62**, 87-92 (2012)
6. D. Mesquita, R. M. Lima, M. A. Flores, C. Marinho-Araujo, and M. Rabelo, Int. J. Ind. Eng. Manag., **6**, 3, 121-131 (2015)
7. R. Nordin, A. A. A. Bakar, N. Zainal, and H. Husain, Procedia - Soc. Behav. Sci., **60**, 271-276 (2012)
8. A. Samer Ezeldin, Z. Omar, R. M. Yasin, R. Hamid, N. M. Tawil, K. Yusoff, and M. S. Rasul, Procedia - Soc. Behav. Sci., **102**, 267-275 (2013)
9. Z. Sari, Procedia - Soc. Behav. Sci., **102**, 560-567 (2013)
10. N. Brahimi, F. Dweiri, I. Al-Syouf, and S. A. Khan, Procedia - Soc. Behav. Sci., **102**, 446-453 (2013)
11. R. M. Lima, D. Mesquita, M. Amorim, G. Jonker, and M. A. Flores, Int. J. Ind. Eng. Manag., **3**, 2, 75-82 (2012)
12. R. Riedel, D. Jentsch, S. Tröger, and E. Müller, Integrating experiential learning into Industrial Engineering curricula – a case study, IFAC Proc., **43**, 326-330 (2010)
13. B. Abaza, I. Paraschiv, M. Spiroiu, and C. Stanciu, Project-Based Pedagogy For A New Product Development, in *INTED2015 Proceedings* (2015)
14. J. W. Volkmann, M. Landherr, D. Lucke, M. Sacco, M. Lickefett, and E. Westkämper, Procedia CIRP, **41**, 632-637 (2016)
15. A. Miller and M. Bures, New Approach to Industrial Engineering Education with the Help of Interactive Tools, Procedia - Soc. Behav. Sci., **174**, pp. 3413–3419, (2015)
16. D. Jelonek, The Development of Software Agents in e-Learning 3.0, in *ICIT 2015 The 7 th International Conference on Information Technology* (2015)
17. P. Maresca, E. Gómez, J. Caja, C. Barajas, and R. Ledesma, Procedia Eng., **132**, 205-212 (2015)
18. N. Jamaluddin, A. Ayob, S. A. Osman, N. T. Kofli, and S. Johar, Procedia - Soc. Behav. Sci., **102**, 648-653 (2013)
19. A. Ayob, S. A. Osman, M. Z. Omar, N. Jamaluddin, N. T. Kofli, and S. Johar, Procedia - Soc. Behav. Sci., **102**, 48–54 (2013)
20. S. Fernandes, D. Mesquita, M. A. Flores, and R. M. Lima, Eur. J. Eng. Educ., **39**, 1, 55-67 (2014)
21. R. M. Lima, P. H. Andersson, and E. Saalman, , Eur. J. Eng. Educ., **42**, 1, 1-4,(2017)
22. M. Palma, I. de los Ríos, and D. Guerrero, Procedia - Soc. Behav. Sci., **46**, 1570-1580, (2012)
23. M. San-Juan, M. D. P. De Tiedra, O. Martín, F. J. Santos, R. López, R. Suárez, and R. Mahillo-Isla, Procedia Eng., **132**, 221-227 (2015)
24. R. M. Lima, D. Mesquita, C. Rocha, M. Rabelo, R. M. Lima, D. Mesquita, C. Rocha, and M. Rabelo, Production, **27** (2017)
25. M. A. L. Domingo, J. A. Albajez, and J. Santolaria, Procedia Eng., **63**, 786-795 (2013)
26. H. Chowdhury, F. Alam, S. K. Biswas, M. T. Islam, and A. K. M. S. Islam, Quality Assurance and Accreditation of Engineering Education in Bangladesh, Procedia Eng., **56**, pp. 864–869, Jan. (2013)
27. R. Ulewicz, System Approach To Assure Quality of Education At the Faculty of Management of Czestochowa University of Technology., Polish J. Manag. Stud., **8**, pp. 259–268, (2013)

Assessment of production processes functioning in the case of air bag production

Krzysztof Mielczarek[1,*], and *Krzysztof* Knop[1]

[1]Czestochowa University of Technology, Faculty of Management, Department of Production Engineering and Safety, Armii Krajowej 19B, 42-200 Czestochowa, Poland

Abstract. In the chapter was presented enterprise producing air bags. described. Toyota's management principles by Jeffrey Liker were pointed out, which refers to production process (second). An innovative BOST questionnaire survey was characterized as a tool for transformation of Toyota's management principles into questions. Survey and research method determined as BOST (the name of the is an acronym created from the first two letters of the name and surname of his creator i.e. Stanislaw Borkowski, professor of technical and economic sciences, the acronym is legally protected). Results of the questionary were introduced. It was made an analysis in the range of the principle 2 of Toyota. It was shown analysis of respondent's characteristics. Next structure of assessments was determined and some statistical graphs were built.

1 Introduction

The air bag is a device of passive protecting passenger's compound from one or two pillows. In the case of a head-on collision are automatically inflated constituting the amortization between absorbing front seats of the driver and the passenger and with the steering wheel or the front part of the passenger cabin. The air bag consists of the generator of gas and the sack made of special material. The charge in the generator of gas is being initiated electrically, through impulse of sensors located in different places of the vehicle. A task of sensors is to recognize where the hitting was started and to activate appropriate air bags. As a result of the slight bump air bags cannot be started. After filling pillows, a gas immediately is escaping in specially prepared holes. Thanks to such a solution correct shock absorption is possible while the body is hitting the pillow. This solution prevents for strangulation of the person who e.g. lost consciousness. We are distinguishing the following kinds of air bags: frontal air bag, the side air bag, the knee air bag, air curtains and the air bag for the center passenger of the back couch. A production process, during the product is coming into existence, is a very essential stage. It decides about final evaluation taken out by the customer. Enterprise, which aspiring to the excellence of its product, realizes the necessity of constant improvement. The product should fulfil the highest requirements in terms of the functionality, the industrial safety and the environmental protection.

* Corresponding author: krzysztof.mielczarek@wz.pcz.pl

2 Presentation of respondent's structure

The personnel rating are the best indicator of the correctness of individual areas in the enterprise. To the purpose of analyzing satisfying the staff, a hierarchy of importance of influencing factors to the correctness of action given area and raising the quality by the vision and satisfying in the company, a BOST questionnaire form was carried out by 30 workers from production department. Amongst selected enterprise the population of respondents was chosen, which consisted from production workers of the examined enterprise, having a contact with production process in the workplace. Studies were marked by a low degree of anonymity, but with large degree of control by the interviewer. Applied questionnaire form - lecture and direct questionnaire - affected on increase in measurement control. It was an intentional treatment, which was supposed to contribute in increase of the correctness in filling surveys by pollsters In Table 1 was presented characteristics of respondents in table set.

Table 1. Respondents features in researched company.

	Feature indication and its characteristics					
	Gender	Educations	Age	Work experience	Mobility	Way of employment
1	Man	<Average	< 30	< 5	1	Normal
2	Woman	Average	31 - 40	6 to 10	2	Transfer
3		Higher I	41 - 50	11 to 15	3	Finance
4		Higher II	51 - 55	15 to 20	4	
5			56 - 60	21 to 25	5	
6			61 - 65	26 to 30	6	
7			> 66	31 to 35		
8				> 36		

On the basis of collected results circular graphs presenting participation of the given feature in the set (Fig. 1). From all respondents, man presents 67% of examined crew (Fig. 1a). The majority of respondents has the university education. In the case of age (Fig. 1c) the most workers were from 31 to 40 years (33% of examined crew). The largest groups constituted persons with the work experience on level 6 - 10 years (38%).

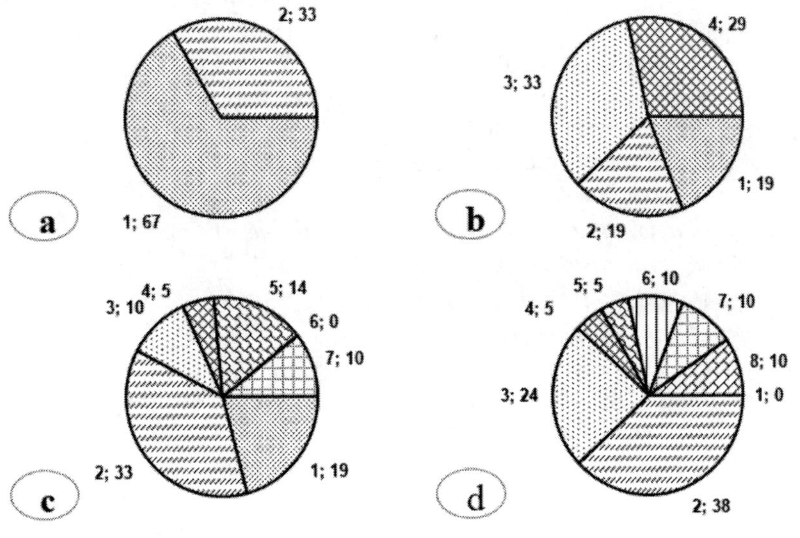

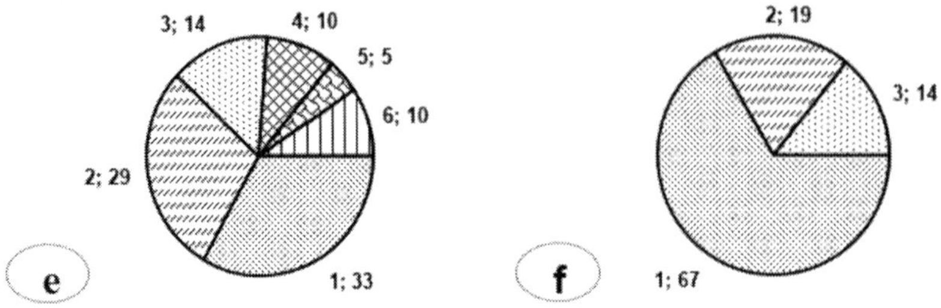

Fig. 1. Percentage (circular charts) characteristics of respondents with consideration: a) gender, b) educations, c) age, d) work experience, e) mobility, f) way of employment.

Analyzing the personal feature of respondents – mobility we observed that persons, for whom the current job is first constituted the majority of respondents. Analyzing way of recruitment (Fig. 1f) we noticed that persons with contract of employment constituted the largest group – (67% of the whole).

3 An evaluation the production process factors according to the BOST method

Toyotarity is a scientific discipline examining human - machine and human - human relationships with the inclusion of a process-based approach, Japanese culture, especially of Toyota, oriented to continuous improvement with the use of knowledge [1]. The BOST method describes Toyota's management principles with its characteristic factors. Valuable complement of carried out research there is interpretation of BOST questionnaire results [2]. The BOST method allows assessing the significance of factors describing the 14 Toyota management principles. It lets better look on the enterprise by eyes of their workers [3-4]. In the purpose to form an opinion it is essential to know the opinion of workers from different ranks in enterprise. BOST is survey where the questions are so well-matched to judge enterprise and its immaterial stores are possible. It was made on the base of 14 principles of the Toyota [5]. The second Toyota management principle is based on the conviction that appropriate process leads to appropriate results [6]. If the process is designed properly, then good results will come automatically [7]. It is necessary that quality is built in the processes already on the initial stage of production [8]. Processes shall be reorganized so that their course is fluent and uniform. Succeeding and improving competitiveness is only possible in self-learning organization and by continuous improvement of an organization culture [9].

Question that was analyzed sounded *„Which factor is the most important in the production process?"* (fill the blanks with 1; 2; 3; 4; 5; 6; 6 – the highest factor). The person who is filling questionnaire is judging the following factors:

- *Continuous system of problem detection* (CP).
- *Stopping production after discovering a quality problem* (PE).
- *Standard tasks, processes, documents* (SZ).
- *Providing authorizations to subordinates* (EU).
- *Application of only reliable technology* (ST).
- *Application of visual control* (SN).

The question has connection with 2 principle of the Toyota (E3 area). In Table 2 a percentage combination of evaluations was presented.

Table 2. Percentage list [%] of importance ratings for factors of E3 area.

Evaluation	Indicating the factors					
	CP	**PE**	**SZ**	**EU**	**ST**	**SW**
1	*0.0*	*4.8*	*28.6*	*14.3*	*19.0*	*33.3*
2	*0.0*	*28.6*	*9.5*	*38.1*	*9.5*	*14.3*
3	*14.3*	*19.0*	*23.8*	*19.0*	*9.5*	*14.3*
4	*19.0*	*23.8*	*23.8*	*14.3*	*14.3*	*4.8*
5	*47.6*	*14.3*	*4.8*	*14.3*	*4.8*	*14.3*
6	*19.0*	*9.5*	*9.5*	*0.0*	*42.9*	*19.0*

On the basis of the results presented in Table 2 the following circle charts have been created (Fig. 2).

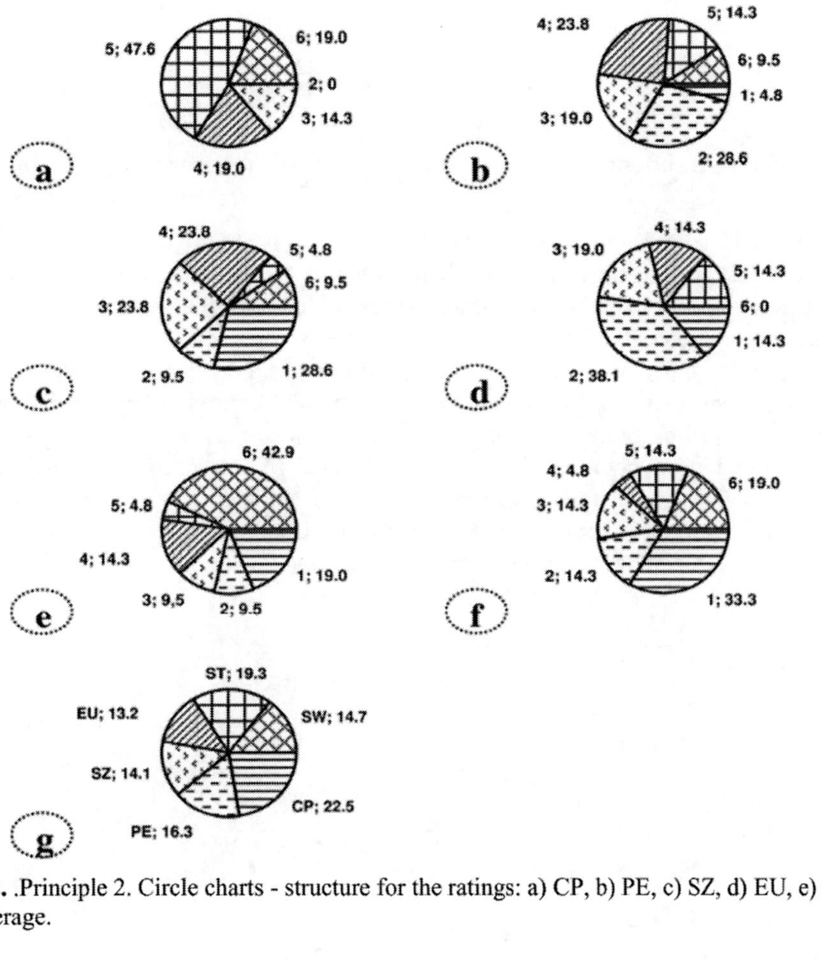

Fig. 2. .Principle 2. Circle charts - structure for the ratings: a) CP, b) PE, c) SZ, d) EU, e) ST, f) SW, g) average.

Continuous system of problem detection (CP) according to 47.6% respondents has a deciding influence on the production process because they granted high rate "5". On the basis of the prepared structure of ratings for separate factors it was found that *Application of only reliable technology* (ST) was the most important factor for 42.9% of respondents. They have awarded that factor the highest rate "6".

On the basis of acquired results the following significance sequences of factors describing the second Toyota management principles have been created in the production process of air bag:

$$CP > ST > PE > SZ > SW > EU \quad (1)$$

From analysis of the range of important factors it results that *Providing authorizations to subordinates* (EU) is not a key factor of the production process (among examined factors it is on the last place in a range of important factors of the production process).

The majority of factors has similar values. For 22.5% of respondents *Continuous system of problem detection* (CP) has high meaning in the organization of the production system. In the case of the factor *Application of only reliable technology* (ST) reached average 19.3%. According to respondents independence *Providing authorizations to subordinates* (EU) with the equal average 13.2% is the least important.

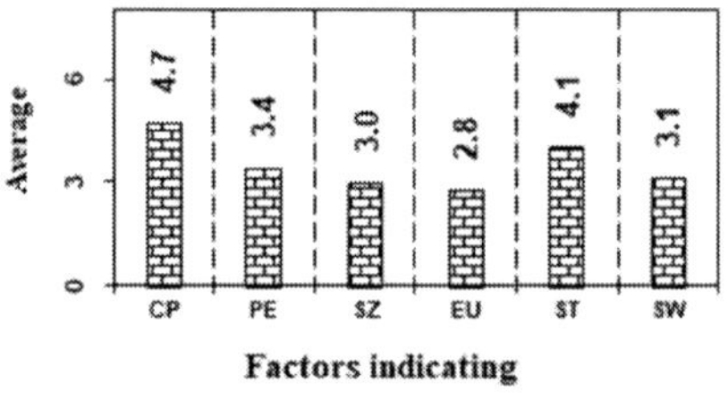

Fig. 3. Average (numerical) structure of the rating describing the second Toyota management principle.

To conclude it is possible, that workers are thinking highly of constant quality improvement, it just goes to show that this factor receiver high assessment. Assessments of remaining factors are distributed quite evenly. Respondents rank the factors in logical way and it proves proper approach of workers towards production process.

Radar graphs (Fig. 4) show importance depicting the percentage disintegration of factors describing the second Toyota management principle. On individual graphs they are arranged in the decreasing order according to the direction of movement of pointers of the clock having begun from hour 12:00, percentage shares of the assigned evaluations by respondents for each factor. Analysis of the relation showed that for some cases the equal percentage participation of factors had appeared. An occurrence of the change of the factor position in the importance series is also happening together with the height of a marking scale.

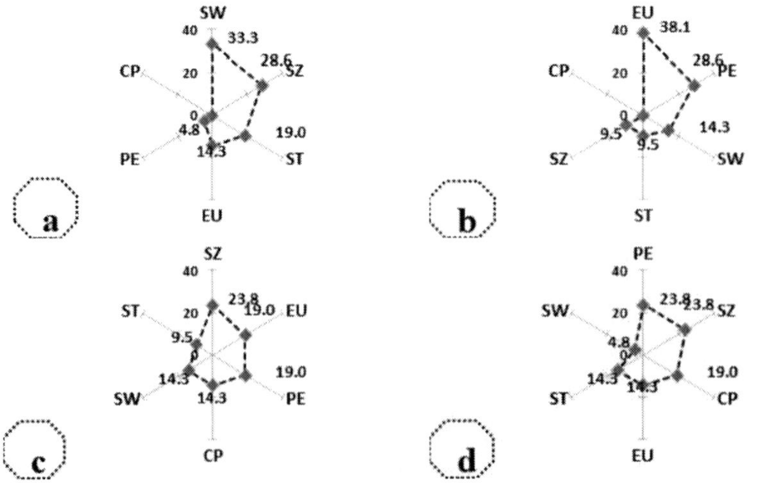

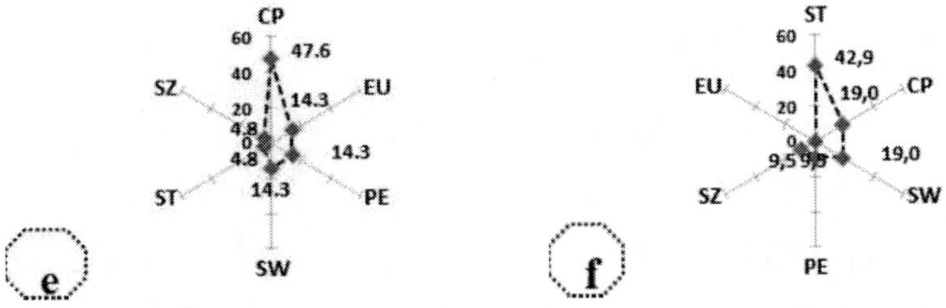

Fig. 4. Radar charts for importance factors of 2 principle of the Toyota: a) „1", b) „2", c) „3", d) „4, e) „5", f) „6".

Factor *Continuous system of problem detection* (CP) appeared on the position dominating only for evaluation „6". For the evaluation „1" and „2" the factor took the last place.

4 Summary

The Toyota management principles which are associated with production process was presented. Innovative BOST questionnaire survey, which are an attempt to convert Toyota's management principles into questions was described. The starting point for changes (improvement) is recording the existing condition. The present situation is definitely known best by participants of the processes implemented in a given enterprise. The analysis of significance of factors describing the second Toyota management principle states that the factor *Continuous system of problem detection* (CP) is the most important. The second position was taken by the following factor – *Application of only reliable technology* (ST). These factors are closely connected with technological aspects of production relating to fluent production rhythm with simultaneous application of reliable methods and technologies. The analysis has revealed diversity in the significance of factors describing the second Toyota management principle. In this way the usefulness of the BOST method has been proved for assessing a production process of goods of high quality requirements.

References

1. J.K. Liker, *The Toyota Way. 14 Principles of management of the world's leading manufacturing company (*Publisher MT Business, Warsaw, 2005)
2. K. Amasaka, *Science TQM, New Quality Management Principle, The Quality Management Strategy of Toyota Introduction* (Bentham Sc., U Arab Emirates, 2012)
3. S. Borkowski, *Toyota's management principles in the questions. The results of BOST (*PTM Publishing House, Warsaw, 2012)
4. S. Borkowski, *Scientific Potential of Toyotarity and BOST Method* (Polish Quality Institute, Warsaw, 2016)
5. K. Knop, Production Engineering Archives, **7/2**, 40-44, (2015)
6. S. Borkowski, J. Rosak-Szyrocka, D. Klimecka-Tatar, M. Jagusiak-Kocik, P. Sygut, 22nd International Conference on Metallurgy and Materials METAL (Brno, 2013)
7. J. Selejdak, Production Engineering Archives, **1/1**, 13-15 (2013)
8. M. Zhuravskaya, E. Morozova, N. Anashkina, M. Ingaldi, Polish Journal of Management Studies, **13/2**, 192-203, (2016)
9. K. Mielczarek, Production Engineering Archives, **8/3**, 32-3, (2015)

Selected programme methods of practical teaching – smart grid course as a case study

Robert Kucęba[1,*]
[1]Institute of Information Management Systems, Czestochowa University of Technology, Armii Krajowej 19B Avenue, Częstochowa, Poland

Abstract. With reference to the subject matter defined in the present paper's title, the methods applied in the processes of formal practical teaching – programme methods, have been distinguished and aggregated on the basis of the research carried out within the ERASMUS+ project "The acceleration method of development of transversal competences in the students' practical training process". As a case study programme methods of formal teaching have been aggregated as well as informatic tools and applications used while developing the simulated by students in laboratory conditions, a virtual management environment of dispersed RES micro-installations, within the Smart Grid course included into the area of prosumer energy. This simulator reflects the functional and logical structure of the Virtual Power Plant, which integrates the prosumer energy sources.

1 Introduction

Formal teaching at higher education institutions (public, non-public and community ones) is usually a formalized institutional form [1, 11], implemented according to programmes that enable to gain qualifications (knowledge, skills and competences), recognized by the legal system as well as on the labour market. Differentiated practical teaching methods are used in the formal teaching processes, which facilitate acquisition of knowledge and skills and their use in professional practice. Application of differentiated practical teaching methods in the teaching process, enables students to evaluate the usefulness of presented teaching content and developed competences in "professional life", also in creating and following the "professional career path". The portfolio of practical teaching methods in formal teaching includes primarily the methods: problem ones, activating ones, exposing ones, programme ones and practical ones. The method which is also frequently applied in formal teaching, not indicated in the distinguished categories of methods, is groupwork (in formalized student teams).

With reference to the subject matter defined in the present paper's title, the cognitive focus concerns the applied in formal processes of practical teaching – programme methods with the use of differentiated informatic tools and applications. E-learning tools are also commonly applied at higher education institutions (both off-line and on-line ones). In case of programme methods used in formal teaching processes, one can observe growing share

* Corresponding author: robertkuceba@wp.pl

of mobile applications use. Basically, programme methods ensure that students can efficiently, frequently without time and space limitations, become acquainted with the teaching content. It should be also indicated that a dynamic growth of applying in laboratory classes programme methods with the use of simulation games can be observed (e.g. MMOG class ones - massively multiplayer online game). Frequently informatic simulators reflecting heterogeneous systems, environments in business practice, are also developed by students. As supplementary to the already mentioned programme methods of teaching the following ones are also used: brainstorming, observation, panel discussion, problem lecture [2].

2 Interdyscyplinarity of prosumer energy

The selection of prosumer energy as a new utilitarian and at the same time interdisciplinary profile in practical teaching is justified by the needs of the labour market without being limited to the European Union ones only. Prosumer energy as a stimulant of local new and simultaneously perspective social "green initiatives" in favour of sustainable development of regions [3, 4]. The justification of selecting prosumer energy in the theoretical grasp is the reference to the third prosumer wave according to A. Toffler. It should be stressed here that A. Toffler rejected the traditional concept of a passive consumer, indicating the development of an active one, involved in the search of the product which best meets their expectations [5]. According to the A. Toffler's theory prosumer energy should be defined as a transformation from energy products acquired from fossil fuels in centralized energy sources – towards energy chains of RES energy value, co-created in economic micro-spaces, generating economic, environmental and social benefits [5]. The new energy value chains stimulate new business models on the energy markets (the new paradigm) – moving away from centralized models towards decentralized ones [6].

Prosumer energy stimulates the demand for a new group of local entrepreneurship, new work posts – and thus, the demand for a new group of specialists and indirectly new interdisciplinary courses/specialties/profiles of studies, and what follows the demand for "professional prosumers". The abovementioned is also justified by the currently observed diffusion of new investments into RES micro-installations, growth of environmental awareness of heterogeneous investors, and thus, their conscious social and environmental responsibility.

In accordance with the aforementioned justification of interdisciplinary –the practical profile of „Prosumer Energy" is confirmed by the convergence of knowledge, skills, social competences in four areas: technical, economic, social and natural – where it seems reasonable to distinguish the environmental one [3, 7].

In particular, in the technical dimension – in the practical teaching processes, attention is paid to diversity of energy sources in local perspectives, their dispersion and managing the dispersed prosumer energy infrastructure (e.g. micro- wind turbines, micro- biogas plants, photovoltaic and photothermal sources) [8]. Their indicated features include technological diversity, non-linear availability and the possibility of energy value growth that is produced, e.g. in RES micro-installations, through integration (also with the energy receivers on the side of prosumers) in the virtual environment of Smart Grid intelligent networks.

In the economic dimension the considered feature include: regional specialisation, pro-environmental investments, pollution reduction including creation of low emission/smog, incorporation of external costs into the prices of fuels and secondary energy, low cost and marginal costs of investments.

In the social dimension in turn, social competences are shaped in the practical teaching processes in the scope of: autonomous energy acquisition, change of energy consumption culture (change of energy consumers behaviours – adjusting own energy needs to the availability of non-linear RES sources) and stimulation of the willingness to pay (Cost-Benefit Analysis) – frequently incurring larger costs resulting from additional external (environmental) costs.

In the environmental dimension the students acquire knowledge and develop skills in the scope of perceiving the relations between prosumer energy diffusion and natural environment pollution reduction. This concerns in particular the ability to analyse, measure, monitor and interpret air pollution indexes.

The abovementioned dimensions confirming the utility and interdisciplinarity of prosumer energy, as well as the demand on the labour market for a new group of specialists, justify the introduction into the didactic offer of Higher Education Institutions a new course or specialty "Prosumer Energy". The course/specialty "Prosumer Energy" has also been included into the educational offer of the universities that cooperate in carrying out the ERASMUS project "The acceleration method of development of transversal competences in the students' practical training process" [2]. It should be indicated here that this course/specialty has been introduced at technical faculties, especially Electrical Faculties.

3 PRACTICAL TEACHING METHODS OF DISPERSED RES MICRO-INSTALLATIONS MANAGEMENT

Taking into consideration the utilitarian and interdisciplinary dimensions of prosumer energy at the Management Faculty (social sciences) of Czestochowa University of Technology/Poland, the Authors of the present paper have started the specialty – "Energy and Environmental Management" within the course Spatial Economy, directly connected with the prosumer energy profile. This specialty comprises issues in the scope of energy and environmental management in territorial government units, as well as business organizations. It prepares substantively and practically specialists in the scope of energy and environmental management (with particular consideration of prosumer energy), in this planning, optimizing and balancing energy needs, designing prosumer energy systems in territorial government units, manufacturing and service providing enterprises or public utility units, e.g. healthcare.

A measurable goal of practical teaching within the subject Smart Grid is consolidation of knowledge, skills and competences in the area of prosumer energy and carrying out own projects reflecting actual conditions of managing these sources and energy receivers on the prosumer side. Smart Grid classes take place in a computer laboratory, primarily with the use of programme methods of practical teaching.

In the present point of the paper the Authors in a synthetic way present the course of project classes within the subject Smart Grid with the use of various practical teaching methods, in particular programme methods and tools. The schedule of the classes is assigned with reference to the designed by the students, in laboratory conditions of the Management Faculty, three layers of Smart Grid [6] – which reflect the structure of the Virtual Power Plant integrating prosumer energy sources [9,10].

STAGE I - SELECTION OF PROSUMER TECHNOLOGIES IN THE PROSUMER TECHNOLOGY LAYER

PRACTICAL TEACHING METHODS: group work, panel discussion or/and brainstorming with the participation of the lecturer.

REQUIREMENTS: theoretical background supplemented with a case study of good practices resulting from the knowledge, skills, competences acquired within the subjects of the specialty: Renewable and Dispersed Energy Systems, Energy Infrastructure Management, as well as Synergetics and Eco-innovations.

IMPLEMENTATION (groups of 3-4 students): The first layer Prosumer Technology (PT) is determined in the actual Smart Grid structure as an operational one. In the first PT layer the students in the basis of a panel discussion or/and brainstorming with the participation of the lecturer select – at least two heterogenous prosumer micro-installations (in the laboratory conditions this usually concerns: micro- photovoltaic sources and micro-wind turbines) according to the criteria: demand for energy at the household where students reside and the local economic potential of RES "carriers" (on the basis of local reports concerning, e.g., wind speed or amount of sunlight). The task of the students at designing the simulator of the PT layer is also selection of electricity receivers (e.g. domestic appliances and intelligent vehicle charging systems: plug-in hybrid electric vehicle –PHEV – only as a design assumption), which can be controlled (e.g. with the use of the IoT technology) adjusting their work schedule to the availability of non-linear RES sources.

STAGE II - SIMULATING PROCESSES IN THE SMART METERING LAYER

PRACTICAL TEACHING METHODS: group work, programme methods: with the use of computer applications (in particular Google Drive, Google Spreadsheets).

REQUIREMENTS: a) theoretical background supplemented with a case study of good practices resulting from knowledge, skills, competences acquired within general majors subjects: Statistics and Econometry, and the specialty subjects: Renewable and Dispersed Energy Systems, Energy Infrastructure Management, as well as Synergetics and Eco-innovations; b) a project of the Prosumer Technology layer.

IMPLEMENTATION: (project teams of 3-4 students): In the second layer of Micro-Smart Grid – an analytic one, which is a simulation of Smart Metering – SM (in real conditions -metering infrastructure – bidirectional intelligent metres), students on the basis of the available power patterns of RES installations, depending on local daily changes of, for example, the amount of sunlight or/and wind speed, create in spreadsheets daily power characteristics of available energy sources – introduced in the first PT layer, e.g., for micro-photovoltaic sources and micro- wind turbines. Another task of designing the SM layer is to elaborate the total daily power characteristics of available RES hybrid (aggregate daily availability of adopted in the project Prosumer RES micro-installations). Due to the limited in the laboratory conditions access to Intelligent Metering Infrastructure, the students perform in the period of seven days readings of energy consumption at one-hour intervals in their households (in real conditions of SM work the readings are performed at intervals not longer than 15 minutes). On the basis of the readings the students prepare in spreadsheets for all the days when the readings were performed, daily, calendar load characteristics (demand for electricity). Another task in simulating the SM layer is preparing the spreadsheet of daily, calendar characteristics of supply – available power of the micro-installation hybrid (the design assumption: repetition of the hybrid availability in subsequent days of measurements) and demand (real demand for energy in the given household) balancing.

STAGE III - DEVELOPING A SIMULATOR OF THE SYSTEM TO MANAGE PROSUMER TECHNOLGY INFRASTRUCTURE

PRACTICAL TEACHING METHODS: group work, programme methods: with the use of computer applications – in the cloud Google Drive: Google Spreadsheets and neural networks generator in the application environment STATISTICA NEURAL NETWORKS PL.

REQUIREMENTS: a) theoretical background supplemented with a case study of good practices resulting from knowledge, skills, competences acquired within general majors subjects: Statistics and Econometry, Economic Forecasting and the specialty subjects: Renewable and Dispersed Energy Systems, Energy Infrastructure Management, as well as Synergetics and Eco-innovations; b) a project of the Prosumer Technology layer; c) the project of Smart Metering layer.

IMPLEMENTATION (coordinated cooperation of all project teams of 3-4 persons. Due to the population of the laboratory groups usually 5 project teams): In the real structure of this Smart Grid layer – Prosumer Technology Infrastructure Management Systems in the virtual environment support management of prosumer energy dispersed resources. This concerns planning, scheduling the work of energy receivers with reference to the availability of RES micro-installations. In order to settle the implementation of the energy and climate package, e.g. 3x20, amounts of generated energy from RES (available power and electricity generated in RES micro-installations) are aggregated in these systems.

The students using spreadsheets determine the indexes of air pollution reduction and reduced use of fossil fuels to produce the same amount of energy in RES micro-installations and thermal power plants. The students using spreadsheets, on the basis of historical supply-demand data, in this balanced one (data aggregated and visualized in the second layer of Smart Metering), design an analytic system, which generates forecasts of periodical availability and demand for electricity, in the adopted prosumer infrastructure. The forecasts are carried out in spreadsheets with the use of linear forecasting models: e.g. with the use of forecasting formulas according to: running average, forecasting models, ARMA and ARIMA models. Additionally, the forecasts are carried out with the use of non-linear models generated in the neural networks environment STATISTICA NEURAL NETWORKS PL4. Non-linear models created by the students in this environment include: *Multilayer Perceptrons* (MLP), *Kohonen Self-Organizing Feature Maps* (KSO), *Probabilistic Neural Networks* (PNN), *Generalized Regression Neural Networks* (GRNN) and *Linear Networks* (LN). The developed forecasting models (linear and non-linear ones) are subject to verification (*ex post*) on the basis of relative and average square errors – generated at the same time intervals. The goal of verifying the accuracy of the forecast is the selection of the final forecasting model.

The last element of the project is cumulating non-linear outputs of available power and the demand for energy generated by all project teams within the laboratory group of the students. A measurable result of such a cumulation is defining the amount of electricity generated in the virtual power plant (cumulated amount of electricity in RES micro-installations, adopted in the PT layer by particular project teams), aggregating the reduction size of the demand/purchase of electricity from the Energy System, as well as determining indexes of air pollution reduction. In particular, in the academic laboratory conditions of the Management Faculty this concerns reduction indexes of CO_2, NO_x.

To sum up, on the basis of the evaluation of practical teaching methods, conducted among the students after the laboratory classes, it was concluded that the applied programme methods (even though the informatic tools portfolio was limited in the laboratory conditions of the University) supplemented with problem methods (e.g. panel discussion, brainstorming) or activating methods, in a synergic way influenced the level and results of the teaching process:

- profound, up-to-date knowledge in the scope of:
  - prosumer energy – RES micro-installations,
  - functions and structure of informatic decentralized systems of energy management;
- acquired skills and competences in the scope of:
  - designing, implementing and optimizing Smart grid networks,
  - generating and analysing the availability of energy sources and loads on the prosumer side,
  - determining and interpreting indexes of greenhouse gases reduction and reduced use of primary fuels in prosumer energy,
  - managing the energy demand and supply on the side of the prosumer,
  - creative thinking and work in project teams.

5 Summary

Diversifying the practical teaching methods, independent creation of procedures, determining the processes (e.g. managing the energy demand and supply in prosumer installations) and their migration to the virtual environment enables students to evaluate the usefulness of presented teaching content and developed competences in "professional life".

This is confirmed by the firs graduates of the specialty "Energy and Environmental Management", directly connected with the profile of prosumer energy. They indicate that the knowledge, acquired skills and competences in the scope of planning, designing, programming the procedures – managing the energy infrastructure, in this dispersed RES micro-installations place them very high on the labour market. On the basis of an analysis of professional career paths of the first graduates of this specialty (it should be stressed here that they are graduates of a non-technological faculty), 54% of the researched population work or participate in professional internships in territorial government units – in the departments of energy management in territorial units. Diversification of practical teaching methods with the consideration of programme methods facilitates creating a "climate" for development of new groups of entrepreneurship, inspires the students to participate actively in didactic activities, establish their professional preferences and combine theoretical and practical knowledge.

References

1. A. Andrzejczak, *Methodology of studying* (2014)
2. The Project The acceleration method of development of transversal competences in the students' practical training process: *„The report concerning applied teaching methods of transversal skills and methods of practical trainings"* (Poznań, 2016)
3. R. Kucęba, L. Kiełtyka, *Present Day Trends of Innovations* (2013)
4. H. Son, T. Y. Kang, H. Kim and J. H. Roh, IEEE Transactions on Smart Grid, **2**, 759-769 (2011)
5. A. Toffler, *The Third Wave* (1980)
6. J. Popczyk, *Dispersed Energy* (2011)
7. D. Niedziółka, *Energetyka Prosumencka. Próba pierwszej konsolidacji* (2014)
8. R. Kucęba and M. Bajor, *Prosumer Energy. First Consolidation Attempt* (2014)
9. R. Kucęba, *Virtual power plant. Chosen aspects of organization and management of dispersed generation subjects* (2011)
10. R. Kucęba, F. Bylok, A. Pabian, M. Zawada, Conference Proceedings *ICSSAM* and *ISEPSS*, 1040-1051 (2014)
11. A. Pacana, R. Ulewicz, Polish Journal of Management Studies **16**, 165-174 (2017)

Improved Education and Trainings in the Field of Fire Protection

Roland Iosif Moraru[1], *Mariusz* Sroka[2,*]

[1]University of Petrosani, Management and Industrial Engineering Department, University Street 20, Romania
[2]Technical University of Czestochowa, Faculty of Management, Armii Krajowej 19B, Poland

Abstract. The article presents the characteristics of the subject "Fire and rescue protection" carried out at the Faculty of Management at the Czestochowa University of Technology in the field of "Work Safety and Hygiene". The modified program consists of lectures and exercises and also presents a practical way of implementing the selected issues in a given building. Next, the methodology of research and the results of questionnaire surveys conducted in the target group of students are presented, referring to the improved content and methods of education and training in the field of fire protection. The article ends with conclusions from the conducted research and the proposed concept of activities for the future.

1 Introduction

Fire protection includes carrying out projects aimed at protecting life, health, property or the environment against fire, natural disaster or other local threat. This is done by preventing the emergence and spread of fire, natural disaster or other local threat, providing manpower and means to fight a fire, natural disaster or other local threat, and carrying out rescue operations. Successful completion of the above tasks requires the functioning of the entire rescue and fire extinguishing system, the medical rescue system and the emergency notification system [1, 2, 3, 4]. It should be highlighted that it is the duty of every natural person, legal person, organization or institution using the environment, building, facility or area to protect them against a fire hazard or other local threat. Detailed tasks, duties and requirements in this area are regulated by relevant regulations, laws, legal acts, standards and instructions [5, 6, 7].

Widely understood fire protection requires an educational system including teaching, education and upbringing [8, 9, 10]. The authors of this study propose changes to the content and teaching methods in the cycle of education of first degree engineering students in the field of health and safety at the subject of fire protection and rescue. The purpose of these changes is to improve the teaching process in the field of fire protection. The research is to answer whether the actions taken, changes and methods are correct and appropriate in the didactic process.

* Corresponding author: mariusz.sroka@wz.pcz.pl

2 General characteristics of the subject: fire protection and rescue

The subject of fire protection and rescue is conducted in the field of occupational safety and health on the sixth semester of engineering studies of the first degree. It is a subject area with a general academic profile of 15 hours of lectures and 15 hours of exercises. The number of ECTS points that students receive after completing the course is 3.

The aim of the course is to familiarize students with the principles of fire protection, the fire protection system in Poland and the rules of conduct in the event of a fire and the principles of undertaking rescue and firefighting activities. It is assumed that the achievement of learning outcomes during the didactic process will allow students to skillfully assess fire risks, to act properly in the event of fire and rescue and firefighting activities.

The curriculum contents of lectures were selected in accordance with the guide prepared by the authors on the subject for a given field of study and the corresponding national qualifications framework. They include:
- Basic concepts and terms related to fire and rescue;
- Legal regulations in the field of fire protection and rescue in Poland;
- Basic concepts, phenomena and processes regarding the burning of materials;
- Equipment and other extinguishing media - purpose and rules of use;
- Technical fire-fighting equipment - (fire-fighting water supply, warning systems, fixed fire extinguishing systems, system installations);
- Causes of fires, identification and assessment of fire hazards;
- Rules of conduct in case of a fire;
- Organization and procedures for the evacuation of buildings;
- Organization of fire protection in public facilities;
- Rescue operations during the release of hazardous materials.

The content of the auditorium exercises was selected in accordance with the guide prepared by the authors on the subject for the given field of study and the corresponding national qualifications framework. They include:
- Basic legal regulations concerning fire and rescue protection in Poland - implementation of the legal acts database;
- Technical conditions of buildings and their location in the context of fire protection - practical exercises on a construction site;
- Technical and organizational conditions for evacuation from buildings - practical exercises on a construction site;
- Fire-fighting equipment and hand-held extinguishing equipment in public facilities - practical exercises on a construction site;
- Rescue operations during emergency events involving hazardous materials - analysis and indication and discussion of rescue procedures.

Particular attention should be paid to the practical form proposed by the authors for the implementation of selected exercises on building objects. The authors depart from the previously strictly theoretical form in favor of a practical local an-site exercise. The improvement of teaching methods in this way gives very positive results, which is confirmed by the research presented below.

3 Research methodology

Questionnaire surveys were carried out on 10 and 12 June 2018. Respondents were students of the Faculty of Management at the Czestochowa University of Technology. The groups in which the survey was conducted were selected on purpose. The conditions for participation

in the research were: studying in the field of occupational safety and health, positive completion of the subject of fire protection and rescue and consent to participate in the research. 34 questionnaires were carried out, of which all questionnaires were classified after eliminating errors, consistency and logic of the answers provided for further analysis. The questionnaire contained 11 questions. Ten questions were closed-ended with the possibility of indicating one answer. One question was an open-ended question. The aim of the conducted research was:
- Determining whether the selection of program content of lectures and exercises on the subject of fire protection and rescue is appropriate;
- Determining the level of understanding of the theoretical and practical knowledge by respondents;
- Evaluation of the practical form of selected exercises;
- Determining the suitability of the information acquired on the subject in terms of increasing the competences and skills of the group of students;
- Determining whether the goals and objectives assumed for the learning outcomes during the didactic classes have been achieved;
- Diagnosing whether the acquired competences will be used in the future work of the surveyed group;
- Getting the answer to the question whether the subject met the expectations of students.

The test results are presented graphically or compiled in tables and analyzed in the chapter below.

4 Analysis of test results

The first and second question was aimed at assessing whether the program content of lectures and exercises on the subject of fire protection and rescue are properly selected according to the respondents. A vast majority of respondents (in total for lectures 91% and for exercises - 97%) evaluate the selection of program content as appropriate or very good. None of the respondents indicated that the choice of program content lectures and exercises was wrong or insufficient. The test results for this part of the survey are presented in Figure 1.

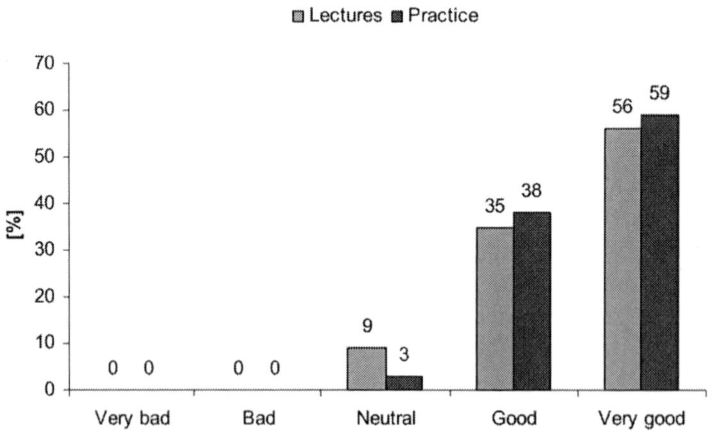

Fig. 1. Evaluation of the selection of program content of lectures and exercises on the subject of fire protection and rescue.

The next two questions concerned the extent to which students improved their theoretical and practical knowledge of the issues discussed in the didactic classes. A vast majority of respondents indicated the improvement of their knowledge as good or very good (in total for theoretical knowledge 88% and for practical knowledge 97%). The understanding was improved sufficiently by theoretical knowledge by 12% of respondents, and 3% by practical knowledge. None of the respondents indicated that the classes did not deepen his knowledge. The results of this part of the survey are summarized in Table 1.

Table 1. Degree of deepening theoretical and practical knowledge in the field of discussed issues.

Degree of deepening knowledge	Theoretical knowledge Indication [%]	Practical knowledge Indication [%]
Not improved	0	0
Sufficient	12	3
Good	38	41
Very good	50	56

The purpose of the next question was to examine how the respondents evaluate the form proposed by the authors to implement selected issues in a practical way on the designated construction site. This question was one of the most important in the conducted research. The percentage of students surveyed positively evaluating this form of classes was 100%. It should be noted that 76% indicate a very positive rating. The test results for this part of the survey are presented in Figure 2.

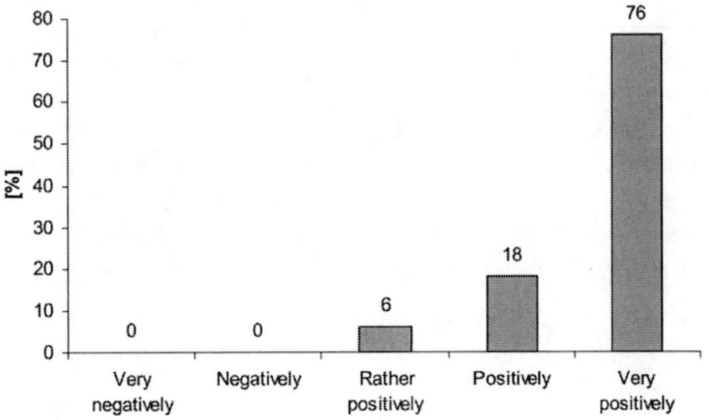

Fig. 2. Assessment of practical classes in the form of a local vision on a construction site.

In the next question, the respondents' task was to assess the suitability of the information acquired on the subject of fire protection and rescue in terms of raising their own competences and skills. 53% of respondents indicated it as very useful and 41% as useful. Only 6% of respondents indicated the answer "moderately useful". It should be noted that none of the respondents indicated the answers: useless and little useful. The test results for this part of the survey are presented in Figure 3.

The next two questions were aimed at getting to know the students' opinions about whether the objectives of the subject were implemented and whether learning outcomes were achieved. This is clearly the answer of 76% of respondents regarding the achievement of goals and 68% of the achievement of the effects of didactic classes. This corresponds to 21% and 29% respectively. None of the respondents indicated "not" and "definitely not". The results of this part of the survey are summarized in Table 2.

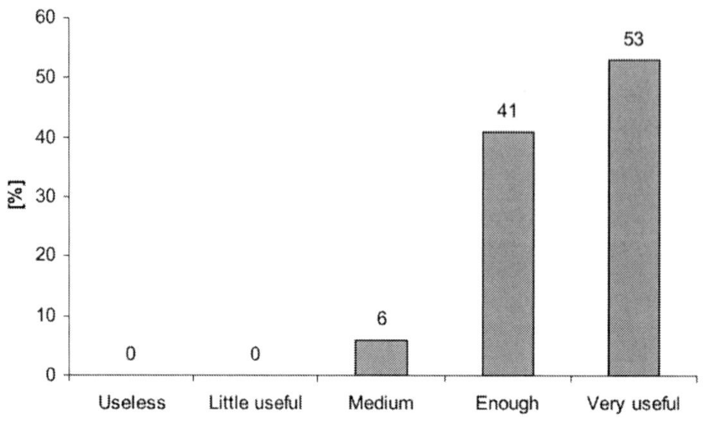

Fig. 3. Usefulness of the acquired knowledge in terms of raising own competences and skills.

Table 2. Realization of goals and achievement of learning outcomes during didactic classes.

Answers	Realization of goals Indication [%]	Achieving the effects Indication [%]
Definitely not	0	0
Not	0	0
Hard to say	3	3
Yes	21	29
Definitely yes	76	68

In the next question, the respondents' task was to indicate whether the acquired competences during the didactic classes will be used in professional work. 55% of the respondents answered that "definitely yes" and 30% that "yes". 15% answered that "it's hard to say". None of the respondents indicated "no" and "definitely not". The results of the survey for this question are presented in Figure 4.

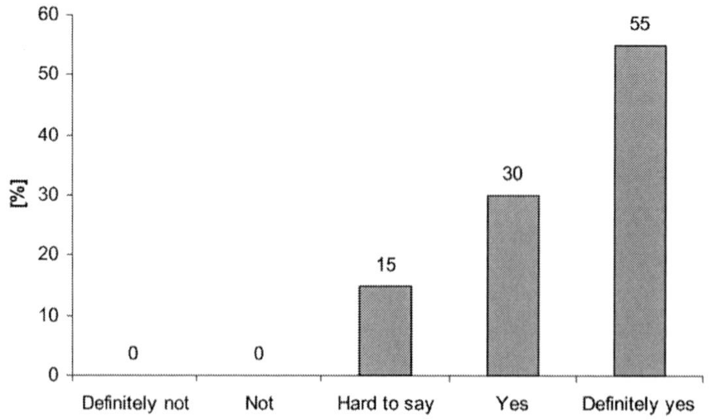

Fig. 4. Use of acquired competences in professional work.

In the last closed question of the questionnaire the respondents' opinion on the fulfillment of expectations regarding the subject of fire protection and rescue was examined. The percentage of respondents indicating that the item met their expectations in to a satisfactory degree is 65%. 26% of respondents indicate the degree of relevance to expectations, and 9% of respondents assess this level sufficiently. None of the participants of the study indicated that the subject did not meet his expectations or did not meet them sufficiently. The results of the survey for this survey question are presented in Figure 5.

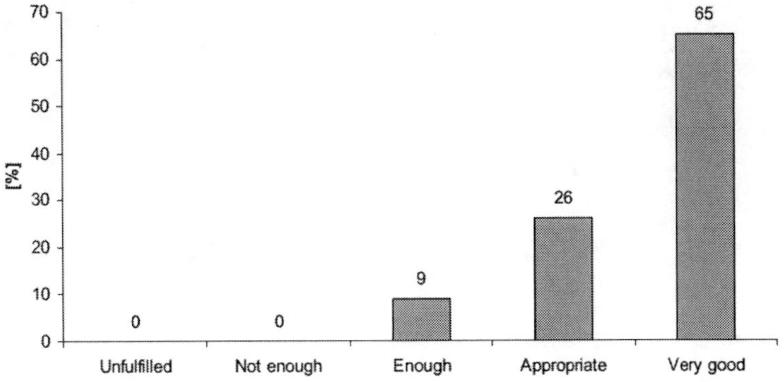

Fig. 5. Degree of meeting respondents' expectations in relation to the subject of fire protection and rescue.

The questionnaire contained one open-ended question, in which the respondents were free to comment and submit their own comments and suggestions regarding the whole training in the field of fire protection. Unfortunately, the answers provided in this part of the questionnaire did not contain a significant research element, therefore, they were not included in this study.

5 Summary

The conducted research clearly indicates that it was appropriate to modify the content of teaching and teaching methods of the subject of fire protection and rescue. Respondents rated highly the correct choice of program content, the degree of theoretical and practical knowledge, the usefulness of this knowledge in terms of raising their own competences and skills as well as usefulness in their professional work. The practical method for the implementation of selected issues was rated particularly highly.

Very important was also the highly satisfactory commitment of the participants during the course and the final grades for the subject.

The intention of the authors in the future is to extend the classes in the field of fire protection and rescue to illustrative, laboratory and simulation methods.

References

1. D. J. Neal, K. T. Stiles, J. V. Salazar, IFSJLM, **11** (2017)
2. W. Lewis, FE, **168**, 4 (2015)
3. R. Holmgren, TP&E, **25**, 2 (2016)
4. M.Sommer, O. Nja, JWL, **23**, 7 (2011)
5. R. Finger, j.r., FE, **169**, 4 (2016)
6. I. Grabara, PJMS, **10**, 1 (2014)
7. K. L. Rossler, A. R. Duvall G. Sankaranarayanan. NE (2018)
8. J. Linder, MFC **49**, 1 (2012)
9. T. Kiurski, FE **163**, 1 (2010)
10. M. Kennedy, AS&U, **85**, 8 (2013)

AUTHOR INDEX